Criminal Justice in England and the United States

Second Edition

David Hirschel, PhD
University of Massachusetts
Lowell, Massachusetts

William Wakefield, PhD
University of Nebraska
Omaha, Nebraska

Scott Sasse, PhD
Dana College
Blair, Nebraska

JONES AND BARTLETT PUBLISHERS

Sudbury, Massachusetts

BOSTON   TORONTO   LONDON   SINGAPORE

Jones and Bartlett Publishers
World Headquarters
40 Tall Pine Drive
Sudbury, MA 01776
978-443-5000
info@jbpub.com
www.jbpub.com

Jones and Bartlett Publishers
Canada
6339 Ormindale Way
Mississauga, Ontario L5V 1J2
Canada

Jones and Bartlett Publishers
International
Barb House, Barb Mews
London W6 7PA
United Kingdom

Jones and Bartlett's books and products are available through most bookstores and online booksellers. To contact Jones and Bartlett Publishers directly, call 800-832-0034, fax 978-443-8000, or visit our website www.jbpub.com.

Substantial discounts on bulk quantities of Jones and Bartlett's publications are available to corporations, professional associations, and other qualified organizations. For details and specific discount information, contact the special sales department at Jones and Bartlett via the above contact information or send an email to specialsales@jbpub.com.

Production Credits
Chief Executive Officer: Clayton Jones
Chief Operating Officer: Don W. Jones, Jr.
President, Higher Education and Professional Publishing: Robert W. Holland, Jr.
V.P., Sales and Marketing: William J. Kane
V.P., Design and Production: Anne Spencer
V.P., Manufacturing and Inventory Control: Therese Connell
Publisher, Public Safety Group: Kimberly Brophy
Acquisitions Editor, Criminal Justice: Jeremy Spiegel
Associate Managing Editor: Amanda Green
Reprints Coordinator/Production Assistant: Amy Browning
Marketing Manager: Wendy Thayer
Manufacturing and Inventory Coordinator: Amy Bacus
Cover and Text Design: Anne Spencer
Cover Images: England: © Gencay M. Emin/ShutterStock, Inc.; U.S.: © iconex/ShutterStock, Inc.;
    Column: © Ron Chapple/Thinkstock/Alamy Images
Chapter Opener Image: © Masterfile
Composition: Auburn Associates, Inc.
Printing and Binding: Malloy, Inc.
Cover Printing: Malloy, Inc.

Library of Congress Cataloging-in-Publication Data

Hirschel, J. David.
    Criminal justice in England and the United States / David Hirschel, William Wakefield, Scott Sasse.
        p. cm.
    Previous ed. published: Westport, Conn.: Praeger, 1995.
    ISBN-13: 978-0-7637-4112-9 (pbk.)
    ISBN-10: 0-7637-4112-4 (pbk.)
    1. Criminal justice, Administration of—Great Britain. 2 Criminal
justice, Administration of—United States.   I. Wakefield, William O.
II. Sasse, Scott. III. Title.
    HV9960.G7H57 2007
    364.941—dc22
                            2007004751

6048

Printed in the United States of America
11 10 09 08 07   10 9 8 7 6 5 4 3 2 1

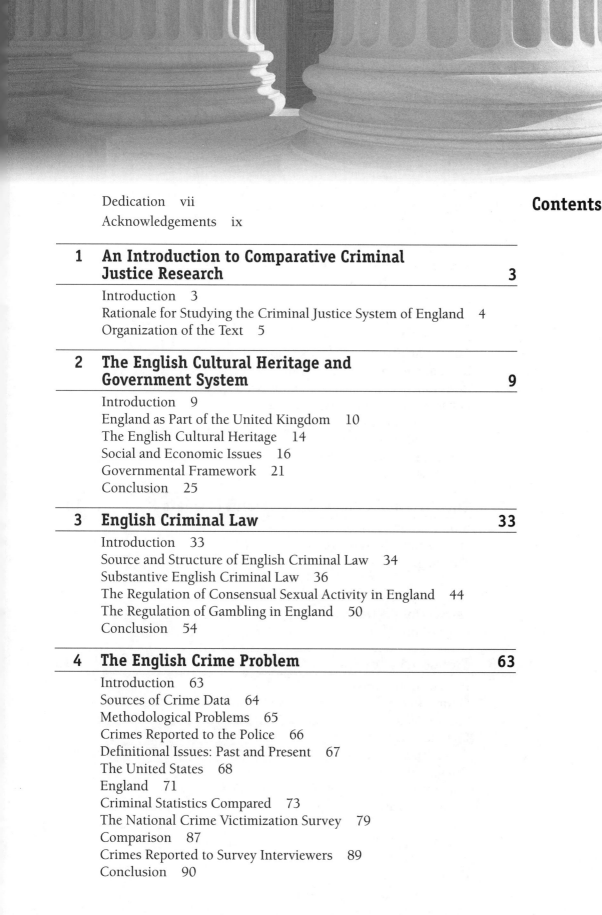

Contents

Dedication   vii
Acknowledgements   ix

1 An Introduction to Comparative Criminal Justice Research   3

Introduction   3
Rationale for Studying the Criminal Justice System of England   4
Organization of the Text   5

2 The English Cultural Heritage and Government System   9

Introduction   9
England as Part of the United Kingdom   10
The English Cultural Heritage   14
Social and Economic Issues   16
Governmental Framework   21
Conclusion   25

3 English Criminal Law   33

Introduction   33
Source and Structure of English Criminal Law   34
Substantive English Criminal Law   36
The Regulation of Consensual Sexual Activity in England   44
The Regulation of Gambling in England   50
Conclusion   54

4 The English Crime Problem   63

Introduction   63
Sources of Crime Data   64
Methodological Problems   65
Crimes Reported to the Police   66
Definitional Issues: Past and Present   67
The United States   68
England   71
Criminal Statistics Compared   73
The National Crime Victimization Survey   79
Comparison   87
Crimes Reported to Survey Interviewers   89
Conclusion   90

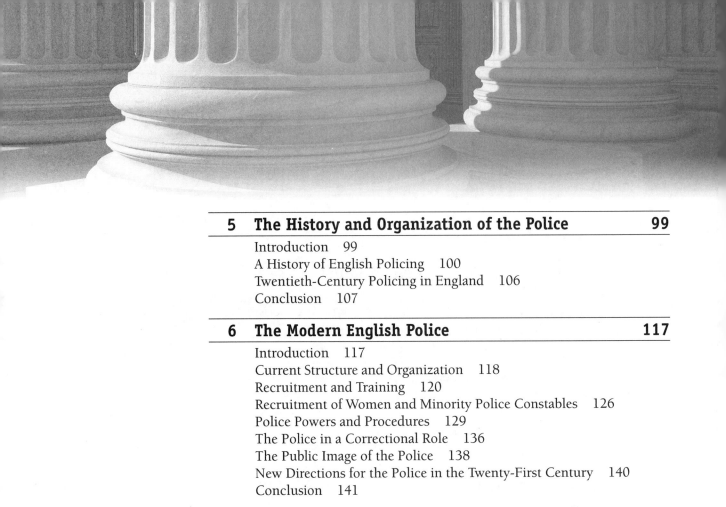

5 The History and Organization of the Police   99

Introduction   99
A History of English Policing   100
Twentieth-Century Policing in England   106
Conclusion   107

6 The Modern English Police   117

Introduction   117
Current Structure and Organization   118
Recruitment and Training   120
Recruitment of Women and Minority Police Constables   126
Police Powers and Procedures   129
The Police in a Correctional Role   136
The Public Image of the Police   138
New Directions for the Police in the Twenty-First Century   140
Conclusion   141

7 The Criminal Courts, Judges, and Lawyers   157

Introduction   157
Structure and Jurisdiction of the Courts   158
The Judges   161
The Lawyers   165
Lawyers for the Prosecution   169
Lawyers for the Defense   173
Conclusion   176

8 The Court Process   185

Introduction   186
Outline of the Court Process   186
Special Features of the English Court Process   187
Pretrial Release   188
Cases Triable-Either-Way   190
Committal Proceedings   192
The Exclusion of Illegally Obtained Evidence   193
Guilty Pleas and Plea Bargaining   195
Juries   197
Sentencing   200
Miscarriages of Justice   208
Conclusion   209

9 The Development of Confinement and Corrections in England   223

Introduction   223
The Early History of Punishment   224
The Move Toward Public Displays of Punishment   228
Examples of Medieval Punishment   229
Early Penal Reformers   230
Eighteenth- and Nineteenth-Century Corrections in England   231
Elizabeth Fry, Reformer   233
Prisons of the Nineteenth Century   234
Alexander Maconochie and Walter Crofton   235
Return to Punishment   236
The Death Penalty in England   237
Conclusion   238

10 The Organization and Operations of Corrections   247

Introduction   247
The Twentieth Century   248
Philosophical Foundations of the Present Corrections System   251
Current Organization and Administration of Corrections
   in England   252
The Prisons of Today   258
Alternatives to Incarceration   267
The National Association for the Care and Resettlement of Offenders
   (NACRO)   272
Conclusion: The Challenges of the Future for Corrections
   in England   273

11 Terrorism   291

Introduction   291
Legislating Against Terrorism   292
Terrorism in England: A Brief History   294
Terrorism Legislation   299
Conclusion   313

12 Juvenile Justice in England   321

Introduction   321
History of Juvenile Justice   322
Changes in Juvenile Justice   323

Juvenile Courts   332
Conclusion   336

13 Evaluating the English Criminal Justice System: Lessons to Be Learned   349

Introduction   349
External Factors Affecting the Operation of the Criminal Justice
  System   352
Differences in the Two Justice Systems   353
Efficiency and Effectiveness in Achieving System Goals   361
"Quo Vadis" England? Where Do You Go from Here?   368
A Final Word   369

Dedication

Dedicated to the memories of:

James C. Kane, Professor
University of Nebraska at Omaha

Dick Hoffman, Lieutenant
Buffalo, New York Police Department

And Paul Chadeyron and Marcus M. Dorfman
Both of whom were English attorneys

For our wives and children:

Fran, Mike, Rob, and Megumi
Ellen, Bill, Eric, and Sean
Sandy, Alexander, and Ryan

Acknowledgements

When one reflects on a project of this magnitude, it quickly becomes apparent that there are many people who have played significant roles in its development. Whether by simply providing advice, suggestions, or substantive contributions to the content of the book, all have had an important part in the development of this endeavor. Ultimately, of course, we take final responsibility for everything in this book, but we would be remiss if we did not specifically acknowledge some special individuals.

Many friends from the criminal justice system in England have provided encouragement and support throughout this project. PC Len Freeman (ret.) of the City of London Police and Inspector Anthony Moore (ret.) of the London Metropolitan Police not only provided information and materials, but were very timely with their advice and encouragement. PC Richard Watson, Sgt. John Girdlestone (ret.), PC Ian Taylor (ret.), Chief Inspector Bernard Hall (ret.), PC Allan Maddox, (ret.), PC David Grant (ret.), and PC Colin Smith (ret.) of the London Metropolitan Police have been most helpful throughout the project. In addition, PC David New, PC Jack Hodge, Sgt. Jeff Ashton, and the rest of the City of London Police staff at the Old Bailey Central Criminal Court provided needed assistance. Over the years, former Assistant Commissioner Brian Hayes of the London Metropolitan Police and former Chief Constable David Williams of the Surrey Constabulary were kind enough to lend assistance and offer advice. Magistrate Valerie West, Marian Sheraton, Mr. Lydiatt, and the staff of the Willesden Magistrates' Court, and Magistrates Quick and Henderson and the staff at the Highbury-Islington Magistrates Court provided access and valuable insight into the workings of the criminal courts. David Povey and Pat Mayhew of the Home Office were kind enough to provide feedback on some of the crime data included in this book. Nick Barnes of the Home Office provided us with much-needed information on the Prison Service Agency. In addition, the staff at Pentonville, Wormwood Scrubs, and Holloway Prisons provided information and insights into the operations of the prison service. Finally, we would like to thank PC Mick Matthews and Ch. Insp. Kevin Bowshear (ret.) who have been invaluable in helping pinpoint similarities and differences between the two countries.

In the United States, the support and help of our colleagues have been of paramount importance. Specifically, we thank Lt. Robert Miller, Eve Buzawa, Ira Hutchison, Richard Lumb, Chris Marshall, Richter Moore (deceased), David Orrick, Sam Walker, George Watson, Tracy Nobling, and BJ Reed. We are also deeply indebted to the painstaking work of Delrae Rippett, who not only read through the whole manuscript and brought many important issues and details to our attention, but also played an instrumental role in producing many of the supplementary educational materials that accompany the text. In addition, we would like to recognize Helen Bentley, Lisa Iannacci, Amy Jobe, and Chandra Mullins for the research and technical assistance they provided. Last, but not least, we appreciate all the hard work of the staff at Jones and Bartlett and in particular our editors, Jeremy Spiegel and Amanda Green, and our production editor, Amy Browning.

Of course, we owe a debt of gratitude to our families, who endured the endless number of times we were unavailable to them, and patiently anticipated the final product with us.

Contextual Factors Affecting Comparative Criminal Justice Research on the Criminal Justice Systems of England and the United States

I

An Introduction to Comparative Criminal Justice Research

Chapter Objectives

After completing this chapter, you will be able to:

- List the benefits that can be obtained from studying the criminal justice system of another country.
- Describe the advantages of focusing on the English criminal justice system.

■ Introduction

This book has been written with a twofold purpose in mind. The first is to provide detailed information on the criminal justice system of England and Wales. The second is to draw comparisons between that criminal justice system and those of the United States, with a view toward policy implications for the administration of criminal justice in the United States.

The book is aimed at a varied audience. It is, however, geared primarily for U.S. readers. It should be useful for the criminal justice professional or for the layperson who is interested in finding out how the English system of criminal justice operates. It should enable students and policymakers alike to question the philosophical underpinnings of their own systems and propose modifications that would help improve those systems. It should provide suitable material for reading on an individual basis, or as a sourcebook for use in the classroom. Finally, it should be a valuable complement for persons undertaking study abroad programs in England.

■ Rationale for Studying the Criminal Justice System of England

On noting the subject matter of this book, a number of questions may immediately be raised by a reader. First, why learn about another criminal justice system? And second, why in particular study the criminal justice system of England?

Some would argue that the acquisition of knowledge is in itself a sufficiently lofty goal. There is, however, much more than the acquisition of knowledge to be advanced in support of studying another system. Such an endeavor also enables one to develop a more complete understanding of one's own system.

Until one looks at another system there is always a tendency to take one's own system for granted, to assume that the way it operates is either the best way or the only way for it to work. Generally, examining another system promotes questions about one's own. Until one observes that another system has a different approach, one might not even consider raising an important question. For example, on discovering that England does not allow counsel to undertake detailed questioning of prospective jurors in criminal trials (the **voir dire** process), a reader might be tempted to ask whether such a modification should be advocated for the United States. Answering this question would involve an examination of the rationale for having voir dire, and a weighing of the factors militating for and against its abolition. Additionally, one would have to question whether England has lost a valuable feature of the trial process that has not been compensated for in some other way.

There are a number of rationales for choosing the English system of criminal justice as the focus of study. First, the United States' systems of criminal justice are for the most part derived from the English system. As a consequence, it is helpful to examine the roots of our current systems to understand more fully why they have been established in the manner they have. Of more than passing interest is the fact that the criminal justice systems of the United States retain some features of the old English criminal justice system that England has since modified. Thus, the states still categorize crimes as felonies and misdemeanors, and generally employ grand juries to investigate allegations of wrongdoing and/or return bills of indictment. In England the division of crimes into felonies and misdemeanors was done away with in 1967. Grand juries were abolished in 1933.

Second, it is often most instructive to examine something that is in some ways similar to and in others different from the object under investigation. In this case, because the criminal justice systems of England and the United States are for the most part derived from the same initial model, there are a vast number of similarities between the systems. However, as has already been demonstrated, there are sufficient differences between the systems to make a comparative study worthwhile. Throughout the course of this book differences between the systems will be highlighted and discussed.

Another advantage that accrues from focusing on the system of a nation with a shared heritage and a similar form of government is that cultural, economic, social, and political differences between the two countries are small enough to allow for the possibility of one country adopting a procedure that has been proven to be highly successful in the other country. History abounds with examples of procedures being transposed from one country to another without due consideration of cultural differences. Thus, for example, it could be suggested that some of the difficulties that African countries have faced are the result of European colonists imposing their systems of law and government without paying proper attention to the indigenous culture. Similarly, lessons that may be drawn from an examination of the legal systems of the former Soviet Union, China, or Japan cannot be applied to the United States without considering the major political and economic differences that have existed between the countries.

On the other hand, one must be careful not to minimize the differences. So that the English criminal justice system can be understood in its cultural, social, economic, and political context, we shall briefly examine the English cultural heritage and governmental system prior to undertaking a detailed examination of that country's criminal justice system.

This book, it should be noted, examines the unified criminal justice system of England and Wales. For the sake of brevity references to England should, unless otherwise indicated, be taken to denote both England and Wales.

It is assumed that the reader possesses some basic knowledge of the criminal justice systems of the United States. Thus, although each chapter contains a description of some aspect of crime or the criminal justice system of England, there is no similar presentation on the United States. However, during the descriptions of England, major differences with the United States are noted. In addition, the discussion sections in each chapter contain detailed references to the United States.

■ Organization of the Text

This book is divided into six parts. In the first part an examination is conducted of the broader context in which the criminal justice system operates. Chapter 2 provides a description of English society and culture, and a discussion of how differences between the societies and cultures of England and the United States might impact on a comparative analysis of the criminal justice systems of the two countries. This is followed by an overview of the English system of government. Chapter 3 contains a discussion of English criminal law, of the types of acts that constitute criminal offenses in England, and of the areas where there are major differences between the substantive criminal laws of England and the United States. In Chapter 4 the English crime situation is examined. This is vitally important to review prior to embarking upon a detailed examination of the English criminal justice system, because differences in both the nature and volume of crime committed in the two nations can impact greatly upon the operation of their respective criminal justice systems.

The second part focuses on law enforcement in England. Chapter 5 traces the development of English policing. Chapter 6 presents the current organization of the English police and discusses the recruitment and training of officers. In addition, police powers and procedures are examined. Other topics include the role of women and minorities in policing and the public image of the police.

After completing an intensive overview and comparative analysis of policing in England and the United States, the book continues by focusing Part III on the judicial and correctional systems of England. Chapter 7 examines the criminal courts and the officials who operate in them—the judges, prosecution and defense lawyers, and clerks. Of particular interest here is the division of the English legal profession into solicitors and barristers, and the transition in the 1980s from essentially police prosecutions to a U.S.-style independent crown prosecution service. Chapter 8 discusses the initial stages of the court process, as well as case resolution and sentencing. Topics covered include an examination of the different stages of the court process, pretrial release, plea bargaining, jury trial, and the sentencing options available to, and used by, English judges.

In Part IV we move on to an examination of the English correctional system. Chapter 9 examines the history of English corrections, and Chapter 10 describes both institutional and community corrections. Part V addresses two issues that were not covered in depth in the first edition of this book: England's response to terrorism (Chapter 11), and the juvenile justice system (Chapter 12). In the sixth part, an overall assessment is presented of the criminal justice systems of the two nations with particular emphasis on the issue of policies and practices that might be transferred from one system to the other.

CHAPTER SPOTLIGHT

The rationale for studying the criminal justice system of England includes:

- The United States' systems of criminal justice are for the most part derived from the English system.
- It is often instructive to examine something that is in some ways similar to and in others different from the object under investigation.

KEY TERMS

Voir dire: Detailed questioning of prospective jurors in criminal trials, which is not allowed in England.

PUTTING IT ALL TOGETHER

1. Why take the time to study the criminal justice system of another country?
2. Why study the criminal justice system of England and Wales?
3. What differences do you expect to find between the English criminal justice system and those in the United States?

The English Cultural Heritage and Government System

2

Chapter Objectives

After completing this chapter, you will be able to:

- List the components of the United Kingdom and outline the United Kingdom's history.
- Describe both the structure of England's system of government and the manner in which it operates.
- Compare and contrast social conditions in England and the United States.
- Compare and contrast economic conditions in England and the United States.
- Compare and contrast the governmental frameworks under which the criminal justice systems of England and the United States operate.

■ Introduction

Prior to embarking upon a detailed examination of the English criminal justice system, it is important to consider factors that are external to the criminal justice system, but which might nonetheless impact upon the operation of the system. Such an investigation is crucial in a cross-cultural analysis because such factors may affect conclusions drawn from the examination of the criminal justice system of one country and sought to be applied to the system of another.

In this chapter we shall examine three sets of factors. The first set deals with the English cultural heritage and involves an investigation of how cultural differences between England and the United States may affect the comparative analysis that is being undertaken in this book. The second set of factors focuses on social and economic differences between the two countries and investigates how these differences may influence comparisons that have been made and conclusions that are to be drawn about the crime problem and crime control issues in England and the United States. The third set is political in nature and focuses on differences in the political and governmental frameworks under which the criminal justice systems of the two countries operate. Before considering these issues, however, we shall examine some basic facts about England and England's history, and its form of government.

■ England as Part of the United Kingdom

The United Kingdom of Great Britain and Northern Ireland consists of England, Wales, Scotland, and Northern Ireland. The United Kingdom covers some 242,910 square kilometers (93,788 square miles), which makes it slightly smaller than the state of Oregon. The population on census day 2001 was 58,836,700, with 49,181,300 (82%) living in England and 2,903,200 (5%) in Wales. The overall population density was 244.2 per square kilometer, with the population density for England registered at 377.1 persons per square kilometer and for Wales at 139.7 persons per square kilometer (*Europa World Handbook*, 2004, p. 4262). This compares with a U.S. population of 284,796,887 and a population density of 29.0 per square kilometer (*Europa World Handbook*, p. 4412). Between 1990 and 2002, the average annual population growth for the United Kingdom was 0.26%, whereas for the U.S. it was 0.98% (Office for Economic Co-operation and Development [OECD], 2005, p. 12).

Unlike the United States, which boasts a "melting pot" of different ethnic heritages, the United Kingdom has a more homogeneous population. Ethnic minorities (non-Whites) account for only 7.9% of the population, with Indians, Pakistanis, and Bangladeshis comprising 44.9% of that minority population (National Statistics Online, 2003a). However, 29% of the inhabitants of London are from minority ethnic groups (Office for National Statistics, 2005, p. 3), and these groups have a far higher proportion of persons under the age of 16 than do Whites (55% versus 19%; Office for National Statistics, 2003, p. 32). In the United States, 25% of the population is classified as non-White (U.S. Bureau of the Census, 2000). Finally, only 8% of the U.K. population was born overseas compared to 12% of the U.S. population (Office for National Statistics, 2005, p. 3).

Historical Background

England possesses a lengthy historical heritage. Vestiges of the past can be observed throughout England today. There is Stonehenge, the configuration of large blocks of stone erected some four thousand years ago. There are numerous Roman ruins including Hadrian's Wall, built around A.D. 122 to keep the marauding Celts and Picts (Scottish tribes) on their side of the border. In the basement of the famous London courthouse, the Old Bailey, one can still see a section of an old Roman wall. There are also cathedrals, inns, and houses that have survived for centuries. What is called old in the United States may be considered to be relatively new in England. Although it is obviously outside the purview of this book to present any kind of detailed exposition of the history of England, a few of the more salient facts merit attention.

For the most part, the English are descended from the Anglo-Saxons who invaded the country in the fifth and sixth centuries and the Danes who came over in the eighth and ninth centuries. For a large part of this period the country was divided into different kingdoms. Even when the country was unified, the central government was weak. This remained so until the Norman invasion of 1066 when William I of Normandy conquered and firmly united the whole country. This was, as the English will proudly remind you, the last time that the country was successfully invaded. Wales, Scotland, and Ireland were not united with England until later.

Wales was not fully subjugated by the Normans and continued to fight with England until 1282, when Edward I finally conquered the country. Wales did not, however, become fully integrated with England until the Act of Union of 1536, which made English law applicable to Wales and allowed Welsh representatives to be admitted to the English Parliament.

Union with Scotland came a short time later. Again, the countries were combined under a common monarch, but initially remained politically separate. On the death of Elizabeth I of England in 1603, James VI of Scotland became James I of England. Apart from their shared monarch, the two countries remained separate, both retaining their own parliaments and systems of justice. In 1707 an Act of Union provided that the two countries should become one by the name of Great Britain, and that there should be only one parliament. The two systems of law were not, however, merged. Even today Scotland retains its own separate and distinct judicial system.

As a result of promises made in the 1997 general election, the Labour Party held referenda in Wales and Scotland on the issue of whether elected national assemblies with limited powers should be established in those two countries. A majority of the electorate in both countries voted for regional assemblies, and such essentially regional assemblies have been established in both Scotland and Wales.

For some centuries the English made incursions into Ireland, alternately waging war and settling there. From the twelfth century the English inhabited a small strip of land known as the "Pale" located around the city of Dublin. In the seventeenth century Oliver Cromwell conquered the whole of Ireland. An Irish parliament did, however, continue to exist. In 1801 Ireland was formally unified with England, and the United Kingdom of Great Britain and Ireland came into existence. Home rule for Ireland, however, remained a volatile issue. In 1921, in a solution that was and still is considered unsatisfactory to many, the country was split into two. The northeastern six counties, with their largely Protestant populations, retained their ties with England. Although ultimately subject to the authority of the Parliament in England, the country was to have a certain degree of autonomy, maintaining its own parliament (later suspended and replaced by a consultative assembly) and its own judicial system. The rest of the island, which was mainly Catholic in composition, became the Irish Free State. However, in 1937 a new constitution was approved that abolished the Irish Free State and established Eire (Gaelic for Ireland) as a new independent nation.

The future of Ireland remains uncertain. Northern Ireland is still part of the United Kingdom, sends representatives to the Parliament in London, and, because of the violence that has occurred, was subject to direct rule from London almost continuously from 1972 to 1999. In the 1990s a series of interrupted peace talks eventually led to the Good Friday settlement of April 10, 1998, and the subsequent Northern Ireland Act of 1998, which provided for local rule and the election of a 108-member assembly with safeguards for minority rights. Elections to the new assembly were held in June 1998, and in December 1999 England transferred authority to the newly elected Northern Ireland government, composed of both Protestants and Catholics. Power sharing by the Protestant and Catholic factions proved difficult, and since the new assembly was established local rule was suspended on a number of occasions, primarily because of the Irish Republican Army's reluctance to disarm. In September 2005 the Irish Republican Army finally did disarm and, after a new agreement was reached in October 2006, a new assembly was elected. On May 8, 2007 this new assembly met and two former bitter enemies, Ian Paisley from the Protestant side, and Martin McGuinness from the Catholic side, were elected First Minister and Deputy First Minister. The establishment of an apparently good working relationship between these two augurs well for the future.

During the sixteenth and seventeenth centuries England developed into a major colonial power, establishing colonies in Africa, Asia, and the Americas. At the height of its power in the late nineteenth century, the British Empire "embrace(d) some eleven million square miles (almost one-fifth of the global land mass) and 372 million people (a quarter of the world's population)" (Haigh, 1985, p. 268). Included in the empire were such countries as present-day Canada, South Africa, Zimbabwe, Nigeria, India, Pakistan, New Zealand, and Australia. Between 1945 and 1985 more than 30 British colonies obtained their independence. Many, however, retain their ties with the mother country as members of the British Commonwealth of Nations.

Government

The form of government that exists in the United Kingdom is a "constitutional monarchy, under democratic parliamentary regime" (Banks, 1987, p. 609). The titular head of state is the **monarch** (king or queen), who inherits the position. The monarch is generally referred to in formal terms as the **Crown**. The government is called **Her (or His) Majesty's Government**, and all government functions are carried out in the name of the Crown. Although the actual power wielded by the monarch has waned considerably over the centuries, the monarch still performs a host of important duties. The sovereign summons and dismisses Parliament, calls upon the leader of the party that has won a general election to form a new government, and appoints all the important state officials, including judges, officers of the armed forces, and representatives to other countries. The sovereign must assent to any bill passed by both houses of Parliament before it becomes law and consent to a declaration of war or the signing of a peace treaty. All of these functions are exercised on the advice of ministers, who have for the most part been elected to Parliament.

The legislative arm of the state is composed of the Crown and the two houses of Parliament: the House of Commons and the House of Lords.

The **House of Commons** is a representative body with each of its 646 (as of 2007) members elected by the voters of a particular district. The members of

Parliament (MPs) are elected either during a general election, when all 646 seats are contested, or during a by-election, when a special election is held for a particular district that has lost its MP through death or resignation. Unlike the situation in the United States, general elections are not held at regular intervals. It is up to the leader of the party in power, subject to the Crown's formal dissolution of the sitting Parliament, to decide when to hold a general election. The only requirement is that a general election must be held within five years of the previous one (Parliament Act 1911, Section 7). Under this framework the government is entitled to call for an election at the time that is most favorable to it remaining in power. The corollary is that the opposition can force a general election by winning a vote of "no confidence" in the government.

Unlike the situation in the United States, there are no primaries before a general election. Indeed, once a general election has been called, there is a period of only three weeks between the dissolution of Parliament and the election. The parties that are not in power always have an official leader and members designated to serve as "shadow" government officials. The main opposition party always has a complete "shadow Cabinet" with party officials holding posts mirroring those held by the government ministers so that should the opposition party gain power, no time would be lost in placing members of that party in the official governmental positions.

The House of Lords, on the other hand, has been a more elite body composed of a mixture of members who inherited the right to sit there and members who were appointed by the Crown. The former category consisted of the thousand or more hereditary peers (dukes, earls, barons, and the like) who still exist in England. The latter category was composed of the 2 archbishops and 24 senior bishops of the Church of England, the life peers who were appointed by the Crown generally in recognition of honorable service to the country, and the **Lords of Appeal in Ordinary**, who were appointed for life by the Crown to perform the judicial duties of the House of Lords.

Although legislation may be introduced in either house, it is generally introduced in the House of Commons. It is required that each bill be read three times in the House of Commons before it is sent to the House of Lords for approval. The House of Lords can send a bill back to the House of Commons with suggested amendments, but the House of Lords has the power only to delay, not to reject, bills passed by the House of Commons.

Since the late 1990s there has been a move to modernize the House of Lords and make it more of a representative body. The House of Lords Act of 1999 took away the automatic right of 660 hereditary peers to sit and vote in that chamber and established a transitional house composed of spiritual and life peers and 92 hereditary peers, most of whom were elected by their own political party. During the 2005 general election the Labour Party promised to remove the remaining 92 hereditary peers from the House of Lords and to introduce additional procedural reforms. The final composition and mode of selection of members of the newly reconstituted House of Lords continues to be the subject of discussion and negotiation. Of concern are such issues as whether a second chamber with mostly elected members would end up challenging the pre-eminence of the House of Commons and whether it is desirable to require representation of non-Christian faiths in the newly constituted chamber.

The **Prime Minister** and the other ministers in the government exercise executive power on behalf of the Crown. This group of ministers is known as the **Cabinet**. Each Cabinet Minister has a certain sphere of responsibility. Thus, there is a minister responsible for financial matters (the **Chancellor of the Exchequer**), another responsible for foreign affairs (the **Foreign Secretary**), and others responsible for such matters as transport, health, and social services. Of prime concern to us in our examination of the administration of justice is the **Home Secretary**, who has the overall responsibility for internal affairs, in particular, the maintenance of law and order. The Prime Minister and the other ministers, who are generally members of the House of Commons (although occasionally members of the House of Lords) are answerable to Parliament and are often called to task for their actions and those of the officials they oversee.

Religion

Unlike the situation in the United States, there is no strict division in England between church and state because the **Church of England** is the established church of the land. The Church of England was established in 1534 when King Henry VIII broke with the Roman Catholic Church. From that time the temporal head of the country, the **sovereign** (also called "Defender of the Faith"), has had the power to appoint archbishops, bishops, and other church officials. The lack of separation of church and state means that there is no national debate over prayer in school and other similar issues.

With greater heterogeneity developing in the composition of the population, the numbers of inhabitants practicing minority religions have increased substantially. It is estimated, for example, that although 71.7% of the population describe themselves as "Christians" and 15% indicate they have no religion, there are currently 1.5 million Muslims, 500,000 Hindus, and 500,000 Sikhs in England (National Statistics Online, 2003b). These religious minorities can exert an impact on local and national matters. With regard to international events, fundamentalist Islamic groups in England have, for example, been the focus of attention following the destruction of the World Trade Center in New York City and the July 2005 London bombings. The infamous "shoe bomber," Richard Reid, is a Londoner who prayed at a mosque in South London.

■ The English Cultural Heritage

England possesses a rich cultural heritage. Over the centuries England has provided the world with literary figures such as Chaucer, Shakespeare, and Dickens; philosophers such as Hobbes and Locke; painters such as Turner and Hogarth; composers such as Elgar and Britten; and architects such as Sir Christopher Wren. England's fine blend of history and culture has led many Americans to admire England in the same fashion that the Romans looked at the Greeks in ancient times.

The English heritage has given rise to a quiet sense of national pride, a feeling that "there'll always be an England." England is the only major European country that succeeded in going through political and economic reforms in the eighteenth and nineteenth centuries without a revolution. In the twentieth century it successfully fought in two world wars. The English do indeed have a long history of coping.

Some critics would argue that the English have a laissez-faire approach that is deficient in that it leaves too much to chance and does not allow for the advantages that accrue from proactive policymaking. This approach was evident in British Prime Minister Neville Chamberlain's dealings with Hitler and was, some would suggest, largely responsible for the economic and worldwide political decline that beset England after the end of World War II in 1945. Whatever the merits of these arguments, it must be noted that a laissez-faire orientation manifests itself in some important ways that impact directly upon the English approach to crime control. Unlike the United States, England has thus far deemed it unnecessary to codify its subjects' rights in a written constitution. Nor, until recently, has it sought to promote the doctrine of separation of powers. These are issues that will be explored in greater detail later in this chapter.

Along with this laissez-faire attitude, the English are said to have developed a greater tolerance than Americans for deviance. The eccentric Englishman is an image depicted in art and the media, both in England and abroad. English television programs routinely contain more bad language and nudity than would be tolerated in the United States, and one of the English national newspapers has been known for its page 3 picture of a semi-nude woman. In a more serious vein, this tolerance of deviance has led the English to take a more permissive legislative approach than the United States to many public order activities such as prostitution, homosexuality, and gambling, topics that will be discussed in some detail in Chapter 3.

In addition to possessing a greater tolerance for deviance, the English are alleged to be less likely than Americans to use violence in their interactions with others. A pervasive belief exists that the United States is one of the more violent, if not the most violent, of the western industrialized nations. For example, there were 886 criminal homicides reported to the police in England and Wales between April 1, 2001, and March 31, 2002, for a rate of 1.7 criminal homicides per 100,000 inhabitants (Home Office, 2002, Table 3.04). In contrast, there were 15,980 such offenses reported to the police in the United States, for a rate of 5.6 per 100,000 inhabitants (U.S. Department of Justice, 2002, p. 19).

The violence apparently inherent in U.S. society is said to be a legacy derived from the hardy immigrants who came over or were brought over to the United States and an outgrowth of the days of the Wild West. In addition, the Civil War resulted in a significant increase in the use of guns by the citizenry (Monkkonen, 1989). The consequences of using violence to settle disputes or achieve other ends are magnified by the continued high rate of gun ownership in the United States. American urban dwellers are far more likely than their English counterparts to own and use firearms (e.g., Kopel, 1990). Although conventional wisdom has estimated that there are firearms in approximately 50% of U.S. households (see, e.g., Kopel, 1992, p. 3), studies conducted in the 1990s have put the figure in the region of 35% to 43% (Cook & Ludwig, 1997). This is, however, still markedly higher than the figure of 4% given for English households (Alpers, 2001). In addition, whereas U.S. law enforcement officers are armed, English police officers do not receive a firearm as part of their standard equipment. Detailed examination of the comparative rates of violence in the two countries, and of the possession and use of firearms, will be undertaken in Chapter 4.

Intertwined with its historical and cultural heritage, England's physical attributes have impacted upon the nature of its crime problem and its crime control policy. As a consequence of its smaller land area, one might expect crime response and crime control coordination issues to be much easier in England than in the United States. Likewise its insular nature has provided a degree of protection against foreign drug traffickers as well as alien armies. This has, however, changed to some extent as a result of completion of the English Channel tunnel between England and France and the greater unification of Europe. As a member of the **European Union (EU)**, a political and economic union consisting of 27 sovereign states, the United Kingdom allows its nationals to move freely to and from member countries and for their nationals to move freely in and out of the United Kingdom.

Particularly noteworthy in terms of crime has been the increase in the human trafficking of illegal aliens seeking to live in the United Kingdom (see, e.g., Sapsted, 2002), and of women who are brought in for the purposes of prostitution (see, e.g., Anti-Slavery, 2005; Arie, 2003; Kelly & Regan, 2000; Taylor, 2005). The men and women trafficked into the United Kingdom for forced labor or prostitution are primarily from Eastern Europe, the Balkans, and East Asia, especially China and Thailand. There is also a problem with trafficking in children, particularly from West Africa. These children often end up as domestic slaves (Home Office, 2006a; U.S. State Department, 2006, p. 253). Measures are being taken to identify potential victims of trafficking at all points of entry into the country, and limited shelter space is being provided for identified victims (Home Office, 2006a, 2006b; U.S. State Department, 2006, p. 253). The law enforcement response has been strengthened by the establishment of the Serious Organized Crime Agency, a national agency entrusted with the task of disrupting organized crime (Serious Organized Crime and Police Act 2005, Sections 1–2).

■ Social and Economic Issues

The nature and composition of the population of England might also be expected to influence the crime situation. The high population density that exists on the island, and in particular in England itself, suggests that the country might be faced with all of the problems that beset urbanized areas, including crime. The greater homogeneity that exists in the English population might, on the other hand, suggest potentially less internal friction than exists in the United States. This situation has, however, been affected by two divergent trends.

First, immigration from the countries of the British Commonwealth, and then later from the European Union, resulted in immigration into the United Kingdom doubling between 1974–1978 and 1984–1988 (Central Statistical Office, 1990, p. 30). Since then the ability of nationals to move freely within European Union member countries has brought about far greater mixing among the different national and ethnic groups, with some nationals even commuting from their home country to work in another country. The U.K. census of 2001 reported that there were 856,200 European Union nationals living in the United Kingdom in 2000 (Office for National Statistics, 2004, p. 22).

Second, there has been a rise of populist nationalist parties in the different sectors of the United Kingdom (in Wales and Scotland as well as Northern Ireland). This resulted, as noted earlier, in some devolution of power from the central government. In addition, the stepped up security measures adopted as a result of the terrorism campaigns waged in England by the provisional wing of the Irish Republican Army provide testimony to the profound changes that can be wrought by the activities of a small faction.

Class Divisions

The United States prides itself on being a classless society. The same claim cannot, however, be advanced with regard to England. The monarch and the rest of the royal family are, in the eyes of most, very different from the common people. In no way do they share the common lot. In addition, the aristocracy has long been part of the English landscape. Even today, some members of this elite class retain the right to participate in the government of the country by sitting in the House of Lords. However, as has been noted earlier, the power of the House of Lords has waned considerably over the centuries and reforms have recently been instituted that may lead to the total abolition of the natural right of any of these designated individuals to sit in that legislative assembly.

Although it is generally accepted that class divisions softened in the twentieth century, it may be suggested that the continued presence of a royal family and an aristocracy presents a vivid picture of how all persons are not in fact born equal and do not enjoy equal opportunity for advancement in society. The impact that the existence of different classes has on the commission of crime is open to debate. It could be argued that such class divisions are helpful in that they provide members of the classes with differential expectations that are in tune with the likely outcome of their lives. This is in contrast to an ostensibly open and fluid society, such as exists in the United States, which may fuel expectations that are not met and as a result generate a higher level of criminal activity. On the other hand, others may argue that the presence of highly visible class divisions accentuates the disparity between the classes and promotes class conflict, including the commission of crime.

Aid for the Underprivileged

Although class divisions do still exist in England, England has developed certain features of socialism that are designed to aid all citizens, and in particular the underprivileged. The National Health Service provides comprehensive health care, generally free of charge, to all those normally resident in the United Kingdom. Emergency medical care is readily accessible to all. Rates of infant mortality are consistently lower in the United Kingdom than in the United States (OECD, 2005, pp. 194–197). Filing for bankruptcy as a result of incurring insurmountable medical bills is unknown in the United Kingdom. In addition, there is extensive public housing and there are widely available unemployment benefits, maternity benefits, child-care benefits, and state pensions.

Many would indeed agree with the proposition that England has provided a far broader and more comprehensive safety net for its underprivileged than has the United States. This assumption is, in fact, supported by available empirical evidence. The Organization for Economic Co-operation and Development (OECD) constructs an indicator of social expenditure. This indicator provides a measure of "the extent to which governments assume responsibility for supporting the standard of living of disadvantaged or vulnerable groups" (OECD, 2005, p. 166). In 2001 the United Kingdom allotted 21.82% of its gross domestic product (GDP) to social expenditures, the United States only 14.78% (OECD, 2005, p. 167). As the data in the report attest, 2001 was a typical year.

The interrelationship between the provision of government support and the commission of crime remains unclear. It might be hypothesized that greater government support of the underprivileged might lead to a lower level of crime committed out of economic desperation. On the other hand, the concept of relative deprivation must still be taken into consideration. A person who by objective standards is considered to be well above the poverty line might still feel deprived when comparing her- or himself to those who are considerably better off, and as a consequence might resort to crime to redress the observed imbalance.

Education

Although England does have a state nursery, elementary, and secondary school system, a small minority of the country's children is educated in independent institutions. Since the 1970s about 7% of the United Kingdom's schoolchildren have received their education in independent (private) schools (Office for National Statistics, 2005, p. 32). This compares to the estimated 10% of U.S. schoolchildren who attend such schools (United States Information Agency, 2003, Chapter 6). Most noticeable in the United Kingdom's education data is the dramatic increase in the past 30 years in the percentage of children under five in nursery schools. While 21% of these children were in school in 1970–1971, 65% were in school in 2003–2004 (Office for National Statistics, 2005, p. 32).

The school leaving age in England is 16. The last 30 years have seen a large increase in the percentage of English school-leavers seeking further education. Although the gap between the percentage of English and U.S. school-leavers seeking further education has narrowed, it still remains substantial. While 16.3% of U.K. school-leavers went into tertiary education in 1991 and 26.1% did in 2001, the comparative figures for the United States were 30.1% and 38.1%, respectively (OECD, 2005, p. 153).

At one time all students accepted by institutions of higher learning in England were eligible for some level of government support, which for the most part was based on the income of the student's parents. This support covered both the cost of tuition and living expenses. The provision of maintenance grants to cover living expenses was, however, discontinued in 1998. The current student financial support system includes a mixture of loans and income-based grants that cover both tuition and living expenses. In 2001–2002, 50% of full-time students did not have to pay any tuition fees and only 35% had to pay the full amount (Higher Education Statistics Agency, 2004, p. 7). Although good comparative information is not readily available, an examination of the data on the cost of higher education in the United States and the provision of financial assistance to students (see, e.g., U.S. Bureau of the Census, 2007a) suggests that low cost further education may be more extensively available in England than in the United States.

Business and Industry

Another feature of a socialist government evident to some extent in England concerns state control of the means of production. Under the leadership of various Labour Party governments, elements of a wide range of industries were nationalized. The government took over responsibility for running parts of the banking, electricity, steel, telecommunications, transport, motor vehicle, shipbuilding, atomic energy, and aircraft industries, among others.

Work conditions and workers' benefits, in both the public and private sectors, have been more regulated and more generous in the United Kingdom than in the United States. U.K. government regulations provide, for example, for paid annual vacation of 4 weeks and for a general limit of 48 hours work a week, including overtime (Department of Trade and Industry, 2003a, pp. 4, 20). As a consequence, the average English worker spends less time at work. In 2001, for example, the average English worker spent about 2 weeks less at work than the average U.S. worker (OECD, 2005, p. 201). Likewise, sick pay provisions and maternity leave are more generous in the United Kingdom than in the United States (Department of Trade and Industry, 2003b; Department for Work and Pensions, 2005).

The Conservative government, however, during the years it was under Margaret Thatcher's, and then John Major's, leadership, engaged in a massive program of privatization of nationalized industries, selling part or all of the government's share in such industries as shipbuilding, civil aviation, railways, and telecommunications. This trend was not reversed by the return of the Labour Party to power in 1997.

In addition to possessing a number of nationalized industries, England has had more representative and far more active trade unions than the United States. Although the percentage of the work force that has membership in a trade union has been declining—29.3% in 2001 (Department of Trade and Industry, 2004, p. 2), down from 34.5% in 1990 (*Whitaker's Almanac*, 1993, p. 688), 43.5 % in 1985, and 55.4% in 1979 (*Europa Year Book*, 1990, p. 2712)—this compares with a figure of 13.5% for union membership in the United States in 2001 (U.S. Bureau of the Census, 2001, p. 412).

Furthermore, English government workers, such as postal employees, are not forbidden by law from undertaking job action, although under the Emergency Powers Acts of 1920 and 1964 the Crown may declare a state of emergency if such worker action results in the deprivation of the essentials of life. Strikes and other forms of job action such as work stoppages and working to rule are, indeed, more prevalent in England than in the United States, though differences in reporting procedures account for some of the differences observed. Thus, between 1985 and 1987 there were 1,016 strikes and lockouts with 16,023 days lost per 100,000 employees in the United Kingdom, as compared to 46 strikes and lockouts with 4,034 days lost per 100,000 employees in the United States (*The Economist*, 1990, p. 200). In 2001 there were 194 strikes and lockouts in the United Kingdom and only 29 in the United States (International Labor Organization, 2005).

As a consequence, industrial action has, in recent years, posed more of a problem for English law enforcement than it has for law enforcement in the United States. The miners' strike of 1984–1985 and the subsequent docklands strike are just two examples of the kind of industrial strife that the English police have had to face. Since the 1980s police officers from one area of the country have, with increasing frequency, been moved to trouble spots elsewhere to help the local police force deal with these problems.

General Economic Conditions

The relationship between these economically related labor issues and the commission of crime is indirect at best. Another economic issue that exerts a more direct effect upon the crime problem involves the general economic conditions that prevail in the United Kingdom and the general supposition that the overall standard of living as measured in material terms is lower in the United Kingdom than in the United States. This supposition is supported by available data that measure the economic conditions of the two nations. Thus, for example, whereas the per capita gross domestic product (GDP) of the United Kingdom in 2001 was $26,627, in the United States it was $35,179 per person (OECD, 2005, p. 24). More vividly, the purchasing power parity ratio of those living in the United Kingdom has been calculated since 1970 consistently to be only about two thirds that of those living in the United States (*The Economist*, 1990; OECD, 2005, p. 85; U.S. Bureau of the Census, 1994, p. 864).

Unemployment rates have also been somewhat higher in the United Kingdom than in the United States. From 1980 through 2000 the standardized unemployment rate was generally 1% to 4% higher in the United Kingdom (OECD, 2005, p. 109; U.S. Bureau of the Census, 1992, p. 838). In 2001, however, the gap closed with the United Kingdom posting an unemployment rate of 5.0% and the United States a rate of 4.7%. In 2002 the unemployment rate of 5.8% in the United States exceeded the United Kingdom's unemployment rate of 5.1% (OECD, 2005, p. 109) and has remained higher, at least through 2004 (U.S. Bureau of the Census, 2007b, p. 852). The United Kingdom, it should be noted, consistently reports a far higher percentage of the unemployed as long-term unemployed (unemployed for 12 months or more). In 2001, for example, whereas 27.8% of the United Kingdom's unemployed were reported to be long-term unemployed, the comparative figure for the United States was only 6.1% (OECD, 2005, p. 113). However, sick pay and unemployment benefits are more generous in the United Kingdom than in the United States. Whereas, for example, unemployment benefits are available for 26 weeks in the United States, comparative benefits in the United Kingdom are available for an indefinite time.

In terms of ownership of material goods, the inhabitants of the United States in general appear to be better off than those of the United Kingdom. Data for 2001 reveal that there were 370.8 automobiles and 11.5 motorcycles per 1,000 population in the United Kingdom, whereas in the United States those figures were 474.8 and 15.3, respectively (U.S. Bureau of the Census, 2003, p. 869). Similarly, whereas in the United Kingdom in 2001 there were 588 telephone main lines and 366 personal computers per 1,000 inhabitants, the comparative figures for the United States were 667 and 625 per 1,000 (U.S. Bureau of the Census, 2003, p. 870). Differences in lifestyle account for much of these observed differences. For example, the greater availability of public transport has made private car ownership more of a luxury than a necessity for the inhabitants of the United Kingdom.

Likewise, with 770 cell-phone subscribers per 1,000 population, as compared to 451 per 1,000 in the United States (U.S. Bureau of the Census, 2003, p. 870), the inhabitants of the United Kingdom have migrated from using land phones to a greater extent than their counterparts in the United States.

These figures do give a general indication of a lower material standard of living in the United Kingdom. They also attest to the fact that there are smaller numbers of some types of material goods subject to theft in the United Kingdom. Whether the lesser availability of these goods makes them more attractive to potential thieves is a matter for empirical testing.

Although the standard of material living may be lower in the United Kingdom, it would be a mistake to conclude that the overall quality of life is not as good as in the United States. It is a question of values. The average inhabitant of the United Kingdom may be materially less well off than her or his counterpart in the United States; however, the average inhabitant of the United Kingdom does spend less time at work. In 2001, for example, the average U.K. worker spent about 2 weeks less at work than the average U.S. worker (OECD, 2005, p. 201). This presumably leaves the average inhabitant of the United Kingdom more time for family and friends and for leisure activities. In addition, the government in the United Kingdom has undertaken to provide a wider safety net for the underprivileged.

■ Governmental Framework

The governmental framework within which the English criminal justice system operates is different from that of the United States in a number of important aspects. These include the existence of a monarchy in England; the fact that England possesses a **unitary governmental system** with no separate state and federal systems; the English doctrine of **Supremacy of Parliament**, which basically asserts that Parliament may pass any law that it likes and that the courts cannot strike down Acts of Parliament; the failure of the English to attempt to emulate the much hallowed U.S. doctrine of separation of powers; and the fact that England has no formal written constitution.

As previously mentioned, the monarch is the formal head of state and as such, at least in theory, wields considerable power. That power has, however, waned considerably over the centuries and has passed through such fundamental documents as the Magna Carta (1215), the Bill of Rights (1689), and the Act of Settlement (1701) to the people, to be exercised on their behalf by Parliament. Although the sovereign can in theory refuse to assent to a bill that has been passed by the House of Commons and the House of Lords, no sovereign has exercised this right since Queen Anne in 1707. Through lack of use this right has in practice been abrogated, and should a present day monarch attempt to exercise it, such action would in all probability promote a national crisis.

A monarch with highly limited powers might appear to some to constitute a financial burden that provides very limited benefit for the state. The amount budgeted for the Queen and her household runs around £9 million a year (Young, 2002, p. 11). This compares with the salary of $400,000 and expense allowance of $179,000 that the President of the United States receives (*World Almanac*, 2003, p. 48).

A modern monarch of this type does, however, fulfill certain very useful functions. The separation between the head of state and the head of government is a division that can be highly beneficial in times of governmental crisis, with the sovereign acting in the role of mediator between two governmental factions.

The existence of a separate head of state also facilitates discussion of governmental policy. Criticism of the head of the government and his or her policies is not taken as an attack upon the state manifesting a lack of patriotism. On the other hand, it is expected that the monarch and the rest of the royal family will not air political views in public. Prince Charles has been criticized for his comments on modern English architecture and the plight of the poor. In November 1988, the Queen herself caused some concern by apparently suggesting that the success of the Scottish National Party in a by-election in Scotland boded ill for the future of the United Kingdom (Greig, 1988, p. 1).

Other issues that must be stressed relate to the utility of the sovereign as a symbol of unity for the country and as a person to preside over important ceremonies. These are important functions served by the sovereign. On the financial side the existence of the monarchy and the attendant royal paraphernalia contribute to tourism. However, although the sovereign may fulfill a number of useful functions, the role of the monarchy in a modern democratic society has been questioned and it has been suggested that the monarchy be abolished.

The fact that England has a unitary system of government with no separate state and federal systems means that England can have a single national policy on any given issue and can with relative ease mount coordinated nationwide efforts to combat any problem. As a consequence, England has a single set of laws and government regulations. Thus, unlike in the United States with federal laws and regulations and 50 separate and distinct sets of state laws and regulations, it is possible in England to state definitively what the law of the land is on a particular issue, and what general regulations govern the operation of a particular type of agency. There is also, in contrast to the United States, a single agency responsible for carrying out a particular function. Thus, there is a single prison system responsible for running the equivalents of the U.S. federal and state prisons and locally administered jails. Whereas the United States entrusts general law enforcement duties to thousands of state and local agencies with largely uncoordinated systems of administration and modes of operation, and often overlapping and conflicting jurisdictions, England has 43 agencies within a single unified system. The problems inherent in a multi-layered system as exists in the United States were highlighted in 2005 by the highly criticized planning for, and response to, the Katrina hurricane.

In the United States great effort has been extended to establish and preserve a balance of power among the three branches of government—the executive, the legislative, and the judicial. Over the years the doctrine of separation of powers has become firmly ingrained in the U.S. psyche, and a delicate system of checks and balances has been established. In England, on the other hand, the doctrine of the Supremacy of Parliament has been espoused, and there has been little attempt over the centuries to establish formal divisions among the three branches of government.

The doctrine of the Supremacy of Parliament contains two aspects. First, no one but Parliament may enact legislation. Second, Parliament may pass any law it desires. If it appears that a statute has been passed in the correct manner, the courts must submit and apply the statute. As Keir and Lawson state: "The principle of parliamentary sovereignty therefore implies that the courts will not intrude into the legislative process, and that an Act of Parliament validly passed under the appropriate procedure and in the accustomed form must be put into effect" (1979, p. 15). Of course, the courts still have to interpret legislation, a function that reserves some power for the courts. Parliament must also be sensitive to the political considerations that in actuality limit its powers, because the members of the most powerful arm of Parliament, the House of Commons, are answerable to the electorate.

Despite these limitations, the doctrine of the Supremacy of Parliament has tremendous practical significance. The case of *Burmah Oil Company v. Lord Advocate* (1965, A.C. 75) is indicative of this doctrine at work. During World War II, in pursuance of a governmental policy of denying resources to the enemy, the British High Command in Burma had ordered the destruction of installations belonging to the Burmah Oil Company. These installations were in fact destroyed the day before the Japanese occupied them. After the war the Burmah Oil Company, which was registered in Scotland, took action at law against the government, claiming damages for the losses suffered. A threshold question to be decided was whether the company could in fact claim compensation. The case reached the House of Lords, the highest court in the land, where it was held that the claimant did have a right to claim compensation. Parliament promptly responded with the War Damage Act of 1965, which stated that:

> *Any right which the subject may have had at common law to compensation from the Crown in respect of lawful acts of damage to, or destruction of, property done by, or under the authority of, the Crown during, or in contemplation of the outbreak of, a war in which the Sovereign was or is engaged, is abolished retrospectively.*

In similar fashion, the Wireless Telegraph Act of 1954 retroactively validated the Post Office practice of charging for licenses, even though the Post Office had recognized that this practice was unauthorized when, in an out-of-court settlement, it had repaid the plaintiff's license and costs in the case of *Davey Paxman & Co. v. Post Office* (*The Times*, 1954). It is unlikely that such legislative interference with the judicial process would be tolerated in the United States.

Along with the doctrine of the Supremacy of Parliament, England has failed to establish a formal separation between the different branches of government. As already indicated, the executive officers (the Prime Minister and the other members of the Cabinet) are also members of the legislature. The possession of what may appear to an American observer to constitute highly conflicting roles has not been particularly problematic to the English. Indeed, they might well suggest that having members of the executive answerable to the legislature is highly beneficial.

There has also traditionally been a lack of separation between the judiciary and the legislature. Indeed, the members of the highest court in the land (the Law Lords of the House of Lords) have also been members of the legislature. The position of Lord Chancellor has been the most anomalous of all. The **Lord Chancellor** has acted in the past as: (1) head of the judiciary, (2) speaker of the House of Lords, and (3) a member of the Cabinet. The Constitutional Reform Act of 2005, however, radically changed this situation. This Act provided for a new Supreme Court of the United Kingdom to take over the judicial role formerly exercised by the Law Lords of the House of Lords with the Lord Chancellor's judicial functions carried out by the President of the new Supreme Court. The Lord Chancellor was stripped of his judicial role as head of the judiciary in April 2006, with that role transferred to the Lord Chief Justice, who will become President of the new Court (Falconer, 2006). At present, it is envisaged that the new Supreme Court will become operational in 2009 (Department of Constitutional Affairs, 2006).

The lack of a formal written constitution has helped Parliament maintain its supremacy, because it means that there is no document that promulgates standards by which courts measure legislative acts. The English have no equivalent of the U.S. Bill of Rights. Such fundamental documents as do exist—such as the Magna Carta (1215), the Bill of Rights (1689), and the Act of Settlement (1701)—do more to delineate and limit the sovereign's powers than to assert citizens' individual rights. Moreover, because of the doctrine of the Supremacy of Parliament, there has been no legal limit to the extent to which Parliament can abridge or abolish rights that in other countries may be regarded as "inalienable." It has been said that the rights of the individual under English law consist of the residue of freedom left after legislative and executive powers have been delineated.

By signing international treaties and agreements England has, however, acknowledged the right of international courts to entertain cases that involve England as a party and that may challenge English laws on the grounds that they infringe upon fundamental rights. Thus, as a member of the United Nations, England submits to the jurisdiction of the World Court in The Hague and, as a member of the European Community, accepts the authority of the European Court of Justice in Luxembourg. As a signatory of the 1950 European Convention for the Protection of Human Rights and Fundamental Freedoms, England has recognized certain fundamental human rights and agreed to allow citizens who claim infringement of those rights to bring cases, via the European Commission of Human Rights, to the European Court of Human Rights in Strasbourg, France. The result has been that cases involving claims of fundamental rights have been decided for the English by a foreign court in France. Such an outcome would not occur in the United States, where the U.S. Supreme Court decides whether a citizen should possess a particular right, such as the right to an abortion (*Roe v. Wade*, 1973) or the right to engage in sexual relations with someone of the same sex (*Bowers v. Hardwick*, 1986; *Lawrence v. Texas*, 2003).

Although the European Commission of Human Rights did little in its early years, in the 1990s it received some 800 cases a year, about a sixth of which came from the United Kingdom (see, e.g., Council of Europe, 1993, p. 29). These cases involved a variety of issues, including complaints about restrictions on the press and inhumane treatment of prisoners. Between 1959 and 1990 the court decided 188 cases, in 130 of which signatories to the treaty were found to have violated provisions of the convention. The United Kingdom was the offending party in 27 of these cases (Berger, 1992, p. 265). In November 1988, for example, in the case of *Brogan and others* the court ruled that police powers to hold suspected terrorists in Northern Ireland for up to 7 days before a court appearance violated the provisions of the convention by denying the right to a prompt appearance before a judicial authority (Berger, pp. 60–64). Success in court did not, however, necessarily guarantee that successful litigants would receive the remedies to which the court had indicated they were entitled. They were still dependent upon Parliament to change the laws and regulations to which the court had objected.

This situation was modified by passage of the Human Rights Act of 1998. This Act of Parliament incorporated the European Convention on Human Rights into English law and provided that English courts and tribunals must take into account any "judgment, decision, declaration or advisory opinion of the European Court of Human Rights" (Section 2[1][a]). English courts and tribunals were instructed that: "So far as it is possible to do so, primary legislation and subordinate legislation must be read and given effect in a way which is compatible with the Convention rights" (Section 3[1]).

As a result of the passage of the Human Rights Act of 1998, inhabitants of the United Kingdom no longer have to take allegations of violations of the European Convention on Human Rights to the European Court of Human Rights in Strasbourg, France. They can pursue their grievances in English courts. However, although the Human Rights Act of 1998 allows English courts to set aside certain laws, the courts still cannot set aside primary legislation. They can merely find that the primary legislation in question is incompatible with the provisions set out by the European Convention on Human Rights and send the offending statutory provision back to Parliament for amendment. In 2004, for example, the House of Lords ruled that allowing the indefinite detention of foreign, but not British, terrorist suspects without being charged violated provisions of the European Convention on Human Rights (*A[FC] and others v. Secretary of State for the Home Department*). Though an initial reaction was to rewrite the regulations to allow similar detention of British terrorist suspects, and the Prime Minister pushed for legislation that would increase the maximum amount of time police could hold suspected terrorists without being charged from 14 days to 90 days, the Prevention of Terrorism Act of 2005 set the maximum amount of such detention at 28 days for both foreign and British suspected terrorists.

■ Conclusion

There are many commonalities between England and the United States. The two countries share both a common language and a common cultural heritage and employ basically similar political and economic systems. The similarities between the two countries should not, however, be allowed to mask differences that may affect both the actual commission of crime and the response to the commission of crime.

Cultural differences arising from England's lengthy historical heritage, present-day social and economic differences, and differences in the political and governmental frameworks under which the two nations operate all need to be taken into consideration. It would be a mistake to ignore the fact that the crime situation and crime control issues may be affected by such factors as England's size and insular nature, the homogeneity of its population, and the existence of a state religion. England's long history of coping with problems characterized by a laissez-faire approach coupled with a certain tolerance of deviance and lack of violence must also be considered. In addition, the presence of a monarchy and distinct social classes, the existence in the economic structure of various socialist features such as nationalized industries and multifaceted social services, and a unitary governmental system that has espoused the supremacy of one of the branches of government and does not employ a written constitution or the doctrine of separation of powers are all differences that should not be overlooked. The extent to which these and other such factors impact upon the commission of crime and the response to crime, and affect the comparative analysis that is being undertaken in this book, will be examined in the following chapters as we embark upon our investigation of the English crime problem and the English criminal justice system.

CHAPTER SPOTLIGHT

- The United Kingdom of Great Britain and Northern Ireland consists of England, Wales, Scotland, and Northern Ireland, with 82% of the population living in England (49,181,300) and 5% in Wales (2,903,200).

- England and Wales are a single legal jurisdiction. The Act of Union (1536) made English law applicable in Wales and allowed Welsh representatives to be admitted to the English Parliament.

- The United Kingdom is a "constitutional monarchy" under a democratic parliamentary regime. The head of state is the Monarch or Crown (king or queen). The government, known as Her or His Majesty's Government, carries out all government functions in the name of the Monarch.

- The government is formed by the leader of the party that wins the most seats in the lower house of Parliament—the House of Commons.

- The House of Commons is made up of 646 directly elected members, each elected from a single member district. The upper house of Parliament is known as the House of Lords. The Monarch, upon the advice of the Prime Minister, mostly appoints members of the Lords. There remain 92 hereditary members, also known as "peers."

- The leader of the party with the most members of the lower house is appointed by the Monarch to be Prime Minister. The Prime Minister appoints other ministers, the most senior of whom form the Cabinet. The Cabinet exercises executive power on behalf of the Crown.

- There is no division of church and state in the United Kingdom. The Church of England is the established church. The Monarch is head of the Church of England and head of state.

- England and Wales have significantly fewer homicides per 100,000 inhabitants than in the United States. Gun ownership is also significantly less than in the United States, and English police officers are still not routinely armed or even trained to use firearms.

- The United Kingdom is a member of the European Union (EU), and citizens of the United Kingdom can move freely around other EU nations, just as citizens of other EU nations can move freely around the United Kingdom.

- Freedom of movement among EU nations has facilitated human trafficking; measures are being taken to combat this problem.

- England has developed a greater tolerance to "deviance," with more permissive attitudes toward prostitution, homosexuality, nudity on television, and gambling than the United States.

- The United Kingdom has a broader safety net for the underprivileged than the United States. This includes a universal healthcare system that is largely free at point of service. The United Kingdom has a lower infant mortality rate than the United States and lower rates of bankruptcy. However, the interrelationship between the provision of services by government and crime is unclear. It may be hypothesized that greater support of the underprivileged may lead to lower levels of crime.

- The United Kingdom has a state education system running from nursery school through higher education. Until 1998 the government provided maintenance grants to students in higher education. Since that time financial support is provided by way of a mixture of grants and loans. However, in 2001–2002 50% of students did not have to pay tuition fees.
- England has a unitary government system. Parliament is supreme, and may pass any law it likes. England can have a single law or policy that applies to the whole nation.
- The United Kingdom has no formal written constitution.
- Constitutional reforms in the United Kingdom have led to a separation of the roles of the Lord Chancellor and creation of a Supreme Court as well as devolved government for Scotland and Wales.
- The United Kingdom is a signatory to the European Convention for the Protection of Human Rights and Fundamental Freedoms. This is incorporated into English law, allowing English and foreign courts some power to challenge the Supremacy of Parliament.

KEY TERMS

Cabinet: A group of ministers who exercise executive power on behalf of the Crown.

Chancellor of the Exchequer: Senior minister responsible for revenue and expenditure (equivalent to the U.S. Treasury Secretary).

Church of England: The established church in England, which is divided into two provinces led by a head. It is part of the Anglican Communion.

European Union (EU): A political and economic union of 27 sovereign European states.

Foreign Secretary: Senior minister responsible for foreign affairs (equivalent to the U.S. Secretary of State).

Her (or His) Majesty's government: The executive formed from the leadership of the party with the largest number of members in the House of Commons.

Home Secretary: Senior minister with overall responsibility for internal affairs, in particular, the maintenance of law and order

House of Commons: The lower house of the legislature, a representative body with each of its 646 members elected by the voters of a particular district.

House of Lords: The upper house of the legislature, composed of a mixture of members who inherited the right to sit there, members who were appointed by the Crown, bishops of the Church of England, and the Law Lords (equivalent to the Justices of the U.S. Supreme Court).

Lord Chancellor: Senior minister who formerly played multiple roles: (1) head of the judiciary; (2) speaker of the House of Lords; and (3) a member of the Cabinet. As of April 2006, the Lord Chancellor is no longer head of the judiciary.

Lords of Appeal in Ordinary: Appointed for life by the Crown to perform the judicial duties of the House of Lords. Law Lords will become the Justices of the Supreme Court of the United Kingdom when this new court becomes operational. It is envisaged that this will occur in 2009.

Monarch/Crown: Titular head of state (king or queen).

Prime Minister: The leader of the party that holds a majority of the seats in the House of Commons.

Sovereign: The temporal head of the country, also called "Defender of the Faith." (See above, *Monarch/Crown.*)

Supremacy of Parliament: Asserts that Parliament may pass any law that it likes and the courts cannot strike down Acts of Parliament.

Unitary governmental system: No separate state and federal systems.

PUTTING IT ALL TOGETHER

1. What are the advantages and disadvantages of having a constitutional monarch?

2. Should all members of the legislature be elected?

3. Should there be some flexibility in the length of time a government stays in power with the government entitled, within specified parameters, to fix the date of a general election and the opposition able to force a general election through a vote of "no confidence"?

4. What kinds of socioeconomic conditions are likely to result in high rates of criminal activity?

5. What is the interrelationship between material prosperity and quality of life?

6. How important is it in a democracy for there to be separation of power, and a system of checks and balances, among the executive, legislative, and judicial branches of government?

REFERENCES

A(FC) and others v. Secretary of State for the Home Department. (2004). UKHL 56.

Alpers, P. (2001). *Firearm-related injury, crime & regulation: Health, justice & compliance data from selected nations.* International Medical Conference on Small Arms, Helsinki, Finland, September 28–30.

Anti-Slavery. (2005). *Trafficking in the UK.* Retrieved from http://www.antislavery.org/homepage/antislavery/traffickinguk.htm.

Arie, S. (2003, November 5). *Janie's secret.* The Guardian. Retrieved from http://www.guardian.co.uk/g2/story/0,3604,1077703,00.html.

Banks, A. (Editor). (1987). *Political handbook of the world: 1987.* Binghamton, NY: CSA.

Berger, V. (1992). *Case law of the European Court of Human Rights.* Vol. 2. Dublin, Ireland: Round Hall.

Bowers. Hardwick. (1986). 478 U.S. 186.

Burmah Oil Company v. Lord Advocate. (1965). A.C. 75.

Central Statistical Office. (1990). *Social trends.* London, England: H.M.S.O.

Cook, P. J., & Ludwig, J. (1997). *Guns in America: Survey on private ownership and use of firearms.* Washington, D.C.: U.S. Department of Justice.

Council of Europe. (1993). *Yearbook of the European Convention on Human Rights: 1988*. London, England: Martinus Nijhoff.

Davey Paxman & Co. v. Post Office. (1954, November 16). *The Times*.

Department of Constitutional Affairs. (2006). *Progress on the Supreme Court and update on the implementation costs*. Retrieved from http://www.dca.gov.uk/pubs/statements/2006/st061017.htm.

Department for Work and Pensions. (2005). *Services and benefits: Statutory sick pay*. Retrieved from http://www.dwp.gov.uk/lifeevent/benefits/statutory_sick_pay.asp.

Department of Trade and Industry. (2003a). *Your guide to the working time regulations: Workers and employers*. London, England: H.M.S.O.

Department of Trade and Industry. (2003b). *Maternity rights*. London, England: H.M.S.O.

Department of Trade and Industry. (2004). *Trade union membership 2003*. London, England: H.M.S.O.

The economist book of vital world statistics: A complete guide to the world in figures. (1990). London, England: Hutchinson.

The Europa world handbook: 2003. (2004). London, England: Europa.

The Europa year book: 1990. (1990). London, England: Europa.

Falconer, Lord. (2006). *Doing law differently*. London, England: Department of Constitutional Affairs.

Greig, G. (1988, November 23). How the party leaders upset the Queen. *Daily Mail*, pp. 1–2.

Haigh, C. (Ed.). (1985). *The Cambridge historical encyclopedia of Great Britain and Ireland*. Cambridge, England: Cambridge University Press.

Hamilton, A. (1985). The *royal handbook*. London, England: Beazley.

Higher Education Statistics Agency. (2004). *Higher education in the United Kingdom*. Retrieved from http://www.lmt.lt/NAUJIENOS/05_10.pdf.

Home Office. (2002). *Criminal statistics: England and Wales 2001*. London, England: H.M.S.O.

Home Office. (2006a). *Tackling human trafficking—Consultation on proposals for a UK action plan*. London, England: H.M.S.O.

Home Office. (2006b). *Human trafficking centre opens*. Retrieved from http://www.homeoffice.gov.uk/about-us/news/human-trafficking-centre?version=1.

International Labor Organization. (2005). *LABORSTA: Internet*. Retrieved from http://laborsta.ilo.org/cgi-bin/brokerv8.exe.

Keir, D. L, & Lawson, F. H. (1979). *Cases in constitutional law* (6th ed.). F. H. Lawson & D. J. Bentley (Editors). Oxford, England: Oxford University Press.

Kelly, L., & Regan, L. (2000). *Stopping traffic: Exploring the extent of, and responses to, trafficking in women for sexual exploitation in the U.K.* London, England: Home Office.

Kopel, D. B. (1990). *Our shoes won't fit your bunions: Should British gun laws be a model for America?* Paper presented at the annual meeting of the Academy of Criminal Justice Sciences.

Kopel, D. B. (1992). *Gun control in Great Britain: Saving lives or constricting liberty?* Chicago: University of Illinois at Chicago.

Lawrence v. Texas. (2003). 537 U.S. 1102.

Monkkonen, E. H. (1989). Diverging homicide rates: England and the United States, 1850–1875. In T. R. Gurr (Ed.), *Violence in America: The history of crime* (pp. 80–101). Newbury Park, CA: Sage.

National Statistics Online. (2003a). *Ethnicity*. Retrieved from http://www.statistics.gov.uk/cci/nugget.asp?id=273.

National Statistics Online. (2003b). *Belonging to a religion*. Retrieved from http://www.statistics.gov.uk/STATBASE/Expodata/Spreadsheets/D7212.xls.

Organization for Economic Co-operation and Development (OECD). (2005). *OECD factbook 2005*. Paris, France: Author.

Office for National Statistics. (2003). *Social trends 2003*. London, England: T.S.O.

Office for National Statistics. (2004). *Social trends 2004*. London, England: T.S.O.

Office for National Statistics. (2005). *Social trends 2005*. London, England: T.S.O.

Roe v. Wade. (1973). 410 U.S. 113.

Sapsted, D. (2002, April 12). Illegal migrants forcing police to ignore rail crime. *The Daily Telegraph*. Retrieved from http://www.telegraph.co.uk/news/main.jhtml?xml=/news/2002/04/12/nrail12.xml.

Taylor, G. (2005). *Evaluation of the victims of trafficking pilot project—POPPY summary findings*. London, England: Home Office Research, Development and Statistics.

U.S. Bureau of the Census. (1992). *Statistical abstract of the United States: 1992*. (112th ed.). Washington, D.C.: U.S. Government Printing Office.

U.S. Bureau of the Census. (1994). *Statistical abstract of the United States: 1994*. (114th ed.). Washington, D.C.: U.S. Government Printing Office.

U.S. Bureau of the Census. (2000). *Profile of selected social characteristics 2000*. Retrieved from http://censtats.census.gov/data/US/01000.pdf.

U.S. Bureau of the Census. (2001). *Statistical abstract of the United States: 2001*. (121st ed.). Washington, D.C.: U.S. Government Printing Office.

U.S. Bureau of the Census. (2003). *Statistical abstract of the United States: 2003*. (123rd ed.). Washington, D.C.: U.S. Government Printing Office.

U.S. Bureau of the Census. (2007a). *Higher education: Finances, fees and staff*. Retrieved from: http://www.census.gov/compendia/statab/education/higher_education_finances_fees_and_staff/.

U.S. Bureau of the Census. (2007b). *Statistical abstract of the United States: 2005*. (126th ed.). Washington, D.C.: U.S. Government Printing Office.

U.S. Department of Justice. (2002). *FBI uniform crime reports: Crime in the United States 2001*. Washington, D.C.: U.S. Government Printing Office.

U.S. State Department. (2006). *Trafficking in persons report June 2006*. Washington, D.C.: U.S. Government Printing Office.

United States Information Agency. (2003). *Portrait of the USA*. Retrieved from http://usinfo.state.gov/usa/infousa/facts/factover/ch6.htm.

Whitaker's almanac 1993. (1993). London, England: J. Whitaker & Sons.

The world almanac: 2003. (2003). New York: World Almanac Books.

Young, R. (2002). *Queen and country: Fifty years on. House of Commons research paper 02/28*. London, England: House of Commons.

Chapter Resources

English Criminal Law

3

Chapter Objectives

After completing this chapter, you will be able to:

- Compare and contrast the different roles legislation and case law have played in the development of English criminal law.
- Describe the general types of offenses covered by English substantive criminal law.
- Describe how English law deals with the issue of entrapment.
- Outline the English approach to consensual adult sexual activity, in particular fornication and adultery, deviant and same-sex sexual relations, and prostitution.
- Outline the English approach to gambling.

■ Introduction

For the most part, the legal systems in operation in the United States today trace their roots back to the English system of law. Thus, as one examines the legal system of any state and compares it with the English legal system, many similarities may be found. With respect to criminal law, this holds true for both **substantive criminal law** (the types of acts that are defined as crimes, such as murder and driving while intoxicated) and **procedural criminal law** (the procedures that are set out for the detection, apprehension, trial, and conviction of suspected offenders, such as "Miranda" rights). In this chapter we will focus on English substantive criminal law and ways in which it differs from U.S. substantive criminal law. Procedural law issues will be covered in subsequent chapters on the police, courts, and corrections.

■ Source and Structure of English Criminal Law

Law is derived from two main sources—statutes and case law. <u>Statutes</u> are acts of the legislative branch of government that declare, command, or prohibit something. They are general in nature and are aimed at the citizenry as a whole, or at a particular defined segment of the citizenry, such as those under the age of 18. <u>Case law</u>, on the other hand, is a judge-made law that emanates from the decision in a particular case, but through the doctrine of precedent has impact beyond the parties involved in that case and the particular facts of that case.

Until fairly recently a high percentage of the law in England was case law. This was an outcome of the historical development of law in England. From early times, law was essentially local in nature. In Anglo-Saxon times the rulers of the different kingdoms were responsible for maintaining peace and order within their realms. The country was divided into counties (also known as shires) with the counties further divided into hundreds. Within the hundreds were the manors administered by lords. Each level had its own system of justice, with the larger units dealing with the more serious issues. Thus, there was the <u>manor court</u> held by the lord of the manor to administer justice among his tenants, the <u>hundred court</u> presided over by the bailiff of the hundred, and the <u>county court</u>, which was run by the sheriff. However, there was a lack of centralization in the system of justice in England. As Maitland points out:

> *resort to any central tribunal, to the king and his wise men seems to have been rare, and this localization of justice must have engendered a variety of local laws. Law was transmitted by oral tradition and the men of one shire would know nothing and care nothing for the tradition of another shire. (1965, p. 4)*

Even after the Norman conquest of 1066, a single law of the land did not exist. However, in the struggle to wrest power away from the nobles, the Crown began to assert the right to settle local disputes. In the twelfth century Henry II divided England into circuits and sent judges around the country to hear and decide cases. Thus began the growth of the <u>common law</u>. Unlike earlier law that had been local in nature, this law was common in the sense that it applied to the whole land. Gradually, more and more of the cases were heard in the King's courts and common law precedents were established for a host of issues.

Case law established the existence of various criminal offenses, and sometimes new offenses were created by extending the boundaries of old offenses. In contrast to the United States, there has been no general codification of the law, although by now most of the law has in a piecemeal manner been enacted into statute.

Most of English criminal law is now found in statutory form. However, different aspects of the criminal law were subjected to legislation at different times. Thus, the offenses of assault and battery were put into statutory form in the Offences against the Person Act of 1861. Larceny and other theft offenses became part of the statutory law in the Theft Act of 1968. Some antiquated common law crimes technically remained in force until fairly recently. Thus until these offenses were abolished by the Criminal Law Act of 1967 (Section 13), it was an offense to eavesdrop or to be a common scold. Where case law has not been superseded by statute, the case law continues in effect. Thus case law still plays a prominent role in English criminal law.

In contrast to the United States, England and Wales have a unitary legal system, in which statutes and case law apply equally throughout the land. It is possible to state authoritatively what the law of the land is, and there are not the kinds of variations in English law that exist among the states in the United States. Law reform is also an easier process. The British Parliament passes laws for England and Wales, and can enact statutes that apply specifically to Scotland or Northern Ireland, though with their own legal jurisdictions and devolved governments, they both enjoy a limited amount of self-government. To ensure that reform of the criminal law is given adequate attention, the Criminal Law Revision Committee was established in 1959 by the Home Secretary "to be a standing committee to examine such aspects of the criminal law of England and Wales as the Home Secretary may from time to time refer to the committee, to consider whether the law requires revision and to make recommendations" (Criminal Law Revision Committee, 1966, p. 2). This committee was extremely active and by the end of 1985 had issued some 17 reports on a variety of topics such as theft, prostitution, and the distinction between felonies and misdemeanors. A great many of the committee's recommendations have been enacted into law. This committee's work is now being conducted by the Criminal Law Policy Unit in the National Offender Management Service Directorate.

As discussed in Chapter 2, the English do not have a formal written constitution. Thus the English courts do not have a written document against which they are to measure the validity of legislative acts. The U.S. model of the courts interpreting and enforcing statutes in the light of constitutional requirements does not apply to England. Provided that a statute has been enacted in the prescribed legislative manner, English courts are bound to apply it. Of course, as a result of British participation in international agreements and treaties, aggrieved parties may be entitled to bring complaints against the British government before foreign courts. Until recently, the European Court of Human Rights in Strasbourg, France, heard many of these complaints.

As discussed in Chapter 2, this situation was modified by passage of the Human Rights Act of 1998. This Act of Parliament incorporated the European Convention on Human Rights into English law and provided that English courts and tribunals must take into account any "judgment, decision, declaration or advisory opinion of the European Court of Human Rights" (Section 2[1][a]). English courts and tribunals were instructed that: "So far as it is possible to do so, primary legislation and subordinate legislation must be read and given effect in a way which is compatible with the Convention rights" (Section 3[1]).

As a result of passage of the Human Rights Act of 1998, inhabitants of the United Kingdom no longer have to take allegations of violations of the European Convention on Human Rights to the European Court of Human Rights in Strasbourg, France. Now, they can pursue their grievances in English courts. However, although English law is subservient to EU directives, and the Human Rights Act of 1998 allows English courts to set aside certain laws, the courts still cannot set aside primary legislation. They can merely find that the primary legislation in question is incompatible with the provisions set out by the European Convention on Human Rights and send the offending statutory provision back to Parliament for amendment.

In the United States, on the other hand, there has always been great concern about limiting the potential for oppression by the government. Mindful of the abuses they had suffered at the hands of the English government, the citizens of the newly established American republic sought to safeguard themselves from oppression by the central government. Formal expression of this sentiment is to be found in the Bill of Rights passed in 1791. Although originally intended to apply only to the federal government, most of the procedural rights contained in the Bill of Rights have through the process of selective incorporation been required by the U.S. Supreme Court to be granted by the states to those living within their various jurisdictions.[1]

As will be discussed in detail in later chapters, this concern with procedural rights has, at least in theory, resulted in Americans obtaining far more protection against the unfair institution and implementation of criminal proceedings than that enjoyed by their English counterparts. Although the development of this aspect of procedural criminal law has provided Americans with markedly more guarantees, in many ways the substantive criminal law became, and still remains, more oppressive in the United States than in England.

■ Substantive English Criminal Law

The functioning of the criminal justice system is naturally affected by the nature and volume of acts that are declared to be criminal. As with any other process, the larger and more complicated the input, the greater the difficulty of ensuring the smooth functioning of the system. Here some major differences may be observed between the criminal justice system of England and those of the United States.

Although the United States inherited from England the division of crimes into felonies and misdemeanors, England no longer has these two categories of crime. The Criminal Law Act of 1967 abolished "all distinctions between felony and misdemeanor" (Section 1) and set up a new division of offenses into arrestable and nonarrestable offenses. **<u>Arrestable offenses</u>** include all offenses, and attempts to commit offenses, for which the sentence is fixed by law (e.g., murder) or for which a convicted defendant may receive a sentence of five years imprisonment or more. Additional specific offenses, such as indecent assault on a woman and taking a motor vehicle without authority, were added to this list by the Police and Criminal Evidence Act 1984 (Section 24). All other offenses are considered **<u>nonarrestable offenses</u>**. Although suspects may be taken into custody without a warrant for both arrestable and nonarrestable offenses, there are wider circumstances under which a warrantless arrest can be made when the offense involved is an arrestable offense.

The significance of the felony/misdemeanor distinction in England had related mainly to the consequences of conviction, because conviction of a felony could result in the forfeiture of property or the imposition of the death penalty. Forfeiture of property had, however, been abolished by the Forfeiture Act of 1870, and the death penalty suspended by the Murder (Abolition of the Death Penalty) Act of 1965 for all offenses except treason (The Treason Act of 1814), piracy on the high seas (The Piracy Act of 1837), and setting fire to Her Majesty's dockyards

(Dockyards Protection Act of 1772). The Crime and Disorder Act of 1998 finally abolished the last capital crimes (Section 36).

In sharp contrast to the United States, there were no procedural consequences of the felony/misdemeanor distinction. The court processing of cases was regulated not by whether they were labeled felonies or misdemeanors, but by whether they were classified as summary or indictable offenses. <u>Summary offenses</u> are offenses such as certain driving and gambling offenses that can be tried only by the lower magistrates' courts. <u>Indictable offenses</u> are offenses such as murder, rape, and robbery that are heard in the higher courts, namely the Crown courts. A number of offenses, such as common assault and most theft offenses, can be tried in either court. Since these are procedural and not substantive issues, they will be discussed in greater detail in Chapter 7.

Serious antisocial acts, such as murder, rape, and robbery, are defined as crimes in both the United States and England. Just as the definitions of these offenses vary from state to state within the United States, so the English definitions vary from those found in U.S. penal codes. By and large, however, the elements of these offenses are very similar to those found in penal codes in the United States. So too are the defenses that may be raised by defendants in both countries.

It is beyond the scope of this book to embark upon a detailed examination of English substantive criminal law. However, it is important for the reader to have a clear indication of the types of antisocial acts that constitute crimes in England, and how those crimes are defined. To this end, three tables are presented to provide a representative picture of the criminal offenses that exist in England and the basic nature of those offenses. It must be stressed that only a sample of English criminal offenses is given in these tables, and that although every attempt has been made to capture the essential nature of each offense in the succinct description that accompanies them, these descriptions are not precise accountings of all the legal elements required for conviction. Indeed, the descriptions focus very much on the physical element (*actus reus*) of the offense with comparatively less attention paid to the mental element (*mens rea*) that is required for conviction. For readers interested in learning more about the exact elements of these offenses, statutory references have been given for those offenses that have been placed on a statutory basis. Where earlier statutory provisions have been replaced, the reference for the most recent provision is given.

From perusing **Table 3-1** the reader can obtain an overview of the types of offenses that constitute crimes against the person under English law. Included are criminal homicide, assault, rape, and sexual assault offenses.

Among the criminal homicide offenses are the common law crimes of murder and manslaughter, and the statutory offenses of complicity in another's suicide (enacted in 1961 when the common law offense of suicide was abolished) and causing death by dangerous driving, an offense enacted to counteract problems arising from jury refusal to convict on manslaughter charges in such circumstances. A variety of offenses, such as infanticide, child destruction, and concealment of birth, involve the death of a newborn child. Although it is an offense for any person to take action to procure the miscarriage of a woman, medically approved abortions are permitted. These are regulated by the 1967 Abortion Act. Many of the assault offenses are to be found in the Offences against the Person Act of 1861. Until 2003, rape and other sexual offenses were, for the most part,

Table 3-1	English Criminal Law Offenses: Offenses Against the Person
Offense	**Essential Nature of Offense**
Murder	Unlawfully kill another with malice aforethought (common law offense).
Manslaughter	Unlawfully kill another without malice aforethought (common law offense).
Letter threatening murder	Send letter threatening to kill a person (Offences against the Person Act 1861, Section 16).
Complicity in another's suicide	Aid, abet, counsel, or procure the suicide of another (Suicide Act of 1961, Section 2).
Death by dangerous driving	Cause death by reckless driving (Road Traffic Act of 1988, Sec. 1; Road Traffic Act of 1991, Section 1).
Infanticide	Mother, while balance of mind disturbed, causes death of child under age of 12 months (Infanticide Act of 1938, Section 1).
Child destruction	Willfully cause a child to die before it has an existence independent of the mother (Infant Life [Preservation] Act of 1929, Section 1).
Procuring illegal abortion	Commit act to procure the miscarriage of a woman (Offences against the Person Act of 1861, Sections 58–59)—exceptions for medically terminated pregnancies (Abortion Act of 1967, Section 1).
Concealment of birth	Secretly dispose of dead body of infant (Offences against the Person Act of 1861, Section 60).
Assault	Cause another to fear the immediate application of unlawful physical violence (common law offense).
Assault occasioning bodily harm	Commit an assault occasioning actual bodily harm (Offences against the Person Act of 1861, Section 47).
Assault on, or obstruction of, a constable	Assault, or willfully obstruct, a constable, or a person assisting a constable, in the execution of a duty (Police Act of 1964, Section 51).
Wounding with intent	Wound with intent to inflict grievous bodily harm (Offences against the Person Act of 1861, Section 18).
Malicious wounding	Unlawfully and maliciously wound another (Offences against the Person Act of 1861, Section 20).
False imprisonment	Without lawful reason inflict bodily restraint on another (common law offense).
Kidnapping	Take or carry away another by force or fraud without consent or lawful excuse (common law offense).
Child abduction	A person connected/not connected with a child under 16 takes or sends the child out of the U.K. without the appropriate consent (Child Abduction Act of 1984, Sections 1–2).
Unlawful administration of poisons	Unlawfully administer poison to another (Offences against the Person Act of 1861, Section 23).
Rape	A person intentionally penetrates the vagina, anus, or mouth of another without consent (Sexual Offences Act of 2003, Section 1).
Sexual assault	Intentional sexual touching of another without consent (Sexual Offences Act of 2003, Section 3).
"Statutory" rape	A person intentionally penetrates the vagina, anus, or mouth of a person under 13 (Sexual Offences Act of 2003, Section 5).
"Statutory" sexual assault	Intentional sexual touching without consent of: (a) a person under 16, (b) a person under 18 with whom the offender is in a position of trust, (c) a family member under 18, or (d) a person with a mental disorder impeding choice. (Sexual Offences Act of 2003, Sections 9, 16, 25, 30).

Table 3-1	English Criminal Law Offenses: Offenses Against the Person, continued
Offense	**Essential Nature of Offense**
Indecent assault	Indecently assault (a) a woman, (b) a man, (c) with intent to commit buggery (Sexual Offences Act of 1956, Sections 14–16).
Indecency with children	Commit an act of gross indecency with a child under the age of 16 (Indecency with Children Act of 1960, Section 1; Criminal Justice and Court Services Act of 2000, Section 39).

set out in the Sexual Offences Act of 1956. The Sexual Offences Act of 2003 broadened the crime of rape, which traditionally only applied to female victims, to include victims of either sex. Although adults are protected against nonconsensual sexual activity, children are given added protection, because they are deemed by law to be incapable of consenting to any such activity. The basic elements of all of these offenses against the person are very similar to those that are to be found in the comparable gender-neutral U.S. statutory provisions.

Table 3-2 outlines the essential nature of offenses against property. The law regarding theft and related offenses was consolidated in the Theft Act of 1968. This

Table 3-2	English Criminal Law Offenses: Offenses Against Property
Offense	**Essential Nature of Offense**
Theft	Dishonestly appropriate property belonging to another with intent to permanently deprive (Theft Act of 1968, Section 1).
Robbery	Steal using force or fear of use of force (Theft Act of 1968, Section 8).
Burglary	Enter building as trespasser with intent to steal, inflict grievous bodily harm, rape, or do unlawful damage (Theft Act of 1968, Section 9).
Aggravated burglary	Above committed with firearm, imitation firearm, weapon of offense, or explosive (Theft Act of 1968, Section 10).
Possession of burglary tools	Possess while not at place of abode articles for use in burglary, theft, or cheat (Theft Act of 1968, Section 25).
Taking conveyance without authority	Take conveyance without consent (temporary deprivation) (Theft Act of 1968, Section 12).
Obtaining property etc. by deception	Dishonestly obtain property, pecuniary advantage, or services by deception (Theft Act of 1968, Sections 15–16; Theft Act of 1978, Section 1).
Blackmail	Make unwarranted demand with menaces (Theft Act of 1968, Section 21).
Handling stolen goods	Handle stolen goods for the benefit of another knowing or believing them to be stolen (Theft Act of 1968, Section 22).
Criminal damage	Destroy or damage property belonging to another—includes arson (Criminal Damage Act of 1971, Section 1).
Forcible entry	Use or threaten violence to obtain entry into occupied premises (Criminal Law Act of 1977, Section 6).
Forgery	Make false instrument with intent that it be accepted as genuine to another's prejudice (Forgery and Counterfeiting Act of 1981, Section 1).
Counterfeiting	Make counterfeit of a currency note or protected coin intending it to be passed off as genuine (Forgery and Counterfeiting Act of 1981, Section 14).

act defines such offenses as theft, robbery, burglary, obtaining property by deception, taking a conveyance without authority, and handling stolen goods. Other offenses against property, such as criminal damage to property, forcible entry, forgery, and counterfeiting, are covered in other statutory provisions. Again, all of these offenses are defined in much the same way as they are in U.S. criminal codes.

Table 3-3 contains a listing of a variety of offenses, broken down into the three general categories of (1) offenses against the State, (2) offenses against public order, and (3) offenses against morality.

Offenses against the State are composed of such offenses as treason, sedition, terrorism, and disclosure of official secrets. These offenses, which are designed to protect the state from subversion and overthrow, include an interesting mixture of antiquated and modern day offenses. On the one hand, there are still in force such offenses as those contained in the provisions of the Treason Act of 1351 and those covered by the common law crime of sedition. On the other hand, there are the recently enacted provisions of the Terrorism Act of 2000, which is covered in detail in Chapter 11.

Offenses against the public order include such offenses as riot, violent disorder, affray, harassment, and incitement to racial or religious hatred, all of which were codified in the Public Order Act of 1986. Other offenses cover a variety of public mischief such as: being drunk and disorderly in public, gaming in a public place, having an offensive weapon in public, being a public nuisance, participating in a bomb hoax, publishing criminal libel, bigamy, perjury, and driving while unfit.

Weapons offenses also have been included in this subcategory. The English have a certification process for the production, sale, and acquisition of firearms. Consequently, it is a criminal offense to manufacture, sell, or transfer firearms unless registered as a firearms dealer, or to possess a firearm without an appropriate certificate. The possession of a firearm in criminal circumstances has been made the subject of a number of distinct offenses. Thus it is, for example, a criminal offense to trespass with a firearm, carry a firearm with criminal intent, or use a firearm to resist arrest.

As in the United States, the English have a problem with motorists who drive while under the influence of alcohol or other drugs. The legislative provisions that cover this issue, which are to be found in the Road Traffic Act of 1972 and the Transport Act of 1981, are very similar to those that exist in the United States.

Offenses against morality focus on crimes that offend moral codes, but ostensibly do not have any direct victims. These criminal offenses are often labeled "victimless" crimes. Included in this subcategory are drug and consensual sex offenses.

The English have a framework of laws classifying drugs into different categories and proscribing their unlawful importation, production, and possession. Although the English do permit heroin to be supplied to registered addicts, this is in fact done far less frequently than might be envisaged. Thus, for example, only 105 (2.1%) of the 5,079 registered addicts in the United Kingdom in 1983 who were being given notifiable drugs for their addiction were receiving heroin. As many as 3,998 (78.7%) of these addicts were being given methadone, while

Table 3-3	English Criminal Law Offenses: Offenses Against the State, Public Order, and Morality

Offense	Essential Nature of Offense
Offenses Against the State	
Treason	Encompass the death of the King, levy war against the King or adhere to the King's enemies in his realm, violate the King's companion, or slay one of the King's ministers or justices (Treason Act of 1351).
Sedition	Make public oral or written statements with seditious intent (common law misdemeanor).
Terrorism	Belong to, support, or wear the uniform of a proscribed organization (Terrorism Act of 2000, Sections 11–13).
Disclosure of official secrets	Disclose without lawful authority information relating to security or intelligence (Official Secrets Act of 1989, Section 1).
Offenses Against Public Order	
Riot	12 or more persons present together threaten unlawful violence for a common purpose so as to cause a reasonable person to fear for his safety (Public Order Act of 1986, Section 1).
Violent disorder	As above but only requires 3 persons (Public Order Act of 1986, Section 2).
Affray	Threaten unlawful violence so as to cause a reasonable person to fear for his safety (Public Order Act of 1986, Section 3).
Harassment	Pursuing a course of conduct amounting to harassment (Protection of Harassment Act of 1997, Sections 1–2).
Incitement of racial or religious hatred	Behave so as to stir up racial or religious hatred (Public Order Act of 1986, Sections 18–23; Anti-terrorism, Crime and Security Act of 2001, Section 39).
Public nuisance (common law offense)	Act unlawfully so as to inflict damage or inconvenience on Her Majesty's subjects.
Drunk and disorderly	While drunk behave disorderly in a public place (Criminal Justice Act of 1967, Section 91).
Gaming in a public place	Take part in gaming in a public place (Gaming Act of 1968, Section 5).
Possession of offensive weapon	Possess an offensive weapon in a public place (Prevention of Crime Act of 1953, Section 1; Offensive Weapons Act of 1996, Sections 1–2).
Bomb hoax	Act to induce a false belief that a bomb will explode (Criminal Law Act of 1977, Section 51). Place substance in a public place with intent to induce belief that it is noxious (Anti-Terrorism, Crime and Security Act of 2001, Section 114).
Criminal libel	Publish defamatory libel (Libel Act of 1843, Section 5); knowing it to be false (heavier penalty, Section 4).
Bigamy	Being married, marry another person (Offences against the Person Act of 1861, Section 57).
Perjury	Willfully make statement known to be untrue while sworn witness in judicial proceeding (Perjury Act of 1911, Section 1).
Unlawful possession of firearms	Produce, acquire, or possess firearms or ammunition without a certificate (Firearms Act of 1968, Section 1).
Unlawful business in firearms	Manufacture, sell, transfer, or repair firearms or ammunition without a dealer's certificate (Firearms Act of 1968, Section 3).

| Table 3-3 | English Criminal Law Offenses: Offenses Against the State, Public Order, and Morality, continued |

Offense	Essential Nature of Offense
Unlawful possession of prohibited weapons	Manufacture, sell, or possess a prohibited weapon (e.g., automatic gun) without official consent (Firearms Act of 1968, Section 5; Firearms [Amendment] Act of 1997, Section 1).
Trespass with firearm	Trespass while carrying a firearm, or imitation firearm (Firearms Act of 1968, Section 20).
Carrying a firearm with criminal intent	Possess a firearm or imitation firearm with intent to commit indictable offense or resist arrest (Firearms Act of 1968, Section 18).
Use firearm to resist arrest	Use a firearm or imitation firearm to resist arrest (Firearms Act of 1968, Section 17).
Driving while unfit	Drive a motor vehicle while unfit to drive through drink or drugs (Road Traffic Act of 1988, Section 4).
	Other offenses of: driving with alcohol concentration above prescribed limit (Section 5); failure to provide a breath test (Section 6).

Offenses Against Morality

Drug Offenses

Offense	Essential Nature of Offense
Import/export drugs	Import or export controlled drugs without license (The Misuse of Drugs Act of 1971, Section 3).
Produce/supply drugs	Produce or supply controlled drugs without authorization (The Misuse of Drugs Act of 1971, Section 4).
Cultivate cannabis	Cultivate any plant of the genus cannabis (The Misuse of Drugs Act of 1971, Section 6).
Unlawful possession of drugs	Unlawfully possess a controlled drug (The Misuse of Drugs of Act of 1971, Section 5).
Permitting drug activities on premises	Occupier or manager of premises knowingly allows certain drug offenses to take place there (The Misuse of Drugs Act of 1971, Section 8).

Sexual Offenses

Offense	Essential Nature of Offense
Public solicitation by prostitute	Common prostitute loiters or solicits in a public place for the purpose of prostitution (Street Offences Act of 1959, Section 1).
Solicitation by "kerb crawling"	A man solicits a woman from a motor vehicle for the purposes of prostitution (Sexual Offences Act of 1985, Section 1).
Causing prostitution	Cause/incite prostitution for gain (Sexual Offences Act of 2003, Section 52).
Living on earnings of prostitution	Control for gain activities related to prostitution (Sexual Offences Act of 2003, Section 53).
Brothel keeping	Keep or manage a brothel (Sexual Offences Act of 1956, Section 33; Sexual Offences Act of 1967, Section 6; Sexual Offences Act of 2003, Section 55).
Exposure	Intentionally expose genitals (Sexual Offences Act of 2003, Section 66).
Incest	Have sex with an adult relative (Sexual Offences Act of 2003, Sections 64–65).
Intercourse with an animal	Intentionally perform act of penetration with a living animal (Sexual Offences Act of 2003, Section 69).
Obscenity	Publish an obscene article (Obscene Publications Act of 1959, Section 2); have an obscene publication for gain (Obscene Publications Act of 1964, Section 1)—but law also provides for the operation of licensed sex shops and cinemas.
Indecent display	Publicly display indecent matter (Indecent Displays [Control] Act of 1981, Section 1).

774 (15.2%) were receiving dipipanone (Home Office, 1984, p. 23). In 1991 virtually only methadone was being prescribed, with 9,521 (98.0%) of the 9,715 registered addicts receiving methadone and only 106 (1.1%) heroin (Home Office, 1992b, Table 3). In 1996, the last year for which figures are available, only 111 (0.06%) of the 18,940 registered addicts were receiving heroin, while 18,776 (99.1%) were being prescribed methadone (Home Office, 1997, Table 3).

The English regulate the publication and dissemination of pornography. However, they take what some might term a rather permissive attitude toward consensual sexual activity between adults. This is an area in which English law differs quite considerably from the laws in the United States, and it will be explored in greater detail later in this chapter.

Criminal defendants in England can assert the same types of defenses as can be raised in the United States. The boundaries of the defenses of self-defense, mistake, intoxication, duress, and necessity are very similar to those found in the different jurisdictions in the United States.

Issues of mental illness on the part of defendants are also dealt with very much as they would be in the United States. A defendant who is deemed incompetent to stand trial will have the case delayed and may be detained for treatment until competence is restored. One difference is that in England it is considered unfair for a defendant to be detained in this fashion unless there is a case to be answered. Thus, the judge may postpone consideration of the issue of competence to stand trial until after the prosecution has presented its case (Criminal Procedure [Insanity] Act of 1964, Section 4; Criminal Procedure [Insanity and Unfitness to Plead] Act of 1991, Section 2).

A defendant who is alleged to have been insane at the time she or he committed the criminal offense will have that issue determined through application of the M'Naghten rules. Until 1991, a verdict of "not guilty by reason of insanity" resulted in the defendant being sent to a mental hospital from which he or she could not be released without the consent of the Home Secretary. Under section 3 of the 1991 Criminal Procedure (Insanity and Unfitness to Plead) Act, however, in cases not involving offenses with fixed sentences (e.g., murder) the court may instead issue a guardianship, or supervision and treatment order, or discharge the defendant. As a result of the Homicide Act of 1957, a defendant charged with murder can plead diminished responsibility, a broader state of mental abnormality than that encompassed in the M'Naghten test of insanity. If the defense is successful, the defendant is found guilty of manslaughter rather than murder, thus avoiding the mandatory life sentence imposed upon conviction of murder.

One defense that has long been recognized in the United States, but until recently has not found any favor in England, is the defense of entrapment. In 1979 in *R. v. Sang* (1979) the House of Lords emphasized that: "It is well settled that the defense of entrapment does not exist in English law" (p. 263). It was noted that in such a case there was no denial that the defendant had committed the *actus reus* and possessed the requisite *mens rea*. The fact that a defendant had been incited to commit a crime by someone acting for the government did not provide a defendant with a defense to the criminal charge. Many crimes are committed at the instigation of others. Their Lordships did comment, however, that the fact that the person inciting commission of the crime was a police officer might justify mitigation of punishment. This situation has changed, however, as a result of the decision in *R. v. Loosely* (2001). In that case the House of Lords held that, although entrapment is not a substantive defense recognized by English law, a court may stay proceedings in a case, or exclude resulting evidence, when entrapment has occurred. The Lords considered that prosecuting an individual for a state-created crime was "an affront to the public conscience" and constituted an abuse of court process (p. 897). It remains to be seen how often proceedings are stayed as a result of entrapment.

By and large, English substantive criminal law is very similar to that found in U.S. penal codes. In the United States, however, there has been a tendency to use the criminal law to criminalize a far wider range of acts. Far more than in England, U.S. criminal law has been used to legislate morality—to give a particular moral position, generally a highly conservative one, the backing of the criminal law. This provides the criminal justice systems of the United States with a larger volume of cases to process. In addition, because offenses against morality generally involve situations with no clear victims, they pose added problems for U.S. law enforcement and prosecution officials. To some, the criminal law in the United States has been used to serve a symbolic, as opposed to an instrumental, function, "symboliz(ing) the public affirmation of social ideals and norms" with this "symbolic function unrelated to its function of influencing behavior through enforcement" (Gusfield, 1968, p. 56). Two examples of how the criminal law has been employed in such a fashion in the United States are gambling and consensual sexual activity. We will examine in some detail the way in which consensual sexual activity and gambling are regulated in England.

■ The Regulation of Consensual Sexual Activity in England

The English approach to consensual sexual activity is fundamentally different from that espoused in the United States. Basically it may be suggested that English law seeks to protect the right of individuals to engage in consensual sexual activity provided that no one else is harmed. A clear statement of the English philosophy is to be found in the report of the Wolfenden Committee, a committee that was set up in 1954 by the British government to examine the laws dealing with prostitution and homosexuality, and to recommend any changes in the laws that were considered desirable. In answering the question of what the role of criminal law should be in this area, the committee stated that:

> It is not, in our view, the function of the law to intervene in the private lives of citizens, or to seek to enforce any particular pattern of behaviour. . . . It follows that we do not believe it to be a function of the law to attempt to cover all the fields of sexual behaviour. Certain forms of sexual behaviour are regarded by many as sinful, morally wrong, or objectionable for reasons of conscience, or of religious or cultural tradition; and such actions may be reprobated on these grounds. (Committee on Homosexual Offences, 1963, para. 14)

More recently, in December 1982, the Criminal Law Revision Committee once again endorsed this approach in its working paper on offenses relating to prostitution and allied offenses (Criminal Law Revision Committee, 1982, p. 3).

The provisions of English criminal law reflect this philosophy. As in the United States, all types of forced sexual activity are proscribed by law and heavily punishable. Thus, forcible rape, which now covers sexual penetration of either a male or female victim, carries a possible maximum penalty of life imprisonment (Sexual Offences Act of 2003, Section 1). Likewise, the law seeks to protect the young and the incompetent. Thus, it is a criminal offense for a person to have sex or engage in sexual activity with a person who is under age 13 (Sexual Offences Act of 2003, Sections 5–7), who has a mental disorder impeding choice (Sexual Offences Act of 2003, Section 30), who is a family member under age 18 (Sexual Offences Act of 2003, Section 25), or who is under age 18 in a position of trust (Sexual Offences Act of 2003, Section 16).

Fornication and Adultery

With regard to persons who are considered old enough and competent to make decisions about their own sexual activities, the English approach differs vastly from the American in legislative practice as well as legislative philosophy. Neither fornication nor adultery is a criminal offense. Provided both parties are legally capable of giving consent, couples who engage in heterosexual premarital or extramarital sexual relations are not violating the criminal law. The one exception is that it is an offense to have sexual relations with a close relative, which is defined as a parent, grandparent, child, grandchild, brother, sister, half-brother, half-sister, uncle, aunt, nephew, or niece (Sexual Offences Act of 2003, Section 65). Even today, both fornication and adultery are prohibited by the laws of a number of U.S. states, though the U.S. Supreme Court decision in *Lawrence v. Texas* (2003), which struck down as unconstitutional a Texas statute that criminalized sexual activity between same-sex partners, renders these statutory provisions constitutionally suspect.

Deviant Sexual Activity

The law in England with regard to prostitution and deviant sexual activity cannot be stated as simply. The general approach is to allow competent adults to exercise their freedom of choice in sexual matters. This general approach is, however, subject to certain exceptions.

Sexual activity may be considered deviant either because of the nature of the parties involved (e.g., they are of the same sex) or because of the nature of the sexual activities in which the couple, whether heterosexual or homosexual, are indulging (e.g., fellatio, cunnilingus, buggery).

English law had long proscribed buggery, whether committed by a man with another man or a woman,[2] and acts of gross indecency (i.e., fellatio, masturbation) between men.[3] Consensual sexual acts committed by adult women (i.e., those over the age of 16) had, however, not been prohibited by the criminal law. As a result of the 1957 Wolfenden Committee report great changes were brought about in the laws affecting consensual sexual activities between adult males. As a result of the Sexual Offences Act of 1967, acts of buggery or gross indecency between men were no longer criminal if (1) they were committed in private, (2) the parties had consented, and (3) the parties had attained the age of 21 (Section 1). These provisions gave adult males far greater sexual freedom than they enjoyed before.

The limitations of this legislation should, however, be noted. First, these legislative provisions did not apply to members of the armed forces (Section 1) or to merchant seamen (Section 2). Second, it still remained a criminal offense for a man to be involved in group homosexual activities because the acts were not committed in private if a third party was present (Section 1). Third, it was still an offense for a man to solicit in public (Sexual Offences Act of 1956, Section 32). And lastly, it was still against the criminal law for a man to procure another man to commit homosexual acts with a third party (Sexual Offences Act of 1967, Section 4).

Two further aspects of this legislation should be noted. First, although consensual acts of buggery between males could no longer be criminal, a consensual act of buggery between a man and a woman still constituted a criminal offense, presumably even if the couple were married to each other. The maximum penalty that could be imposed upon conviction of this offense was life imprisonment (Sexual Offences Act of 1956, Section 37). This was an incredibly harsh sentence for consensual activities that, if committed by two males, would not have even constituted a criminal offense. Second, although a woman of 16 could consent to homosexual activities, a man could not do so until he was 21. These anomalies were finally rectified by the Sexual Offences (Amendment) Act of 2000.

Since 1967 the laws relating to consensual deviate sexual activities have been further liberalized. The age of consent has been lowered, first from 21 to 18 by the Criminal Justice and Public Order Act of 1994 (Section 145), and then from 18 to 16 by the Sexual Offences (Amendment) Act of 2000 (Section 1). In addition, the Criminal Justice and Order Act of 1994 extended the provisions of the 1967 Act to members of the armed forces and merchant seamen, exempting them from criminal liability for homosexual acts committed in private, although still allowing such acts to lead to dismissal (Section 146).

The 1967 Sexual Offences Act had significantly liberalized the laws dealing with male homosexual activity, and had presumably left homosexuals freer to pursue their activities, provided they did so in a discreet manner. One might, as a consequence, have expected a decrease in the number of homosexual offenses reported and prosecuted. However, as one researcher has documented, after the passage of this legislation "the recorded incidence of the offence of indecency between males [has] approximately doubled, and the number of persons prosecuted for the offence [has] trebled" (Walmslay, 1978, p. 400). Admittedly, there had been a marked decrease in the number of offenses of indecency between males in the 11 years prior to the passage of the Sexual Offences Act of 1967, and the number of offenses recorded in the period under study (1968 through 1976) never reached the peak set in 1955.[4] This does not, however, explain the upswing in the number of cases reported after passage of the 1967 legislation. Among the possible hypotheses for this phenomenon, the one most strongly advocated was that the 1967 Act itself had resolved previous uncertainty about the laws dealing with homosexual activities by delineating that when such acts were conducted in private they should be legal, but when performed in public or with a person deemed incapable of giving consent they should be illegal. This "provided the police with an up-to-date basis on which action could more confidently be taken against those involved in homosexual acts in public" (Walmsley, 1978, p. 405). This increased level of police action was facilitated by a change in criminal procedure that allowed summary trial for such offenses (Sexual Offences Act of 1967, Section 9[2]). All this tends to suggest that the police may be more willing and better able to enforce a law that does not attempt to place a blanket proscription on consensual activity, but instead outlines what aspects of that activity will not be tolerated.

Prostitution

English legal provisions dealing with prostitution follow the pattern observed in the other types of consensual sexual activity. Prostitution, which has traditionally been defined as occurring when "a woman offers her body for purposes amounting to common lewdness for payment in return" is not itself illegal (Per Darling, J. in *De Munck*, 1918, p. 637). The Wolfenden Committee acknowledged the futility of using the criminal law to proscribe such behavior when it stated that:

> *Prostitution is a social fact deplorable in the eyes of moralists, sociologists and, we believe, the great majority of ordinary people. But it has persisted in many civilizations throughout many centuries, and the failure of attempts to stamp it out by repressive legislation shows that it cannot be eradicated through the agency of the criminal law. (Committee on Homosexual Offences, 1963, para. 225)*

The criminal law is not, then, used to prohibit prostitution. It is, however, used to protect the young and the incompetent, to regulate the manner in which the business of prostitution is conducted, and to prevent prostitutes from constituting a public nuisance.

As mentioned previously, the young and the incompetent are protected against sexual exploitation by legal provisions that make it illegal to have sexual relations with such persons. It makes no difference that they are offered payment for their sexual services. In addition, it is an offense for a person to cause or incite a person under 18 to become a prostitute (Sexual Offences Act of 2003, Section 48) or to arrange or facilitate child prostitution (Section 50). Likewise, it is an offense to allow a person under 16 to be in a brothel (Children and Young Persons Act of 1933, Section 3; Children Act of 1989, Section 2).

The law does, in fact, take a rather negative attitude toward the involvement of third parties in the process of prostitution. Although the legislators have taken the approach that it is permissible for a competent adult woman to engage in prostitution, they have made "illegal the activities of those who promoted prostitution and encouraged women to become prostitutes, and exploited prostitution for their own financial advantage" (Criminal Law Revision Committee, 1982, p. 2). Thus, it is a criminal offense for a person to cause or incite prostitution for gain (Sexual Offences Act of 2003, Section 52); to "control any of the activities of another person relating to that person's prostitution" (Sexual Offences Act of 2003, Section 53); to keep, manage, or act or assist in the management of a brothel (Sexual Offences Act of 1956, Section 33); for a landlord to allow his premises to be used as a brothel (Sexual Offences Act of 1956, Section 34); and for the tenant or occupier of any premises knowingly to permit the whole or part of the premises to be used either as a brothel (Section 36) or for the purposes of habitual prostitution (Sexual Offences Act of 1956, Section 35).

The general approach embodied in the English law dealing with prostitution is, then, basically civil libertarian in nature, allowing the prostitute to ply her trade, provided this is done in a manner that does not constitute a public nuisance, and provided that third parties are not involved in promoting, or reaping the benefits of, the trade.

There is, also, perhaps a certain gentleness to be discerned in the law towards the prostitute. As the Criminal Law Revision Committee pointed out in 1982:

> *However much the public may disapprove of the prostitute, the law, we think, must always remember the human nature of prostitutes. They have to have somewhere to live and if they are not to outrage decency by following their trade in public they must have somewhere to carry it on. (p. 7)*

This philosophy is to be seen underlying the committee's proposal to narrow the meaning of the term *brothel* so that "the law should allow two prostitutes who share a home to work there as prostitutes without the risk of being prosecuted for keeping or managing the establishment" (p. 21). The reason advanced for this proposal was that the committee had "received much evidence to the effect that a relaxation of the law on the lines suggested might, without detriment to the public, make a contribution to the security of individual prostitutes (and their children) and their quality of life" (p. 21). The current problem of women being trafficked into the United Kingdom to work in brothels and massage parlors (see, e.g., Kelly & Regan, 2000) does, however, appear to militate against the legalization of these establishments.

The prostitutes themselves may run afoul of the law when, like any other citizens, they breach any of its provisions, such as engaging in disorderly conduct. Only in one area have their professional activities received special attention. Because public solicitation for prostitution has been considered to constitute a public nuisance, such behavior has long been proscribed by the criminal law, whether engaged in by the prostitute herself or by someone on her behalf. Under the present legislative provisions it is an offense "for a common prostitute to loiter or solicit in a street or a public place for the purpose of prostitution" (Street Offences Act of 1959, Section 1). A similar provision had made it illegal "for a man persistently to solicit or importune in a public place for immoral purposes" (Sexual Offences Act of 1956, Section 32). Although it was generally believed that this provision and the earlier laws it replaced were aimed at penalizing the prostitute's pimp as he solicited for clients, it is clear that it was used mainly to deal with homosexual soliciting (Criminal Law Revision Committee, 1982, p. 36; Grey, 1975, p. 333).

Male prostitution is, in fact, a topic that until recently had not been dealt with in as great detail by the criminal law as female prostitution. The legal situation has, however, always been very similar. Since the passage of the Sexual Offences Act of 1967, consensual homosexual acts carried out in private by competent adult males have not been illegal. When these acts are performed "for payment in return" they are still legal. However, as already noted, it has been against the law for a man to solicit in public and for a man to procure another man to commit homosexual acts with a third party. In addition, it was an offense under the Sexual Offences Act of 1967 for a man or a woman knowingly to live wholly or in part on the earnings of prostitution of another man (Section 5). In addition, the Sexual Offences Act of 1967 broadened the term *brothel* to encompass the use of a place for homosexual activities (Sexual Offences Act of 1967, Section 6). The gender-neutral language of the Sexual Offences Act of 2003 includes both males and females in its provisions prohibiting third parties from profiting from the activities of prostitutes (Sections 52–53).

One further concern that has been raised is that the law has focused on the solicitation activities of the prostitute while failing to pay adequate attention to similar actions on the part of potential customers. Particularly troublesome was the nuisance caused by men cruising around in automobiles soliciting women for sexual purposes. The Criminal Law Revision Committee suggested in 1982 that such activities be circumscribed by law (pp. 39–41), and this recommendation was enacted into law in Sections 1(4) and 4(1) of the Sexual Offences Act of 1985. The passage and enforcement of this legislation resulted in a reduction in the volume of traffic and prostitution and prostitution-related activities in areas known for prostitution (Matthews, 1993). Such displacement as occurred was to areas such as "shopping streets, which residents found less objectionable" (Matthews, 1993, p. iii).

Because prostitution is legal, prostitutes are free to advertise their services. However, if an advertisement is unduly graphic, and thus likely to "deprave or corrupt," it could contravene provisions of the Obscene Publications Act of 1959. A popular method that prostitutes in London and some other cities, such as Brighton and Norwich, use to advertise their service has been to place business cards in telephone booths. It has been estimated by British Telecom that about 14 million cards advertising some 350 to 400 businesses are placed in telephone boxes each year (Home Office, 2004, p. 66). Not surprisingly, this has been perceived to constitute a nuisance, and as a result the Criminal Justice and Police Act of 2001 made it a criminal offense to place in a public telephone box a card that advertises the services of a prostitute (Section 46).

The English continue to be concerned about certain aspects of prostitution, in particular protecting young and vulnerable people from being ensnared in prostitution, the problems caused by street prostitution, and the interrelationship between prostitution and drug use. Starting in 2000 the government has funded a number of research projects designed to develop effective strategies to reduce the number of young people engaged in prostitution, alleviate the crime and disorder associated with street prostitution, and assist those engaged in prostitution in leaving the trade (Hester & Westmarland, 2004, p. v).

In 2004, the government published *Paying the Price: A Consultation Paper on Prostitution* (Home Office, 2004). This paper highlighted the problems caused by prostitution for both individuals and communities and stressed the need for a "clear and coherent strategy" to reduce these problems (p. 7). More specifically, the paper focused on closing the pathways taken into prostitution by the young and vulnerable, helping and supporting those abused by prostitution, taking firm action against those who groom or coerce young people into prostitution and against pimps and drug dealers, lessening the impact of prostitution on neighborhoods, and placing more emphasis on curbing the demand side of prostitution. By raising questions at the end of each chapter, the paper sought comments on the issues that had been raised.

In January 2006, the government issued its report *A Coordinated Prostitution Strategy* (Home Office, 2006). This report discussed the 861 responses received to the consultation paper and recommended a strategy that sought to "challenge the view that street prostitution is inevitable," "achieve an overall reduction in street prostitution," "improve the safety and quality of life in communities affected by prostitution," and "reduce all forms of commercial sexual exploitation" (Home Office, 2006, p. 1).

The report specifically rejected the approach of adopting managed areas where no arrests are made for prostitution offenses, although other laws, such as drug laws, are enforced (p. 8). It stressed the need for local assessments of the problems posed by prostitution and a coordinated response to alleviate those problems. To help localities address their problems, the government has set up an interactive self-help site (www.together.gov.uk). The proposed approach includes a focus on "prevention" by raising awareness of the realities of prostitution; "tackling the demand" by disrupting street sex markets, in particular by enforcing the law against curb crawling; "developing routes out" of prostitution by providing housing, health services, drug treatment, and advice on education, training, and employment; "ensuring justice" by bringing to justice individuals who exploit and/or abuse individuals through prostitution; and "tackling off-street prostitution" as well as street prostitution in order to take action against all forms of commercial sexual exploitation (Home Office, 2006, pp. 2–12).

■ The Regulation of Gambling in England

Gambling is a general term that refers to "the act of playing at games, particularly games of chance, for money or other valuable stake, or the agreement to risk or hazard such stake on an uncertain event" (Corpus Juris Secundum, 1943, Vol. 38, p. 51). Although gambling is often used interchangeably with the older term *gaming*, English law uses the word *gaming* in its narrower sense of meaning "playing a game of chance for a prize" (Gambling Act of 2005, Section 6).

Gambling takes on many forms. It encompasses card games, betting on the outcome of sporting or other events, playing slot machines or bingo, and buying a raffle or lottery ticket. It can be conducted in just about any location: at places specifically designated for such activities, such as racetracks and casinos; at places not so designated, such as pool halls and social clubs; and more generally on the street, in the home, and over the Internet. Gambling can occur between two or more friends who simply bet a sum of money on a particular outcome, or it can be conducted by third parties who take a percentage of the money wagered in order to cover administrative expenses and earn some profit.

As is the case with consensual sexual activity, the English have not depended on criminal law to regulate gambling. Again, a more liberal attitude can be discerned toward behavior that is considered immoral in some quarters. Social gambling is perfectly legal in England. Commercialized gambling, which is freely available and highly visible, is subject to governmental control.

The nature of this governmental control, however, has tended to differ from that imposed in the United States. To some extent this may be the result of a different underlying philosophy. In England there is greater official acceptance of gambling. The framework under which commercialized gambling operates has exhibited greater concern for protecting the individual player, and less concern than is manifested in the United States with the government making money through such ventures. As one commentator has noted "the paradigm of legal intervention in Britain" has been the "player protection" model, a model that "aims to minimize the impact of gaming on the individual player" (Miers, 1980, p. 170). Other commentators have observed that "the British are relatively uninterested in the revenue potential of gambling taxes" (Beare & Hampton, 1984, p. 160). For, as the authors go on to explain, "British gambling laws are [therefore] not intended to provide new tax revenues. The primary purpose of the legislation is to control the types and amount of gambling and thereby reduce the harmful effects upon British society" (p. 160).

Corroboration of this attitude can be found in numerous English governmental reports. The 1949 Royal Commission on Betting, Lotteries, and Gaming, for example, stated that:

> *The objective of gambling legislation should be to interfere as little as possible with individual liberty to take part in the various forms of gambling but to impose such restrictions as are desirable and practicable to discourage or prevent excess. (p. 55)*

Some 29 years later another Royal Commission explicitly endorsed this philosophy as being "appropriate . . . in a liberal democracy" (Royal Commission on Gambling, 1978, p. 4). Although precautions were to be taken against whetting the appetite for gambling, it was recognized that the existence of an "unstimulated demand" for any type of illegal gambling should result in serious consideration of its legalization (p. 378).

Since the 1990s, however, this focus on protecting players has undergone some change. The 1993 National Lottery etc. Act established the National Lottery, which from its inception in 1994 through 2005 generated "48 billion [pounds] in ticket sales, raised 16 billion [pounds] for good causes, and transformed Britain with 180,000 grants" (National Lottery Commission, 2005, p. 1). The English have clearly recognized the appeal of using state-sponsored gambling to generate revenue for the state. However, protecting individuals from the harms incurred through gambling and ensuring that gambling is conducted in a fair manner are still of great concern to the English. The Gambling Act of 2005, which revamped the English law on gambling, set the following as its objectives:

> *(a) preventing gambling from being a source of crime or disorder, being associated with crime or disorder or being used to support crime,*
> *(b) ensuring that gambling is conducted in a fair and open way, and*
> *(c) protecting children and other vulnerable persons from being harmed or exploited by gambling. (Gambling Act of 2005, Section 1)*

As noted above, social gambling is legal in England, with that legitimacy acknowledged by statute (formerly the Betting, Gaming and Lotteries Act of 1963, Section 32[1], now the Gambling Act of 2005, Section 296). Thus, there is no prohibition against adults engaging in friendly bets (betting) or playing games of chance for winnings in money or money's worth (gaming), provided that the gambling takes place in private and there is no charge for participation (Gambling Act of 2005, Schedule 15). Even juveniles can participate in social gambling (formerly the Betting, Gaming and Lotteries Act of 1963, Sections 32[2]–32[3], now the Gambling Act of 2005, Section 46).

Commercialized gambling is likewise freely available, though subject to governmental control. Here, as with consensual sexual crimes, there are provisions protecting the young. Thus, it is an offense to have a commercial betting transaction with a person under the age of 18 or to send materials promoting betting to a person under 18 (formerly the Betting, Gaming and Lotteries Act of 1963, Sections 21–22, now the Gambling Act of 2005, Section 46). Certain forms of gambling, such as the football pools and the lottery, are, however, exempted from these provisions.

To a citizen of the United States, the pervasiveness of commercialized gambling in England may appear overwhelming. Not only is there betting at the numerous horse race and greyhound tracks dotted throughout the country, gambling at licensed casinos and clubs, and wagering at a multitude of off-track betting offices, but there are also slot machines in public houses (pubs), games on piers that provide small monetary prizes to contestants of all ages, commercial football (soccer) pools circulating in the home and workplace, and odds given in newspapers and accepted in betting shops for all kinds of sporting and nonsporting events. Bets can even be placed on who will win the general election or what name will be given to the newest member of the royal family. Nothing, it appears, is sacrosanct.

This subjective impression is validated by more objective research data. Thus, 94% of the respondents in a 1977 national survey of persons aged 18 and over stated that they engaged in some form of gambling. For most (63%) gambling was an occasional activity. Those who gambled regularly indulged in betting on the football (soccer) pools (35%) and/or horse or dog racing (9%). Only 0.1% said that they visited a casino regularly (Royal Commission on Gambling, 1978, p. 11). A 2004 national survey showed that both the prevalence and the pattern of gambling had changed since 1977. Only 71% of those surveyed said that they had gambled in the past 12 months. In addition, only 5% said that they bet on the football pools, while 61% had participated in the National Lottery. Interestingly, only 1% said they had bet on the Internet. Among regular gamblers the most popular forms of gambling were the lottery and the football pools (Creigh-Tyte & Lepper, 2004).

Internet gambling is treated in much the same way as other forms of "remote" gambling, which is defined under the Gambling Act of 2005 as including the use of the telephone, Internet, radio, television, or "any other kind of electronic or other technology for facilitating communication" (Section 4). This is in stark contrast to the United States, where President Bush recently signed into law the Gambling Enforcement Act of 2006, which seeks to make online gambling illegal by prohibiting the use of credit cards, checks, or electronic fund transfers for betting.

Although all of this commercialized gambling is freely available in England, it is all subject to government control and supervision. Until passage of the Gambling Act of 2005, a variety of licensing authorities and supervisory boards regulated the different forms of gambling. Thus, the Horserace Betting Levy Board was responsible for approving racecourses where betting takes place (British Information Services, 1974, p. 2), while the Horserace Totalisator Board was responsible for authorizing pool betting on races at the racetracks (Betting, Gaming and Lotteries Act of 1963, Section 4). The bookmakers who took the bets had to obtain permits issued by the local licensing authority, a committee of the local licensing justices (Section 1). Local authorities were also responsible for licensing greyhound tracks (Royal Commission on Gambling, 1978, p. 105), authorizing public lotteries for charitable purposes (Lotteries Act of 1976, Section 5 & Sched. 1), and registering and supervising the promoters of football (soccer) pools (Betting, Gaming and Lotteries Act of 1963, Section 4). Finally, gaming machines required a license issued by the Board of Customs and Excise (British Information Services, 1974, p. 13) and came under the general regulation of the Gaming Board.

Until passage of the Gambling Act of 2005, the Gaming Board was the body with overall responsibility for overseeing gaming in England. It had been created to remedy the unsatisfactory conditions that had arisen as a result of the ill-drafted 1960 Betting and Gaming Act, which had legalized gaming premises and had unintentionally spawned the birth of virtually uncontrolled legal gaming.

As a result of the 1968 Gaming Act, government control over gaming was strengthened considerably. Gaming premises could be established only in certain permitted areas. Responsibility for granting the certificates required of owners and employees of gaming premises, and of persons supplying or selling gaming machines, rested with the Gaming Board. The actual operating licenses were issued by the local licensing authorities, who received advice and guidance from the Gaming Board. The Gaming Board also assisted the police in enforcement of the law through its staff of trained inspectors, who enjoyed the right of access to licensed premises. The Gaming Board showed the strength of its authority in the late 1970s when it revoked the licenses of some major corporations (including Playboy) for violations of operating provisions.

As indicated above, the Gambling Act of 2005 totally revamped the manner in which gambling is regulated in England. A 2001 government report had recognized the need to update and simplify the regulation of gambling in Britain (Department for Culture, Media and Sport, 2001), and government proposals were set out in a white paper issued in 2002 (Department for Culture, Media and Sport, 2002). In that white paper, the government stressed the need to bring "all operators of commercial gambling within a single system of licensing and regulation" in order to promote "fairness and efficiency"(p. 7).

Under the Gambling Act of 2005 a single authority, the Gambling Commission, has been established to oversee all forms of commercial gambling except for the National Lottery and spread betting, which are to remain under the province of the National Lottery Commission and the Financial Services Authority, respectively. The Gambling Commission is responsible for regulating such diverse forms of gambling as casino gaming, horse and greyhound racing, the football pools, bingo, lotteries, gaming machines, and remote gambling (gambling using such technology as the Internet, telephone, or television). The commission is to issue codes of practice for gambling and provide guidance to local authorities who are responsible for licensing gambling premises. The commission is also responsible for issuing operating licenses to those who provide the public with these forms of gambling. Conditions may be placed on these licenses, and violations of license conditions may result in administrative and/or criminal sanctions. There is a Gambling Appeals Tribunal authorized to hear appeals from decisions of the Gambling Commission. Although the Gambling Act was passed in 2005, full implementation of the Act is taking some time.

From the above rather succinct and condensed description of gambling in England, it may be seen that both social and commercial gambling are permitted by law. Although commercial gambling is subject to government regulation, it is generally the more potentially exploitive forms of gambling that receive greater regulation and stricter regulation. As a result of the far tighter restrictions imposed by the 1968 Gaming Act, "the number of clubs offering hard gaming was reduced from over 1,000 in 1967 to about 120 in 1970" (British Information Services, 1974, p. 8).

Overall, because of the legal structure under which gambling operates, the English criminal justice system is faced with fewer legal issues than the systems in the United States. Sports betting, for example, does not present the same problems in England as it does in the United States. Betting on football (soccer) pools has long existed without problem. As the Royal Commission on Gambling acknowledged: "The football pools are a well established feature of the British scene, and are not only popular but widely accepted as a harmless activity" (1978, p. 128).

After betting shops were legalized, it was reported that "illegal bookmaking has been virtually unknown" (Beare & Hampton, 1984, p. 193). The introduction of, and subsequent increases in, betting taxes, however, led to a slight resurgence in illegal bookmaking (Royal Commission on Gambling, 1978, pp. 24–25). This fact underscores the need for legal gambling to remain competitive with potential illegal gambling, and serves as a warning to governments of the danger inherent in overtaxing legal gambling winnings.

Examination of available criminal justice statistics supports these claims. Thus, for example, from 1979 to 1985 less than 400, and from 1986 to 1990 less than 200, and from 1991 through 2003 less than 100 persons a year were found guilty or cautioned for indictable betting or gaming offenses (Home Office, 1992a, p. 119; Home Office, 2004, p. 46). And these offenses included violations of license provisions by license holders.

■ Conclusion

To a great extent, then, English substantive criminal law is very much like that found in U.S. penal codes. This is hardly surprising because nearly all of these codes trace their roots back to English law. In some ways, however, English law has developed in a rather different fashion.

Historically, case law played a major role in the development of English law. Although most of English criminal law has been placed on a statutory basis, case law is still a more prominent feature of the legal landscape in England than in the United States. Unless case law has been superseded by statute it continues in effect, and quite a few common law offenses still exist, such as murder, manslaughter, and false imprisonment (see Table 3-1 earlier in the chapter).

Although the criminal laws of the United States derive from England, the division of crimes into felonies and misdemeanors was abolished in England in 1967. Offenses are now classified as arrestable or nonarrestable offenses, with broader arrest powers accompanying the former category. The court processing of a case is determined by whether the case involves a summary or indictable offense. Some offenses may be tried in either way. This is a versatile feature of the English court process that merits further attention, and will be examined in greater detail in Chapter 8.

The type and nature of both the criminal charges that may be brought against a defendant and the defenses that may be raised by a defendant are substantially the same in England as in the United States. One defense that the English still do not recognize is the defense of entrapment. The fact that the defendant was incited by someone working for the government to commit a crime does not negate the findings that the defendant committed the *actus reus* and possessed the requisite *mens rea*. As a consequence, the English believe that the defendant should be held liable for the wrongful act that has been committed. However, incitement by a government official may justify mitigation of the punishment or, since the *Loosely* decision in 2001, a stay of proceedings. This failure to provide a defendant a substantive defense to a criminal charge as a result of questionable actions on the part of government officials is a distinctive feature of English legal practice. It is open to question whether such an approach should be adopted in the United States.

Although by and large English and U.S. criminal law provisions are very similar, there are some major differences between them in the laws regulating consensual sexual activities and gambling. The criminalization of more of these types of activities in the United States means that the criminal justice systems of the United States have a wider sphere of human behavior to control. This wider area of responsibility produces additional problems for the criminal justice systems of the United States. Many of these problems, such as the difficulty of obtaining evidence to sustain convictions in such cases and the added danger of police corruption, will be explored in later chapters.

There is, indeed, a considerable difference that should not be underestimated between having restrictive laws on the books that are rarely enforced and disregarded with impunity, and having more permissive laws that are enforced with some regularity. Although criminal law is to a great extent used in the United States to promulgate a conservative moral position on consensual sexual activities and gambling, its lack of enforcement and disregard by a large segment of the public brings its effectiveness into question. Although some may extol the symbolic function of the law, it can be argued that a law that is broken consistently with almost total impunity harms a moral position more than it helps. On those rare occasions when such a law is enforced there is likely to be a feeling of outrage at the selective enforcement of the law. One can understand the need for strong procedural safeguards for the citizenry when it is faced with oppressive substantive criminal law.

Criminal law has not been used in England to the same extent as it is in the United States as a mechanism for promulgating a conservative moral position on consensual sexual activities and gambling. Neither fornication nor adultery is unlawful. Consensual homosexual activity between competent adults is permissible. Prostitution is not per se illegal, although a number of activities associated with prostitution are proscribed by criminal law. Gambling is subject to regulation rather than prohibition. Competent adults in England have, then, more freedom than their counterparts in the United States to indulge in vices. By not criminalizing the activities of a large segment of the population, the English have not demanded as much of the criminal law, nor have they overburdened the criminal justice system in the same way that has been done in the United States. The English experience with liberalization of laws governing homosexual activities has indicated that a more narrowly written and more narrowly focused law may allow for more concentrated efforts on suppressing the more harmful aspects of forbidden activity. By allowing the citizenry greater sexual freedom and liberating them from the threat of criminal prosecution, the English have perhaps contributed to their psychological well-being, and enhanced their overall quality of life.

The implications of this are important for policymakers in the United States. By delineating under what circumstances consensual sexual activity and gambling will not be tolerated, rather than applying a blanket proscription of all such behavior, the English have achieved greater congruence between the dictates of their laws and popular practice, and have provided the criminal justice system more manageable laws to enforce. The extent to which all of this actually impacts upon the operation of the criminal justice system will be examined in later chapters.

CHAPTER SPOTLIGHT

- English law is derived from two main sources: (1) statute—law written and passed by the legislature, and (2) case law—law created by the decisions of judges in particular cases.

- Case law also is known as the "common law." This dates from the twelfth century when Henry II sent judges to hear cases on circuits, one judge moving between a number of different courts. This ensured that the laws around England were consistent or "common."

- Case law was able to create new offenses and expand old offenses without intervention from the legislative branch of government.

- Unlike in the United States, English law is not codified in a general code, although now most law is found in statutory form.

- Where case law has not been replaced by statute, case law remains in effect.

- England and Wales is a single "unitary" legal jurisdiction. All laws apply evenly across the jurisdiction, unlike in the United States where laws vary among the states.

- The United Kingdom does not have a single written document as a constitution. The courts do not have to enforce laws in light of the constitution, as is the case in the United States.

- The United Kingdom is a signatory of the European Convention on Human Rights (ECHR). The courts can declare that primary legislation is "incompatible" with the ECHR and ask the legislature to change it. The courts may strike down secondary legislation.

- Since 1967, English law no longer divides offenses into felonies and misdemeanors. There are now arrestable and nonarrestable offenses. An arrestable offense is one where the sentence, fixed by law, is five years or more in prison. There is no fixed sentence for nonarrestable offenses.

- Unlike in the United States, the courts' processing of cases is not determined by whether they are felonies or misdemeanors but by whether they are summary or indictable offenses. Summary offenses can only be tried by the lower courts, before a bench, in a magistrate's court. Indictable offenses, such as murder, rape, or robbery, are heard in the higher courts, the Crown courts, before a jury. Some offenses can be tried either way, which means they can be tried in either court.

- For the most part, English substantive criminal law offenses are very much like the criminal offenses found in U.S. criminal codes. There are offenses against the person, against property, and against the state, against public order, and against morality.

- English law makes it an offense to manufacture, sell, or transfer a firearm without a certificate. It is also an offense to possess a firearm without a certificate.

- English law has a framework for classifying illegal drugs (class A, class B, etc.) and laws proscribing their importation, production, or possession.

- English law takes a more permissive attitude to consensual sexual activity between adults. The age of consent for both heterosexual and homosexual activity is 16.

- English law provides the same defenses as in the United States, such as self-defense, mistake, intoxication, and duress. The boundaries of these are similar to those found in U.S. jurisdictions.
- The defense of entrapment does not exist in U.K. law. However, evidence obtained by entrapment may be excluded during trial.
- English law has been used less than U.S. law to legislate morality. Gambling and prostitution are, for example, not illegal, although disorderly conduct and public solicitation for prostitution are.
- Because prostitution is legal, it is legal for prostitutes to advertise their services, so long as the advertisements are not unduly explicit.
- Gambling, "playing a game of chance for a prize," is legal and regulated in England and Wales. The Gambling Commission oversees all gambling (except the National Lottery) and issues operating licenses to those who provide the public with gambling.

KEY TERMS

Arrestable offenses: Include all offenses, and attempts to commit offenses, for which the sentence is fixed by law (e.g., murder) or for which a convicted defendant may receive a sentence of five years imprisonment or more.

Case law: Judge-made law that emanates from the decision in a particular case, but which through the doctrine of precedent has impact beyond the parties involved in that case and the particular facts of that case and is binding on lower courts in subsequent cases.

Common law: Case law that is applied to the whole land.

County court: Run by the sheriff.

Hundred court: Presided over by the bailiff of the hundred.

Indictable offenses: Offenses, such as murder, rape, and robbery, that are heard in the higher courts, namely the Crown courts, before a jury.

Manor court: Held by the lord of the manor to administer justice among his tenants.

Nonarrestable offenses: All other offenses that are not classified as arrestable offenses.

Offenses against morality: Include such offenses as drug use and prostitution. These offenses generally involve situations with no clear victims.

Offenses against the public order: Include such offenses as riot, violent disorder, affray, harassment, and incitement to racial or religious hatred, all of which were codified in the Public Order Act of 1986.

Offenses against the state: Composed of such offenses as treason, sedition, terrorism, and disclosure of official secrets.

Procedural criminal law: The procedures that are set out for the detection, apprehension, trial, and conviction of suspected offenders (e.g., Miranda rights).

Statutes: Acts of the legislative branch of government that declare, command, or prohibit something.

Substantive criminal law: The types of acts, such as murder and driving while intoxicated, that are defined as crimes.

Summary offenses: Offenses such as certain driving and gambling offenses that can be tried only by the lower magistrates' courts in a "bench trial" without a jury.

PUTTING IT ALL TOGETHER

- Should all criminal law offenses and defenses be fully set out in statutory form?
- Should a defendant who proves that he or she was entrapped into committing a criminal offense be acquitted of the criminal offense he or she committed?
- What types of adult consensual activities should be prohibited by the criminal law? What should be the basis for deciding this?
- Should your state adopt the English model of legalized prostitution? Would this approach be better or worse than the Nevada system of state sanctioned houses of prostitution?
- What types of gambling activities should be prohibited by the criminal law? What should be the basis for deciding this?

ENDNOTES

1. *Mapp v. Ohio*, 367 U.S. 643 (1961) (protection against unreasonable search and seizure); *Malloy v. Hogan*, 378 U.S. 1 (1964) (privilege against self-incrimination); *Benton v. Maryland*, 395 U.S. 784 (1969) (protection against double jeopardy); *Gideon v. Wainwright*, 372 U.S. 335 (1963) (right to assistance of counsel); *Klopfer v. North Carolina*, 386 U.S. 213 (1967) (right to speedy trial); *Duncan v. Louisiana*, 391 U.S. 145 (1968) (right to trial by jury); *Pointer v. Texas*, 380 U.S. 400 (1965) (right to confrontation of witnesses).

2. In 1553, in the reign of Henry VIII, buggery, which was originally an ecclesiastic offense, was made a secular offense punishable by death.

3. Acts of gross indecency committed in private between consenting parties first became criminal offenses in 1885 (Criminal Law Enactment Act of 1885, Section II).

4. There were 2,322 offenses recorded in 1955. This fell to 840 in 1967. Between 1967 and 1973 the number of offenses known rose from 840 to 1,567. The average yearly number in the years 1973 through 1976 was just under 1,660 (Walmsley, 1978, p. 400). From 1977 to 1987 the number of offenses ranged from a high of 1,706 in 1978 to a low of 857 in 1985. The yearly average was 1,243 (Home Office, 1988, p. 34). In 1988 there were 1,306 such offenses recorded, in 1989 2,022, in 1990 1,159, and in 1991 965 (Home Office, 1992a, p. 47).

REFERENCES

Beare, M. E., & Hampton, H. (1984). *Legalized gambling: An overview.* Ottawa, Canada: Ministry of the Solicitor General.

British Information Services. (1974). *Control of gambling in Britain.* London, England: Central Office of Information.

Committee on Homosexual Offenses and Prostitution. (1963). *The Wolfenden report.* New York: Stein & Day.

Creigh-Tyte, S., & Lepper, J. (2004). *Gender differences in participation in, and attitudes towards, gambling in the UK: Results from the 2004 NOP survey.* London, England: Department for Culture, Media and Sport.

Criminal Law Revision Committee. (1966). *Eighth report: Theft and related offenses.* London, England: H.M.S.O.

Criminal Law Revision Committee. (1982). *Working paper on offences relating to prostitution and allied offences.* London, England: H.M.S.O.

De Munck. (1918). 1 K.B. 635.

Department for Culture, Media and Sport. (2001). *Gambling review report.* London, England: H.M.S.O.

Department for Culture, Media and Sport. (2002). *A safe bet for success.* London, England: H.M.S.O.

Grey, A. (1975). Sexual Law Reform Society Working Party report. *Criminal Law Review*, 323–335.

Gusfield, J. R. (1968). On legislating morals: The symbolic process of designating deviance. *California Law Review, 56*, 54–73.

Hester, M., & Westmarland, N. (2004). *Tackling street prostitution: Towards an holistic approach.* London, England: Home Office Research Unit.

Home Office. (1984). *Statistics on the misuse of drugs in the United Kingdom, 1983.* Surbiton, Surrey, England: Home Office Statistical Department.

Home Office. (1988). *Criminal statistics: England and Wales 1987.* London, England: H.M.S.O.

Home Office. (1992a). *Criminal statistics: England and Wales 1991.* London, England: H.M.S.O.

Home Office. (1992b). *Statistics of drug addicts notified to the Home Office, United Kingdom, 1991.* Croydon, Surrey, England: Home Office Research and Statistics Department.

Home Office. (1997). *Statistics of drug addicts notified to the Home Office, United Kingdom, 1996.* Croydon, Surrey, England: Home Office Research and Statistics Department.

Home Office. (2004). *Paying the price: A consultation paper on prostitution.* London, England: H.M.S.O.

Home Office. (2006). *A coordinated prostitution strategy and a summary of responses to paying the price.* London, England: H.M.S.O.

Kelly, L., & Regan, L. (2000). *Stopping traffic: Exploring the extent of, and responses to, trafficking in women for sexual exploitation in the U.K.* London, England: Home Office.

Lawrence v. Texas. (2003). 537 U.S. 1044.

Maitland, F. W. (1965). *The constitutional history of England.* Cambridge, England: Cambridge University Press.

Matthews, R. (1993). *Kerb-crawling, prostitution and multi-agency policing.* London, England: Home Office Police Department.

Miers, D. (1980) Eighteenth century gaming: Implications for modern casino control. In J. A. Inciardi & C. E. Faupel (Eds.), *History and crime: Implications for criminal justice policy* (pp. 169–192). Beverly Hills, CA: Sage.

National Lottery Commission. (2005). *A lottery for the future: Shaping the structure of the third competition.* London, England: H.M.S.O.

R. v. Loosely. (2001). 4 All. E. R. 897.

R. v. Sang. (1979). 3 W.L.R. 263.

Royal Commission on Gambling. (1978). *Final report.* London, England: H.M.S.O.

Walmsley, R. (1978). Indecency between males and the Sexual Offences Act 1967. *Criminal Law Review*, 400–407.

Chapter Resources

The English Crime Problem

4

It is vital to measure crime accurately if we are to tackle it effectively.

—The Right Honorable David Blunkett MP, Home Secretary, July 2001
(National Crime Recording Standard)

Chapter Objectives

After completing this chapter, you will be able to:

- Describe the differences, similarities, and methodological problems of the English and U.S. methods for crime data collection.
- Compare and contrast the trends in crime rates for violent and property crimes for England and the United States in the past and the present.
- Outline why there were problems with the old methods of collecting crime statistics and how these methods were changed in both countries.
- Explain the effects of new crime recording standards in the United States and England.
- Explain how the victim reporting of crimes varies between the United States and England.
- Discuss how the gap between English and U.S. violent and property crime rates is narrowing.

Introduction

Before embarking upon a comparison of the criminal justice systems of England and the United States it is important to investigate thoroughly the nature and volume of criminal offenses committed in the two nations, because differences on either of these dimensions are likely to impact heavily on the operation of the criminal justice systems. In Chapter 3, we saw that the more serious criminal offenses such as murder, forcible rape, robbery, and burglary are defined in a similar fashion in both nations. Indeed, the same kinds of wrongful acts against persons and property are considered to constitute criminal offenses. This is so whether we are dealing with more serious offenses or lesser ones. However, in the realm of so-called victimless crimes, in particular consensual sexual and gambling activities, we observed that Americans have been more inclined than the English to use the criminal law to regulate these activities. Thus in the United States we entrust a far wider range of human activities to the criminal law and its agencies (law enforcement, the courts, and corrections) for regulation than do the English. As a consequence, we might expect the criminal justice agencies in the United States to be far more overburdened than their English counterparts.

But what about the extent of general criminal offenses against both person and property? We have already stated that similar wrongful acts are against the criminal laws of both nations. How do the volume and nature of these types of offenses compare? In this chapter we will compare English and U.S. crime data.

■ Sources of Crime Data

Many sources of data are available for comparing crime rates in different nations. We are interested in the actual numbers of offenses committed by individuals. This information might, at least in theory, be accessed by **self-report studies** in which those sampled are asked whether they have engaged in particular types of criminal activity. However, in practice, few respondents are likely to give much, if any, information that is going to incriminate them. This is particularly likely to be the case if the information involves serious offenses and/or this information is not already known by the authorities. As a consequence, self-report studies have not been regularly conducted on large samples, and when such studies have been conducted they generally have not investigated the commission of serious offenses.

A second potential source of information is people who have been the victims of criminal offenses (**victimization surveys**). Although such information is by definition not available for so-called victimless crimes (drug activities, consensual sex, and gambling offenses, for example) and for murder, it can be obtained for a variety of offenses against both person and property. Although victims are a potentially rich source of information, there are a number of problems with victimization surveys. First, because it is generally not feasible, or even desirable, to survey the whole population in which the surveyors are interested, only a certain percentage of that population is likely to be sampled. As a consequence, victimization figures derived from victim surveys are not actual figures but estimates based on the samples drawn from the population, and are subject to the limitations of the sampling procedures employed. Second, not everyone has been the victim of a criminal offense, and few are likely to have been the victims of serious criminal offenses such as forcible rape and robbery. Thus, large samples are required to obtain accurate estimates for infrequently committed offenses, and even with large samples, estimates for those offenses may occasionally be unreliable. Third, all information gathered is obtained from the victims without third-party verification and is subject to victim definition of the situation and victim memory, including both backward and forward telescoping into the time period covered by the survey.

Despite these disadvantages, highly useful information may be obtained from crime victim surveys, which can provide valid and reliable measures of certain types of crime, such as assault, theft, and burglary. Most significantly, these surveys offer information about crimes not reported to the police, or as they are also known, the "dark figures" of crime. They provide information about why the victim did or did not report the crime to the police and supply extra information about the victim(s), the offender(s), the incident(s), and the consequences of the incident(s). Collecting all this additional information gives a different perspective and a more complete picture of both the extent and nature of crime than can be obtained from police-generated data. Consequently, and although they may need to be viewed with some caution, victim surveys have been a popular mechanism for obtaining crime data, and as you will see, in recent years confidence in the results has increased.

Victimization surveys have been conducted annually in the United States by the National Crime Victimization Survey (NCVS) since 1973. In England, the **British Crime Survey (BCS)** has been conducted in 1981, 1983, 1987, 1991, 1993, 1995, 1997, 1999, 2001/2002, 2002/2003, 2003/2004. Later in this chapter, we will examine in more detail problems specifically related to these two sets of victimization surveys and how they are combined with police-recorded data.

Because it is not possible to obtain comprehensive data on all crimes from either self-report or victimization studies, it is necessary to take a step back from the data in which we are really interested (crimes committed) and examine data that we can readily obtain—crimes reported to the police. Although these data have gone through one set of filters—the decision to bring the offenses to the notice of the police—information is available on all reported acts that violate the criminal law. Provided the offense has been reported to the police, information on that offense is available from police records, although these are somewhat limited in detail. Unfortunately, although the English publish national statistics each year on all offenses reported to the police in the annual publication **Criminal Statistics: England and Wales,** in the United States such information has been provided in the Federal Bureau of Investigation (FBI) Uniform Crime Reports (UCR) for only the eight index-one offenses (murder, forcible rape, robbery, aggravated assault, burglary, larceny, motor vehicle theft, and arson). These eight offenses do, however, represent a significant proportion of the more serious offenses committed against the person and against property. Although the FBI began the National Incident-Based Reporting System (NIBRS) in the early 1990s, which, as discussed later in this chapter, encompasses a wider range of offenses, it has still been adopted by only a minority of the police jurisdictions in the United States.

■ Methodological Problems

Before examining the relevant data, we must acknowledge that there are methodological problems in comparing similar types of crime data, such as police data or victimization data, from two nations. First, it is highly unlikely that offenses will be defined in precisely the same fashion in both nations, particularly for less serious crimes. Second, it is unlikely that the method of classifying and counting procedures will be exactly the same. Third, it is unlikely that the sampling procedures employed or the sample surveyed in victimization studies will be the same. All of these issues are likely to affect both the data obtained and the interpretation of those data.

In the early 1990s, both the English and Americans began to alter radically the process of how crime data had been collected, with the goal of improving the validity and reliability of their crime statistics. In the following sections we discuss the previous methods of defining, counting, and collecting crime data in England and the United States; the reasons for the changes; the new data collection paradigms within the respective countries; and how the changes will hopefully provide a more accurate and more complete picture of the crime situation in both England and the United States.

■ Crimes Reported to the Police

In this section, we will examine data on crimes reported to the police in England and the United States. The data sources used for these analyses are the revamped Home Office's Criminal Statistics: England and Wales (CSEW), which now includes the British Crime Survey (BCS); the FBI Uniform Crime Reports (UCR); and the National Crime Victimization Survey (NCVS). Before presenting the data, however, it is necessary to examine more closely the manner in which the different crimes have been defined, classified, and counted in these data sources. It is also necessary to discuss the validity and reliability issues of the data as well as the recent evolution of these data sets.

To promote further the ideal of uniform reporting, both the CSEW and UCR established from their inception detailed procedures for classifying and counting criminal incidents. Both systems established hierarchical classifications of offenses with generally only the most serious of multiple offenses attached to a particular incident being counted. In general, both the attempted and completed offense have been included in the figures given for a particular offense category.

In the UCR system, if a female has been raped and then murdered, only the murder is counted because the rape is subsumed under the murder according to the hierarchical classification rule. However, if there were multiple victims of index crimes against persons (i.e., five persons murdered, raped, or the victims of aggravated assault at the same time), all five offenses are counted.

With crimes against property, a distinct operation with multiple victims (e.g., five persons robbed at the same time, or items belonging to five people stolen from one room at the same time) is treated as one offense. However, each theft of a motor vehicle is recorded as a separate offense. Although a separate offense is counted for the burglary of each house, condominium, or other distinct structure, burglaries of rooms in hotels and lodging houses where the clientele was transient are recorded as only a single offense. Finally, although arson is classified as a property offense and thus each arson is counted as a single incident regardless of the number of victims, it has been exempted from the hierarchy rule, with all arsons reported "regardless of their commission in conjunction with another Crime Index offense" (Federal Bureau of Investigation, 1981, p. 36).

In England, the approach has been basically the same. The counting and classifying of criminal incidents have been based on hierarchical classification procedures with "only the most serious . . . offence counted where several offences were committed in one incident" (Home Office, 1989, p. 183). Thus, as in the United States, in a rape-murder only the murder is counted. Because only one set of criminal law provisions was involved, the English were able to use the established potential maximum penalties to determine the hierarchical order of the offenses, and do not, as in the United States, leave it to the collecting agency (the FBI) to establish that order.

Again in similar fashion to the United States, in situations involving "offences of violence against the person and sexual offences where there was more than one victim . . . one offence is usually counted for each victim" (Home Office, 1989, p. 183). Multiple victims in robberies generally constitute only one robbery incident. Thus, if five people are simultaneously threatened and robbed, only one offense of robbery is counted. However, if injuries are inflicted, in addition to a robbery being recorded for one of the victims, offenses of violence against the person are counted for each injured victim (Home Office, 1992, p. 66).

With regard to property offenses, an important distinction has been drawn between "collective" and "personal" protection. Although collective protection "applies to property left in a place having some characteristic making it safe to leave it there—e.g., property belonging to several persons in a changing room" and results in only one offense being recorded no matter how many victims there were, personal protection "applies when no such characteristic exists" and results in an offense being counted for each victim (Home Office, 1992, p. 110). Although the basic counting rules for burglary have been similar in England and the United States, the English count each burglary of an occupied hotel room as a separate offense. Again, as in the United States, each theft of a motor vehicle is recorded as a separate offense. Unlike the situation with the UCR, arson has been included within the hierarchy rule. Finally, in contrast to the United States, the English crime counts include offenses that a convicted offender asks to be "taken into consideration" at sentencing, provided that the recording police force have sufficient grounds to believe that the offense was committed.

■ Definitional Issues: Past and Present

Although the definitions used in the English CSEW are the legal definitions of those offenses in England, the definitions employed in the UCR did not conform to the legal definitions of any particular jurisdiction. Rather, they were the reporting definitions drawn up by the FBI to ensure consistent reporting of offenses by police departments in states with different legal definitions of criminal offenses. Thus, in theory at least, in the United States as in England, there was thought to be acceptable uniformity in the categorization and counting of different types of criminal incidents.

However, in the mid-1980s in the United States and England, serious debate developed concerning the accuracy of the respective national crime statistics and the benefits and liabilities of the current data collection systems. The rather large statistical discrepancies between police data and victim surveys (the U.S. NCVS and the British Crime Survey) drew attention. In both countries, the victim survey crime frequencies and rates were consistently higher than the police-collected statistics due in part to the manner in which the respective data collection systems operated. Another concern was a possible lack of uniformity in the manner in which the police in different agencies defined and counted criminal incidents. A major concern raised with the UCR was that it only provided summary data, and those summary data were available for only a few offenses. To deal with these issues, the respective people and agencies on each side of the Atlantic sought different solutions, but both with the goal of improving the accuracy, reliability, and validity of crime statistics.

■ The United States

The Uniform Crime Report (UCR)/National Incident-Based Reporting System (NIBRS)

The mid 1980s witnessed both government agency and academic dissatisfaction with the UCR. There was neither sufficiently complete nor sufficiently detailed information on crimes for an accurate depiction of crime. In reference to validity issues, or how well or accurately crime statistics reflect reality, there were a number of serious issues. First, a count of crimes reported to the police was provided for only the eight Part I offenses. For the Part II offenses only the numbers of arrests were reported for each offense. Second, the hierarchy rule eliminated many more crimes from being recorded at all, as in the murder-rape situation discussed previously. Concerning reliability issues, definitions of crimes varied from agency to agency, and a compounding problem was that not all agencies submitted reports. Consequently, these measurement problems pressured advisory groups to explore the continued feasibility of the current measurement of crime. They made several recommendations in the summer of 1985 with the ultimate goal of converting how crimes statistics were recorded from summary, as they had been recorded starting in the 1930s, to incident recording.

Advisory groups recommended that definitions of index crimes be revised and that other offenses that might be added to the report be identified. Additionally, all crime definitions, new and old, required refining, and the data elements needed to convert to incident recording required development (National Atlas, p. 4). What evolved was the National Incident-Based Reporting System (NIBRS). The NIBRS collects data on each single incident and arrest within 22 crime categories and 46 crimes (known as "A" offenses). "B" offenses consist of eight offenses and an "all other offenses" category. **Table 4-1** provides a comparison of the two systems. Information on each incident, all victims, all offenders, all arrestees, and property is now collected. In addition, the hierarchy rule no longer applies, so now all crimes in each criminal incident are counted, as opposed to

Table 4-1 Comparison of UCR and NIBRS Offenses

UCR

Part I (Index) Offenses (Offenses and arrests are in hierarchical order)	Part II Offenses (Arrests only are reported for the following)
Murder	Curfew and loitering law violations
Forcible rape	Disorderly conduct
Robbery	Driving under the influence
Aggravated assault	Drug abuse violations
Burglary—breaking or entering	Drunkenness
Motor vehicle theft	Embezzlement
Larceny	Forgery and counterfeiting
Arson (not subject to the hierarchy rule)	Fraud
	Gambling
	Liquor laws
	Offenses against family and children
	Other assaults
	Prostitution and commercial vice
	Sex offenses (except forcible rape and prostitution)

NIBRS

Group A (Offenses and arrests are reported for the following, for which a hierarchy does not apply)	Group B (Arrests only are reported for the following)
Arson	Bad checks
Assault Offenses	Curfew/loitering/vagrancy
Bribery	Disorderly conduct
Burglary/breaking and entering	Driving under the influence
Counterfeiting/forgery	Nonviolent family offenses
Destruction/damage/vandalism of property	Peeping Tom
Drug/narcotic offenses	Runaways
Embezzlement	Trespassing
Extortion/blackmail	All other offenses
Fraud offenses	
Gambling offenses	
Homicide offenses	
Kidnapping/abduction	
Larceny/theft offenses	
Motor vehicle theft	
Pornography/obscene material	
Prostitution offenses	
Robbery	

Source: Rantall & Edwards 2000, p. 3.

only the most serious as was done in the past. As one can imagine, in moving from summary to incident recording, the size of the data sets required to record such numbers of crimes would grow exponentially to accommodate the new recording scheme.

As may be expected, the process of changing from recording the summary UCR offenses to an incident-based system of recording crime has been and continues to be slow and deliberate. In order for all these new changes to be effective, substantial law enforcement input and cooperation are very much needed to provide such detailed information. As one might imagine, there was some reluctance by law enforcement to embrace this new system due to the size of the data sets, the amounts of data required to be gathered, and the effort needed to gather the data elements. Thus, there has not been a great rush to convert to the new format. Data started being collected in this format in January 1989, and as of 2004, 5,271 police agencies in approximately half the states (24) submitted data to the NIBRS (Federal Bureau of Investigation).

With the goal of increasing the amount of information collected for each crime by moving to incident crime reporting rather than tallying total numbers of certain crimes from each law enforcement agency, there have been benefits and liabilities. A benefit is greater reliability in that the FBI is now more comprehensive on how crime is recorded and classified. Also, computers and audits verify how each agency/state is collecting data, thus increasing both the reliability and validity of data. To ensure data and collections methods are sound, quality assurance reviews are conducted every 3 years for those states that utilize the NIBRS (National Atlas, p. 6). Apparent problems are that the computerized data files are massive, requiring a great deal of effort to collect data, and the recording of these data is expensive because of the need for new computer systems or systems upgrades for those who do record the data on a consistent basis.

To summarize, reporting is voluntary, expensive, and expansive so it is not surprising that law enforcement agencies have not been adopting this scheme too quickly or in great numbers. The changes have been a slow process rather than a sudden implementation, and this will most likely continue to be the case in the future. Hopefully a clearer idea of the extent of crime will be achieved as more police agencies adopt the new reporting scheme.

The National Crime Victimization Survey (NCVS)

The NCVS has been conducted since 1972 by the U.S. Census Bureau with the purpose of recording those crimes not reported to the police by victims or those "dark figures" of crime that may never be recorded by the UCR. The survey reports data on a yearly basis utilizing consistent procedures with a nationally representative sample of 50,000 households that includes 100,000 persons ages 12 and older. Crimes included in the NCVS are rape, robbery, aggravated assault, simple assault, purse snatching/pocket picking, burglary, theft, and motor vehicle theft. As discussed previously, rich demographic information is collected on the victims, the offenders, the consequences of the victimization for the victim, and the probabilities for victimization for each crime.

Just as the Uniform Crime Reports underwent changes, in 1989, the NCVS underwent a redesign to broaden its coverage and the depth of that coverage and to increase the validity or the accuracy of the crime statistics. The first report to utilize the new measures was available in October 1994. To this improved end, better measures of sexual assault crimes and domestic violence were pursued by the inclusion of more direct survey questions with the idea that today, victims' reluctance to discuss crimes of such a sensitive nature is not what it was in the past. Improvements were also sought in helping respondents recall victimizations (U.S. Bureau of Justice, 1995, p. 1), and the effort is now made by NCVS representatives to help victims discriminate between different types of potential crimes, particularly in domestic situations. Thus, potentially, all crimes in a given incident may be recorded. The results of these efforts have been a greater depth of knowledge and a wider range and larger number of crimes reported and recorded, particularly sexually oriented and assault crimes. Thus, with the questioning and collecting methods improved in both the police-collected and victim-driven statistics, a more complete picture of overall crime in the United States can possibly emerge.

■ England

Crime in England and Wales/British Crime Survey (BCS)

After working through their own evaluative processes, the English also recognized validity and reliability issues with their police-collected statistics. Also acknowledged was the divergence between crime survey rates and the police-collected statistics, with the former showing that only approximately 40% of crimes are reported to the police (Simmons, Legg, & Hosking, 2003, p. 2). Several reforms were suggested with the goal of improving the comparability of the police and victim-driven statistics and their reliability and validity in England and Wales.

After nearly 20 years of debate that culminated in a recommendation made in the Home Office Review of Crime Statistics (Simmons, 2002), a new scheme was embraced that combined the results from the British Crime Survey and police-recorded crime in a single report in order to give a more complete depiction of trends and patterns in crime statistics (Simmons et al., 2003, p. v). Definitions of particular crimes were also altered and refined to improve comparisons between the two sources. In 2001/2002, the first crime volume was produced utilizing data from both the British Crime Survey and the police-recorded statistics (Simmons et al., p. v). To further aid in the comparability between surveys and police statistics, the **National Crime Recording Standard (NCRS)** was adopted by all police agencies in England and Wales in April 2002 with the goals of increasing the reliability and comparability of English crime data, and ". . . to increase the ability to compare performance between police agencies and to provide a better service to the victims and the general public" (Simmons et al., p. 3). It was thought that this approach would provide more complete data that could ultimately lead to a better understanding of crime and therefore more effective police interventions (Simmons et al., p. 3).

To these ends, the police recording standards were broadened, counting rules were altered, and a victim-centered approach was adopted for collecting all crime data. The result was a shift in how police report crime. Before the adoption of the NCRS, the police had to satisfy themselves that a crime had been committed, and only when they were satisfied that a crime had been committed would that offense be recorded in police statistics. Now an allegation of victimization is enough for a criminal offense to be recorded. The reason for this is that calls for service must be met, and victims therefore become the driving force behind crime statistics. Thus, this victim-centered approach was advocated to increase the number of crimes coming to the attention of the police, to gain a clearer picture of crime, to "... provide a better measure of the service demanded from the police, and ... to promote public confidence in the police service as a whole" (Simmons et al., 2003, p. 3). This approach also advanced the goals of assuring victims that crimes reported by them were properly recorded and that future crimes could be prevented. Specifically, the NCRS states that:

1. *All reports of incidents, whether from victims, witnesses or third parties and whether crime related or not, will result in the registration of an incident report by the police.*

2. *The incident is considered a crime if, in all probability, the circumstances amounted to a crime as defined by law and,*

3. *If there is no credible evidence to the contrary.*

4. *Once recorded, a crime would remain recorded unless there was additional verifiable information to disprove that a crime had occurred. (Home Office, 2006, Annex A, Section 2.3)*

From a statistical perspective, a few problems arise under this scheme. First, those driving crime statistics (victims) have less experience than the police in interpreting the validity of an occurrence as criminal, thus widening the net of crimes that will be counted in the tally. Second, the standard as to what may be interpreted as criminal is less stringent, which may further compound the problems inherent in the redefinition of some crimes by the NCRS. In other words, the act is considered criminal until proven otherwise, which may temporarily artificially inflate crime statistics and/or may cause misclassification of criminal acts. Third, in the time that may be required to determine whether a crime is founded, the most recent crime statistics—specifically, the quarterly crime reports generated by the Home Office—may not be entirely accurate. Given these problems, crime data would be of limited use to law enforcement and other agencies in the short run. These difficulties have been acknowledged by the British government, and the conclusion was that, "Auditing should be consistent and ongoing at the local and national levels and at all stages of the process with the goals of identifying and addressing weaknesses in crime recording practices" (Simmons et al., 2003, p. 4).

Therefore, it is hypothesized that until several years of data have been collected, the short-run impact that the new victim-centered approach to counting crimes has on the individual crimes and their statistics cannot be effectively perceived (Simmons et al., p. xi). The reasoning behind this is based in large part on practicality and from moving from old to new. From a practical perspective, and despite the supporting processes mentioned above, police data systems were designed for information management rather than the monitoring of how the NCRS changes affected crime levels. Further, as discussed previously, police cannot always accurately differentiate types of incidents in some cases due to the altering of criminal definitions (Simmons et al., p. 12). For example, under the NCRS definitional heading of "burglary" are four burglary subheadings based on the location where the burglary may have occurred and also on the intent of the burglar while committing the crime. Potentially, one crime may fall into two or more categories. The victim-oriented approach also may confound categorization and ultimately statistical accuracy.

The results of these changes have already become apparent. After the implementation of the NCRS, crimes counted on a national scale increased by 10% (Simmons et al., 2003, p. x), but if the NCRS was not included in the picture, crime decreased overall by 3% from 2001/2002 to 2002/2003 (Simmons et al., p. v). Thus far, it appears implementation of the NCRS and the new counting rules has had the most impact on the violent crime statistics, and it is most likely that the recording of crime statistics from this victim-oriented approach will show continued increases in this type of crime until the "bugs" are worked out and/or until the "fail-safes" have taken hold (Clegg, Finney, & Thorpe, 2005).

These fail-safes and mechanisms for monitoring the changes effected by the NCRS are already having an impact. In 2004, the Audit Commission concluded that most of the police agencies were applying the NCRS, and as a result there was an improvement in the quality of data. Compliance was noted to be "weakest for violent crimes," but the commission suggested that "any changes in or increases in violence figures should be due to the new recording standards rather than real increases in these types of crimes." Indeed, these new figures would provide "more accurate statistics in the serious violence crimes categories due to victims' tendency to report those crimes when seriously assaulted or injured" (Clegg, Finney, & Thorpe, 2005, p. 4).

One may expect that improvements in crime statistics will be realized far more quickly in England simply because the English police force is nationalized rather than compartmentalized as it is in the United States. Also, the changes in England are mandatory as opposed to being more "voluntary" as they are in the United States. Ground will also be gained towards statistical clarity more quickly in England because the effort has been made to not only combine survey data and police statistics in one report, but also to make meaningful comparisons where possible. As to the total impact of the NCRS, time will tell. Until these changes are fully implemented, the accuracy of crime statistics will be affected to a certain extent.

■ Criminal Statistics Compared

The terms used to categorize criminal offenses in the new reporting systems and the definitions of those terms in two countries naturally differ. In constructing the tables that follow, considerable effort has been made to ensure that the comparable categories are as similar as possible. The U.S. crime categories and definitions are taken from the revamped UCR and NCVS, and also provide U.S. population figures. The English categories likewise are taken from their most recent annual reports that take into account crime statistics before and after the National Crime Recording Standard and the utilization of the British Crime Survey in the annual reports.

Both the FBI and the Home Office have changed their methods, definitions, and the like on how crimes will be counted, so there are several challenges for us as to how these respective data sets will be compared country to country and how the crimes counted under the new paradigms may be compared with the old. For example, the National Crime Victimization Survey and the British Crime Survey cover different age ranges (respondents older than 12 and older than 16, respectively), and variations in offense definitions still exist in spite of the respective improvements. To these ends, in the following section the crime statistics selected for discussion will be those that are the most comparable among the police-recorded statistics, the British Crime Survey, and those crime statistics generated by the new UCR and the NCVS.

According to Thorpe (2004), the crimes that overlap most comparatively between English police-recorded statistics and the British Crime Survey are vandalism, burglary, vehicle-related theft, bicycle theft, theft from the person, robbery, common assault, wounding, and vehicle interference (p. 1). Obviously, if only these crimes were used for our discussion, a clear picture of the crime problems in both nations would not be as fruitful because violent crimes such as rape and homicide would not be addressed. Therefore, for the sake of consistency and the ability to compare statistics over time, the violent crimes discussed here will be murder and nonnegligent manslaughter, robbery, rape, and aggravated assault. Property crimes in our discussion will consist of burglary, larceny/theft, and motor vehicle theft. The totals as well as rates for both types of crimes also will be reported when available. The reasoning behind this choice of offenses is that these offense statistics have been shown to yield valid and reliable measures over time and from country to country. They are the crimes most likely to be reported and recorded, about which there is much consistent data, and also the least to be affected by newly created definitions or new strategies in data collection.

The Data: The United States

The number of offenses reported to the police in the United States for the four violent offenses and their totals for the years 1984 through 2003 are presented in **Table 4-2,** and the rates per 100,000 population are presented in **Table 4-3.** The number of offenses reported to the police in the United States for the three property offenses for the years 1984 through 2003 is presented in **Table 4-4;** the rates per 100,000 population are presented in **Table 4-5.**

An examination of Table 4-2 reveals a marked decrease over the years in the numbers of all violent offenses, with particular emphasis on the decrease in crime from 1994–2003. In that time period, violent offenses decreased by 25.6%. The most profound decreases occurred with robbery (–33.2%), homicide (–29.3%), and aggravated assault (–22.9%).

In Table 4-3, these figures are adjusted to account for population growth. As seen in Table 4-3, there is also a marked decrease in the rate of violent crimes,

| Table 4-2 | Uniform Crime Report: Violent Crime in the United States by Volume and Rate, 1984–2003 | | | | |

| Population[1] | Violent Crime | Violent Crime | | | |
		Murder and Nonnegligent Manslaughter	Forcible Rape	Robbery	Aggravated Assault
Population by Year	*Number of Offenses*				
1984: 235,824,902	1,273,282	18,692	84,233	485,008	685,349
1985: 237,923,795	1,327,767	18,976	87,671	497,874	723,246
1986: 240,132,887	1,489,169	20,613	91,459	542,775	834,322
1987: 242,288,918	1,483,999	20,096	91,111	517,704	855,088
1988: 244,498,982	1,566,221	20,675	92,486	542,968	910,092
1989: 246,819,230	1,646,037	21,500	94,504	578,326	951,707
1990: 249,464,396	1,820,127	23,438	102,555	639,271	1,054,863
1991: 252,153,092	1,911,767	24,703	106,593	687,732	1,092,739
1992: 255,029,699	1,932,274	23,760	109,062	672,478	1,126,974
1993: 257,782,608	1,926,017	24,526	106,014	659,870	1,135,607
1994: 260,327,021	1,857,670	23,326	102,216	618,949	1,113,179
1995: 262,803,276	1,798,792	21,606	97,470	580,509	1,099,207
1996: 265,228,572	1,688,540	19,645	96,252	535,594	1,037,049
1997: 267,783,607	1,636,096	18,208	96,153	498,534	1,023,201
1998: 270,248,003	1,533,887	16,974	93,144	447,186	976,583
1999: 272,690,813	1,426,044	15,522	89,411	409,371	911,740
2000: 281,421,906	1,425,486	15,586	90,178	408,016	911,706
2001: 285,317,559[2]	1,439,480	16,037	90,863	423,557	909,023
2002: 287,973,924[3]	1,423,677	16,229	95,235	420,806	891,407
2003: 290,809,777	1,381,259	16,503	93,433	413,402	857,921
Percent Change, Number of Offenses					
2003/2002	−3.0	+1.7	−1.9	−1.8	−3.8
2003/1999	−3.1	+6.3	+4.5	+1.0	−5.9
2003/1994	−25.6	−29.3	−8.6	−33.2	−22.9
2003/1984	−8.5	−11.7	+10.9	−14.8	+25.2

Source: Uniform Crime Report, retrieved from: http://www.fbi.gov/ucr/cius_03/xl/03tbl01.xls

[1]Populations are U.S. Census Bureau provisional estimates as of July 1 for each year except 1990 and 2000, which are decennial census counts.

[2]The murder and nonnegligent homicides that occurred as a result of the events of September 11, 2001, are not included in this table.

[3]The 2002 crime figures have been adjusted.

Note: Although arson data are included in the trend and clearance tables, sufficient data are not available to estimate totals for this offense.

Table 4-3	Uniform Crime Report: Violent Crime in the United States by Volume and Rate, 1984–2003, Rate per 100,000 Inhabitants				
		Violent Crime			
Population[1]	Violent Crime	Murder and Nonnegligent Manslaughter	Forcible Rape	Robbery	Aggravated Assault
Year					
1984	539.9	7.9	35.7	205.7	290.6
1985	558.1	8.0	36.8	209.3	304.0
1986	620.1	8.6	38.1	226.0	347.4
1987	612.5	8.3	37.6	213.7	352.9
1988	640.6	8.5	37.8	222.1	372.2
1989	666.9	8.7	38.3	234.3	385.6
1990	729.6	9.4	41.1	256.3	422.9
1991	758.2	9.8	42.3	272.7	433.4
1992	757.7	9.3	42.8	263.7	441.9
1993	747.1	9.5	41.1	256.0	440.5
1994	713.6	9.0	39.3	237.8	427.6
1995	684.5	8.2	37.1	220.9	418.3
1996	636.6	7.4	36.3	201.9	391.0
1997	611.0	6.8	35.9	186.2	382.1
1998	567.6	6.3	34.5	165.5	361.4
1999	523.0	5.7	32.8	150.1	334.3
2000	506.5	5.5	32.0	145.0	324.0
2001[2]	504.5	5.6	31.8	148.5	318.6
2002[3]	494.4	5.6	33.1	146.1	309.5
2003	475.0	5.7	32.1	142.2	295.0
Percent Change, Rate per 100,000 Inhabitants					
2003/2002	−3.9	+0.7	−2.8	−2.7	−4.7
2003/1999	−9.2	−0.3	−2.0	−5.3	−11.8
2003/1994	−33.4	−36.7	−18.2	−40.2	−31.0
2003/1984	−12.0	−27.8	−10.1	−30.9	+1.5

Source: Uniform Crime Report, retrieved from: http://www.fbi.gov/ucr/cius_03/xl/03tbl01.xls

which had a peak in 1991 at a rate of 758.2 per 100,000 population and steadily decreased thereafter to a rate of 475 in 2003. Between 1994 and 2003, violent crime decreased by 33.4%. This decrease in the rate of reported violent offenses was fairly evenly distributed across all crimes with the exception of forcible rape. Although there was a gradual decrease in the reported rates of forcible rape between 1997 and 2001, rates started increasing again in 2002. With a decrease of 40.2% between 1994 and 2003, the rates of robbery showed the greatest de-

| | | **Property Crime** | | |
| | | | Larceny- | Motor Vehicle |
Population[1]	Property Crime	Burglary	Theft	Theft
Population by Year				
1984: 235,824,902	10,608,473	2,984,434	6,591,874	1,032,165
1985: 237,923,795	11,102,590	3,073,348	6,926,380	1,102,862
1986: 240,132,887	11,722,700	3,241,410	7,257,153	1,224,137
1987: 242,288,918	12,024,709	3,236,184	7,499,851	1,288,674
1988: 244,498,982	12,356,865	3,218,077	7,705,872	1,432,916
1989: 246,819,230	12,605,412	3,168,170	7,872,442	1,564,800
1990: 249,464,396	12,655,486	3,073,909	7,945,670	1,635,907
1991: 252,153,092	12,961,116	3,157,150	8,142,228	1,661,738
1992: 255,029,699	12,505,917	2,979,884	7,915,199	1,610,834
1993: 257,782,608	12,218,777	2,834,808	7,820,909	1,563,060
1994: 260,327,021	12,131,873	2,712,774	7,879,812	1,539,287
1995: 262,803,276	12,063,935	2,593,784	7,997,710	1,472,441
1996: 265,228,572	11,805,323	2,506,400	7,904,685	1,394,238
1997: 267,783,607	11,558,475	2,460,526	7,743,760	1,354,189
1998: 270,248,003	10,951,827	2,332,735	7,376,311	1,242,781
1999: 272,690,813	10,208,334	2,100,739	6,955,520	1,152,075
2000: 281,421,906	10,182,584	2,050,992	6,971,590	1,160,002
2001: 285,317,559[2]	10,437,189	2,116,531	7,092,267	1,228,391
2002: 287,973,924[3]	10,455,277	2,151,252	7,057,379	1,246,646
2003: 290,809,777	10,435,523	2,153,464	7,021,588	1,260,471
Percent Change, Number of Offenses				
2003/2002	−0.2	+0.1	−0.5	+1.1
2003/1999	+2.2	+2.5	+0.9	+9.4
2003/1994	−14.0	−20.6	−10.9	−18.1
2003/1984	−1.7	−27.8	+6.5	+22.1

Table 4-4 — Uniform Crime Report: Property Crime in the United States by Volume and Rate, 1984–2003

Source: Uniform Crime Report, retrieved from: http://www.fbi.gov/ucr/cius_03/xl/03tbl01.xls

[1]Populations are U.S. Census Bureau provisional estimates as of July 1 for each year except 1990 and 2000, which are decennial census counts.

[2]The murder and nonnegligent homicides that occurred as a result of the events of September 11, 2001, are not included in this table.

[3]The 2002 crime figures have been adjusted.

Note: Although arson data are included in the trend and clearance tables, sufficient data are not available to estimate totals for this offense.

crease in reported crimes. And, despite the great publicity given to criminal homicide, the rates for this offense have shown an overall decrease, with a slight increase in recent years. The peak rate of 9.8 per 100,000 population occurred in 1991.

| Table 4-5 | Uniform Crime Report: Property Crime in the United States by Volume and Rate, 1984–2003, Rate per 100,000 Inhabitants | | | |

| | | Property Crime | | |
Population[1]	Property Crime	Burglary	Larceny-Theft	Motor Vehicle Theft
Year				
1984	4,498.5	1,265.5	2,795.2	437.7
1985	4,666.4	1,291.7	2,911.2	463.5
1986	4,881.8	1,349.8	3,022.1	509.8
1987	4,963.0	1,335.7	3,095.4	531.9
1988	5,054.0	1,316.2	3,151.7	586.1
1989	5,107.1	1,283.6	3,189.6	634.0
1990	5,073.1	1,232.2	3,185.1	655.8
1991	5,140.2	1,252.1	3,229.1	659.0
1992	4,903.7	1,168.4	3,103.6	631.6
1993	4,740.0	1,099.7	3,033.9	606.3
1994	4,660.2	1,042.1	3,026.9	591.3
1995	4,590.5	987.0	3,043.2	560.3
1996	4,451.0	945.0	2,980.3	525.7
1997	4,316.3	918.8	2,891.8	505.7
1998	4,052.5	863.2	2,729.5	459.9
1999	3,743.6	770.4	2,550.7	422.5
2000	3,618.3	728.8	2,477.3	412.2
2001[2]	3,658.1	741.8	2,485.7	430.5
2002[3]	3,630.6	747.0	2,450.7	432.9
2003	3,588.4	740.5	2,414.5	433.4
Percent Change, Rate per 100,000 Inhabitants				
2003/2002	−1.2	−0.9	−1.5	+0.1
2003/1999	−4.1	−3.9	−5.3	+2.6
2003/1994	−23.0	−28.9	−20.2	−26.7
2003/1984	−20	−41	−13.6	−0.1

Source: Uniform Crime Report, retrieved from: http://www.fbi.gov/ucr/cius_03/xl/03tbl01.xls

[1]Populations are U.S. Census Bureau provisional estimates as of July 1 for each year except 1990 and 2000, which are decennial census counts.

[2]The murder and nonnegligent homicides that occurred as a result of the events of September 11, 2001, are not included in this table.

[3]The 2002 crime figures have been adjusted.

Like the rates of violent crimes, the rates of all U.S. property crimes decreased from 1984 to 2003, and in most cases the decreases were substantial. The offense that decreased the most dramatically was burglary (−28.9%), from a rate of 126.5 per 100,000 in 1994 to a rate of 740.5 per 100,000 in 2003. Concerning motor vehicle theft, which returned a rate of 433.4 per 100,000 for 2003 and a decrease of 18.1% between 1994 and 2003, recent years have witnessed small but steady increases in this crime; the larceny rate, which had the smallest decrease (–10.9%), fluctuated the most in this time period.

■ The National Crime Victimization Survey

As **Table 4-6** indicates, and as was reflected with the UCR data, overall violent crime peaked in the mid 1990s, with the rate being 51.2 per 1,000 in 1994, and thereafter dropped off rather precipitously to 22.3 per 1,000 in 2003. The offense of simple assaults consistently had the highest rates, with the peak being 31.5 per 1,000 in 1991 and a low of 14.6 in 2003. The offense of rape (which in the NCVS includes both female and male victims) had the lowest rates with a peak of 2.5 per 1,000 in 1984 and a low of 0.5 per 1,000 in 2003. Although robbery and aggravated assaults increased slightly from 2002–2003, rape and simple assaults continued with this decreasing trend, as occurred with these offenses in the UCR.

In reference to the property victimization rates found in **Table 4-7,** the NCVS indicates that total property crime rates have on average steadily declined at the amazing average rate of 12.4% per year, and that the rate more than halved between 1984 and 2003. Further, burglary and theft have progressively declined in this time period, and motor vehicle theft increases fluctuated before peaking in 1991 at 22.2% and thereafter declined to the low rate of 9.0% in 2003. The average decrease for theft per year was 9.6% and for burglary was 2.5% per year. Thus, as it was for violent and property crimes alike in the UCR, the decreases for the NCVS have been profound, all-encompassing, and dramatic.

England: Crime in England and Wales Report/ British Crime Survey

The following tables and data and discussion have been generated from the 2003/2004 version of the annual report, Crime in England and Wales, which also now reports data from the British Crime Survey. This report was utilized because it was the newest version available at the time of this writing, and it is furthest removed from the changes and implementation of the NCRS and the utilization of the BCS in the reporting of crime statistics. Also, because there is occasionally small variation in the crime report numbers due to audits that may find victim reports as founded or unfounded, using the latest statistics available should help alleviate that problem. This 2003/2004 report is also being used in the hope that some of the "growing pains" and variation in crime statistics that may occur in recording crime from agency to agency, such as varying quality, varying interpretations on definitions, and the like, may have been alleviated.

Beyond the audits and quality assurances mentioned earlier in the chapter, because of the shifts in crime recording discussed previously and other past changes, some data in the cells of the tables may be missing. For example, starting in 1995, males were included in crime statistics as victims of rape, and so the statistics for rape were adjusted accordingly with rape now differentiated by gender. Assaults are now only those that did not cause bodily injury. Those assaults that do cause injury are placed into different categories of "wounding," depending upon the extent of the harm to the victim. These gaps in the tables

Table 4-6	National Crime Victimization Survey Violent Crime Trends, 1984–2003

Adjusted Violent Victimization Rates

Number of Victimizations per 1,000 Population Age 12 and Over

Year	Total Violent Crime	Rape	Robbery	Aggravated Assault	Simple Assault
1984	46.4	2.5	5.8	10.8	27.2
1985	45.2	1.9	5.1	10.3	27.9
1986	42.0	1.7	5.1	9.8	25.3
1987	44.0	2.0	5.3	10.0	26.7
1988	44.1	1.7	5.3	10.8	26.3
1989	43.3	1.8	5.4	10.3	25.8
1990	44.1	1.7	5.7	9.8	26.9
1991	48.8	2.2	5.9	9.9	30.6
1992	47.9	1.8	6.1	11.1	28.9
1993	49.1	1.6	6.0	12.0	29.4
1994	51.2	1.4	6.3	11.9	31.5
1995	46.1	1.2	5.4	9.5	29.9
1996	41.6	0.9	5.2	8.8	26.6
1997	38.8	0.9	4.3	8.6	24.9
1998	36.0	0.9	4.0	7.5	23.5
1999	32.1	0.9	3.6	6.7	20.8
2000	27.4	0.6	3.2	5.7	17.8
2001	24.7	0.6	2.8	5.3	15.9
2002	22.8	0.7	2.2	4.3	15.5
2003	22.3	0.5	2.5	4.6	14.6
Percent Change					
2003/2002	−2.2	−14.3	−56.9	+7.1	−46.3
2003/1999	−30.5	−44.4	−30.1	−31.3	−29.8
2003/1994	−28.9	−104.2	−60.3	−61.3	−53.7
2003/1984	−51.9	−80.0	−56.9	−57.4	−46.3

Source: National Crime Victimization Survey retrieved from: http://www.ojp.usdoj.gov/bjs/glance/tables/viortrdtab.htm

Note: 1973–1991 data were adjusted to make data comparable to data after the redesign. Estimates for 1993 and beyond are based on collection year, whereas earlier estimates are based on data year. Rape does not include sexual assault.

are simply due to the aforementioned broadening, narrowing, or reclassification of offenses and/or that data may have been shifted to other categories or new offenses, leaving certain cells empty. In some cases, data from the Home Office Recorded Crime Statistics 1898–2002/2003 Excel file (http://www.homeoffice.gov.uk/rds/pdfs/100years.xls) were utilized to fill in the blanks where necessary and possible.

Table 4-7	National Crime Victimization Survey Property Crime Trends, 1984–2003			
	Adjusted Property Victimization Rates			
	Number of Victimizations per 1,000 Households			
Year	Total Property Crime	Burglary	Theft	Motor Vehicle Theft
1984	399.2	76.9	307.1	15.2
1985	385.4	75.2	296.0	14.2
1986	372.7	73.8	284.0	15.0
1987	379.6	74.6	289.0	16.0
1988	378.4	74.3	286.7	17.5
1989	373.4	67.7	286.5	19.2
1990	348.9	64.5	263.8	20.6
1991	353.7	64.6	266.8	22.2
1992	325.3	58.6	248.2	18.5
1993	318.9	58.2	241.7	19.0
1994	310.2	56.3	235.1	18.8
1995	290.5	49.3	224.3	16.9
1996	266.4	47.2	205.7	13.5
1997	248.3	44.6	189.9	13.8
1998	217.4	38.5	168.1	10.8
1999	198.0	34.1	153.9	10.0
2000	178.1	31.8	137.7	8.6
2001	166.9	28.7	129.0	9.2
2002	159.0	27.7	122.3	9.0
2003	163.2	29.8	124.4	9.0
Percent Change				
2003/2002	−12.6	+7.1	+1.7	0.0
2003/1999	−17.6	−12.6	−19.2	−10.0
2003/1994	−47.3	−47.1	−47.1	−52.1
2003/1984	−59.1	−61.2	−59.5	−40.1

Source: National Crime Victimization Survey, retrieved from: http://www.ojp.usdoj.gov/bjs/glance/tables/proptrdtab.htm

Note: 1973–1991 data were adjusted to make data comparable to data after the redesign. Estimates for 1993 and beyond are based on collection year whereas earlier estimates are based on data year.

In order to gain perspective on the overall crime problem in England, and to fully take advantage of a report that utilizes police-recorded and victim surveys data, general statistics will be explored first, followed by those that address individual crimes. The police recorded 5.9 million crimes in 2003/2004, which was an overall increase of 1% from the previous 2002/2003 report. The BCS estimated approximately 11.7 million crimes, which is a decrease of approximately 5% for this measure (Dodd, Nicholas, Povey and Walker, 2004: 8). As it was in the United States in the mid 1990s, there was a peak in the numbers and rates of victimizations for most crimes in Britain followed by declines for many crimes, particularly property crimes.

The specific numbers and rates for comparable offenses reported to the police in England and the BCS rates of crime are presented in **Tables 4-8, 4-9,** and **4-10.** An examination of Table 4-8 reveals marked increases in crimes counted during the latter 2 years and, as many sources have indicated, this increase is due to the implementation of the new National Crime Reporting Standard. In fact, for 2002/2003, the impact of the new recording practices of the NCRS generated an overall "artificial" increase in crime of 10% rather than the increase being due to genuine crime (Simmons et al., 2003, p. xi). Of general interest is that although the crime number declines have slowed in recent years, every category in the BCS noted sharp declines in both violent and property crime rates since 1995. Specifically, the BCS has noted a 39% decrease in crime since 1995 (Dodd, Nicholas, Povey and Walker, 2004: 12).

Violent Crimes Defined: Police-Recorded/British Crime Survey

"Violent crime," as defined by the new classifications and the 2003/2004 report, is now organized into three categories: violence against the person, sexual offenses, and robbery.

"Violence" against the person is composed of:

- "Homicide, threat, or conspiracy to commit murder
- Serious wounding intentionally inflicted
- Less serious wounding, which includes less serious injury and with actual bodily harm or grievous bodily harm without intent
- Common assault, which is now defined as an assault with no injury
- Harassment and possession of weapons
- BCS overlap: "wounding and common assault" (Dodd, Nicholas, Povey and Walker, 2004: Dodd, Nicholas, Povey and Walker, 2004: 71)

"Sexual offenses" is composed of:

- **Buggery** (sodomy or bestiality)
- Indecent assaults on males
- Indecency between males
- Rape—female
- Rape—male
- Rape
- Indecent assaults on females
- Unlawful sexual intercourse with girl under 13
- Unlawful sexual intercourse with girl under 16
- Householder permitting defilement of girls
- Incest
- **Kerb crawling** (soliciting from a motor vehicle for the purposes of prostitution) and procuration
- Abduction
- Bigamy
- Soliciting or importuning by a man

Table 4-8	Home Office Recorded Crime Statistics, 1984–2002/2003, Violent Crimes								
	Total Violent Crime	Homicide Murder Manslaughter Infanticide	Rape— Female	Rape— Male	Rape	Robbery (all)	Wounding	Common Assault (some minor injury)	Common Assault (no injury)
Year									
1984	159,299	621			1,433	24,890	5,276		
1985	170,650	616			1,842	27,463	5,885		
1986	178,203	661			2,288	30,020	6,616		
1987	198,829	688			2,471	32,633	7,942		
1988	216,214	624			2,855	31,437	8,678		
1989	239,858	641			3,305	33,16	8,926		
1990	249,904	669			3,391	36,195	8,920		
1991	265,085	725			4,045	45,323	9,408		
1992	284,199	687			4,142	52,89	10,741		
1993	294,231	670			4,589	57,845	10,701		
1994	310,332	726			5,032	60,007	11,033		
1995	310,936	745	4,986	150		68,07	10,445		
1996	344,766	679	5,759	231		74,035	12,169		
1997	347,064	739	6,281	347		63,072	12,531		
1997/1998	352,873	748	6,523	375		62,652	12,833		
1998/1999 (old rules)	331,843	750	7,139	502		66,172	13,960		
1998/1999 (new rules)	605,797	750	7,132	504		66,83	14,006	151,469	
1999/2000	703,107	766	7,809	600		84,277	15,135	189,783	
2000/2001	733,387	850	7,929	664		95,154	15,662	203,427	
2001/2002	813,271	891	8,990	730		121,370	16,556	226,472	
2002/2003	991,800	1,048	11,441	852		108,045	17,882	..	234,244
2002/2003 and 2003/2004	1,109,017	853	12,354	893		101,195	19,358	..	237,701
Percent Change Between									
2002/2003 and 2003/2004	+12	−18	+8	+5		−6	+8	..	+1

Source: http://www.homeoffice.gov.uk/rds/pdfs/100years.xls

- Abuse of position of trust
- Gross indecency with a child
- BCS overlap: According to the Crime in England and Wales 2003/2004 report, the number of sexual offenses that came to the attention of the BCS was too small to provide reliable estimates and was therefore not included in the report (Dodd, Nicholas, Povey and Walker, 2004: 78).

"Robbery" is composed of:

- Robberies of persons
- Robberies of businesses
- BCS overlap: mugging/snatch theft, robbery, and attempted robbery (Dodd, Nicholas, Povey and Walker, 2004: 71)

According to the Crime in England and Wales 2003/2004 annual report, violent crime comprised 23% of all BCS crimes and 19% of police-recorded crime

Table 4-9	Police Recorded Crime Statistics, 1984–2003/2004, Property Crimes			
	Total Property[1]	Total Burglary	Theft or Unauthorized Taking of Motor Vehicle	Total Theft and Handling Stolen Goods
Year				
1984		892,923	344,806	1,807,981
1985		866,697	367,426	1,884,069
1986		931,620	411,060	2,003,873
1987		900,104	389,576	2,052,005
1988		817,792	366,713	1,931,274
1989		825,930	393,399	2,012,760
1990		1,006,813	494,209	2,374,409
1991	4,976,466	1,219,464	581,901	2,761,119
1992		1,355,274	585,501	2,851,638
1993		1,369,584	592,660	2,751,901
1994		1,256,682	536,579	2,564,608
1995	4,738,600	1,239,484	502,280	2,452,109
1996	4,636,028	1,164,583	485,695	2,383,946
1997	4,191,467	1,015,075	399,208	2,164,952
1997/1998	4,131,483	988,432	392,381	2,144,973
1998/1999 (old rules)	4,086,694	951,878	381,080	2,126,718
1998/1999 (new rules)	4,303,712	953,184	381,709	2,191,439
1999/2000	4,410,543	906,468	364,270	2,223,620
2000/2001	4,260,810	836,027	328,037	2,145,372
2001/2002	4,524,827	878,547	316,321	2,267,063
2002/2003	4,693,395	888,951	305,618	2,365,535
2003/2004	4,610,310	818,642	279,111	1,378,972
Percent Change Between				
2002/2003 and 2003/2004	−2	−8	−9	−1

[1]From Crime in England and Wales 2003/2004.

Table 4-10	Trends in BCS Victimization Rates per 10,000 Adults/Households and Percentage Changes, Property and Violent Crimes						
	Total BCS Violence	Common Assault w/ Some Minor Injuries	Wounding	Robbery	Burglary	Theft of Vehicles	Total Theft from the Person
Year							
1981	558	362	131	42	409	156	112
1991	651	432	154	45	678	257	108
1995	1,046	718	225	83	835	241	167
1997	897	599	196	82	752	175	152
1999	832	563	157	98	588	153	154
2001/2002 Interviews	669	412	155	85	441	144	144
2002/2003 Interviews	665	404	168	72	439	126	164
2003/2004 Interviews	640	391	155	67	422	108	147
% change 1995 to 2003/2004	−39	−46	−31	−20	−49	−55	−12
% change 1997 to 2003/2004	−29	−35	−21	−18	−44	−38	−3
% change 1999 to 2003/2004	−23	−31	−2	−32	−28	−29	−5
% change 2001/2002 to 2003/2004	−4	−5	−1	−21	−4	−25	+2
% change 2002/2003 to 2003/2004	−4	−3	−8	−7	−4	−14	−10

Source: Simmons et al. (2005), p. 22.

Note: Total property rates cannot be calculated because they are based on household rates and personal rates, which cannot be combined (Simmons et al., 2005: p. 22.)

(Dodd et al., 2005, p. 11). The British Crime Survey estimated 2,708,000 violent incidents occurred against adults in England and Wales (Dodd, Nicholas, Povey and Walker, 2004: 67), which, overall, is a decrease of 36% in crime since 1995 (Dodd, Nicholas, Povey and Walker, 2004: 67). However, this year's BCS report also noted that violent crime has stabilized (Dodd, Nicholas, Povey and Walker, 2004: 9). The police recorded 1,109,017 violent incidents, which is a 12% increase since 2002/2003, partly attributed to the implementation of the NCRS (Dodd, Nicholas, Povey and Walker, 2004: 67), increases in the reporting of crime by the general public, increases in police activity, and improvements in recording crime (Dodd, Nicholas, Povey and Walker, 2004: 69).

Table 4-8 displays the homicide statistics for the time period of 1984–2004. Although there was a definite increase in the number of homicides, with a spike in the numbers in 2002/2003, overall, the increase was sporadic with a moderate decrease from 2002/2003 to 2003/2004. Of interest in this decrease was the impact the homicides by serial killer Harold Shipman had on the statistics. Shipman was a doctor imprisoned for the poisoning murder of 15 of his patients in 2000. With Shipman's murders (172) included in the homicide statistics, homicides diminished 18% from 2002/2003 to 2003/2004, but if his murders are excluded from the final tally, homicides decreased approximately 2% (Dodd, Nicholas, Povey and Walker, 2004: 70). This situation demonstrates a stark difference between U.S. and British homicide statistics. It is highly unlikely that one person's crimes would have such an impact on the U.S. homicide rates. Given that Shipman murdered a minimum of 215 of his patients, and because that number is still rising as his crimes are being investigated, his effect on homicide statistics should be seen for years to come.

There are 17 different offenses in the category of sexual offenses, but for our discussion, total numbers for rape of a male, female, and the category comprising both will be addressed. Table 4-8 indicates that rape has steadily increased since 1984, and with the new recording standard implemented in 1995 whereby males are now counted as victims of rape, the increase since that time applied for both males (5%) and females (8%), with an overall increase of 7% for both genders (Dodd, Nicholas, Povey and Walker, 2004: p. 78). Females comprised 93% percent of the victims (Dodd, Nicholas, Povey and Walker, 2004: p. 78).

Robbery of a business and a person is included in the totals in Table 4-8. During the time period under study, there was a steady increase in robberies with two peaks followed by sharp declines occurring in 1996 and 2001/2002, the latter of which may be partly attributed to the implementation of the NCRS. Ninety percent of the robberies were of personal property, while the balance was robberies of businesses (Dodd, Nicholas, Povey and Walker, 2004: p. 79). The BCS also recorded increases in the robbery rates between 1981 and its 2003/2004 interviews, with a peak in 1999. Following this peak, rates declined significantly (−32%). In the Crime in England and Wales 2003/2004 report, the conclusion was that for this time period, robbery numbers were too low to provide reliable estimates, although the report did note that there was a general decline for the last decade (Dodd, Nicholas, Povey and Walker, 2004: 79).

As with most of the crimes we have explored, the crimes associated with assault have undergone redefining, refining, and delineation. As of April 2002, police-recorded reports defined assaults as those with no injury to the victim. Included in this category are assaults on constables and race-motivated assaults. Assaults with minor injuries are now classified as "less serious wounding" (Dodd, Nicholas, Povey and Walker, 2004: p. 72) whereas assaults with more serious physical consequences for the victim are classified as "wounding." The BCS definitions were altered for comparative purposes in the BCS 2002/2003 interviews (Dodd, Nicholas, Povey and Walker, 2004: p. 34). The BCS definition of assault includes both those assaults with no injury and those involving minimal injury. "Wounding" is also now reflective of police-recorded definitions.

Common assaults account for a large percentage of recorded crime. In the BCS, common assaults comprised 61% of all violent crimes. Assaults comprised 24% of their violent crime totals for police-recorded crime (Dodd, Nicholas, Povey and Walker, 2004: p. 67). As is noted in the report, this difference in percentages is indicative of the low reporting rate of this crime (Dodd, Nicholas, Povey and Walker, 2004: p. 72). In Table 4-8, police-recorded statistics indicate very steady increases in all categories of assault, particularly wounding, and although the data on the newly formed common assault definitions (with and without injury) have only been recorded for a few years, the same upward pattern is apparent. BCS rates for assaults and woundings demonstrate a stark difference. In Table 4-8, sharp declines in victimization rates are the standard, particularly after the peak in 1995. Since that year, common assaults with minor injuries declined 46% while woundings decreased 31%. As we have seen with other crimes, the explanation for this divergence has again been attributed to the impact rule changes have had in counting this crime (Dodd, Nicholas, Povey and Walker, 2004: p. 39).

Property Offenses

Property offenses comprise 78% of both the BCS and police-recorded crime (Dodd, Nicholas, Povey and Walker, 2004: p. 7). Both the BCS and police-reported crime have noted decreases in property crime since 1995. Specifically, the BCS has reported a decrease of 46% between 1995 and 2003/2004 (Simmons et al., 2003, p. 45). Concerning burglary, the results are mixed. According to BCS statistics, the rates for domestic burglary have leveled off whereas police-recorded statistics noted an 8% drop from 2002/2003 and 2003/2004 (Dodd, Nicholas, Povey and Walker, 2004: p. 47). Burglaries in and out of a dwelling dropped 8% for the BCS and 9% for the police-recorded statistics (Dodd, Nicholas, Povey and Walker, 2004: p. 45). Vehicle-related theft also dropped for both measures. The BCS recorded a 10% decrease whereas police-recorded statistics noted a 9% drop in the time period from 2002/2003 and 2003/2004 (Dodd, Nicholas, Povey and Walker, 2004: p. 45). The theft of the actual vehicles declined by 14% and 9% for the BCS and police-recorded measures, respectively. Lastly, both measures recording theft from a person have noted peaks and valleys in the numbers and rates, but have overall seen declines.

■ Comparison

A comparison of the trends of reported serious crime in England and the United States from 1984 through 2003/2004 shows some variation in the rates of reported crimes in both countries. Although the overall rate of violent crime and property crime in the United States has decreased for both police-recorded crime and victim surveys (UCR violent crime down 33.4% and property crime down 23% from 1994 to 2003, and violent and property crime nearly cut in half in the same time period according to the NCVS), England has shown increases in police-recorded violent crime while posting fluctuating rates for property crimes. British victim surveys have recorded decreases in both violent and property crime for the same time period. And although much has been stated concerning the effects the changes in definitions and counting practices have had on the crime rates in England, one may note that the increases in violent crimes were already occurring before the implementation of the new counting schemes, continuing a trend that had already existed in the 1980s.

Even with the decreases in most crime measures recently, historically the rates of crimes in England were consistently lower than those in the United States. However, times are changing, and in 1996, English crime rates surpassed the U.S. rates for assault, burglary, and motor vehicle theft (Langan & Farrington, 1996). With other crimes, once large disparities between the countries have narrowed. An example is in the rates of criminal homicide. Whereas the rate in England was only 25% of the rate in the United States in 1999 (1.4 vs. 5.7 per 100,000 population), in 1992 the rate was 1.3 versus 9.3 per 100,000 which was 14% of the U.S. rate. The rate of reported rape offenses, which was comparable to the homicide rate, has climbed in recent years and is now 88% of that experienced in the United States (28.9 vs. 32.8) compared to 18.9% (8.1 vs. 42.8) in 1992. And although the English police-recorded robbery rate has fallen from 2002 to 2004, there was a substantial increase between 1997 and 2001/2002. The rate that formerly constituted 39.2% of the U.S. rate (103.2 vs. 263.6) in 1992 was exceeding the U.S. rate by 9% in 2000 (U.S. 145 versus 160 per 100,000 in England; Smith, 2003, p. 14).

In reference to assault, in 2003 aggravated assault comprised the largest proportion of all violent crime at 62.1% (U.S. Department of Justice, 2004, p. 11) in the United States, whereas in England, common assaults made up a comparable 61% and 24% of BCS and police-recorded statistics, respectively (Simmons et al., 2005, p. 72). Although much variation exists in the manner in which this crime is defined and delineated in the respective countries, it has been fairly well established that the assault rates in England have been consistently higher than those of the United States since the early 1980s. In 2002, according to victims' surveys, the U.S. levels of assault were less than half of the English rates, with statistics being 198 per 10,000 (U.S. Department of Justice, 2003a, p. 14) versus 404 per 10,000.

It is clear that the reported rates of violent offenses are becoming much more comparable between the United States and England. However, crimes of violence are far more likely to be committed in the United States with the use of a firearm. Overall, firearms were involved in 26.8% of the total murders, robberies, and aggravated assaults collectively during 2000 in the United States (U.S. Department of Justice, 2003b, pp. 1–8). Whereas an average of 71.1% of the criminal homicides, 42.1% of the robberies, and 19% of the aggravated assaults reported to the police in the United States in 2002 were committed with a firearm, only 9% of the criminal homicides, 17% of the robberies, and 33% of the violence against the person crimes reported to the police in England were committed with a firearm (Povey, 2005, pp. 28–33). In fact, firearms are used in only 0.4% of all crimes in England, or once for every 250 crimes; of these, 57% involved air weapons (Povey, p. 28).

For property crimes, the situation is similar. The rise in the rate of property crimes in England has seen the English rate catch up and surpass (for the first time, in 1982) the U.S. rate. By 1992, the overall U.S. property crime rate was less than half of the English rate (4,903.7 vs. 10,310). In 2003/2004, the U.S. burglary rate, which from 1982 onwards was consistently the lower of the two, was 41% of the English rate, while the motor vehicle theft rate, which was lower for all 20 years under examination, is 79% of the English rate. This latter statistic must be taken with a grain of salt, because the English police-recorded statistics have delineated three separate crimes that may occur with a vehicle. If the total recorded vehicle-related crime rate is compared with the UCR rate, the U.S. rate is 49% of England's. Thus, although the United States would appear to have a significantly greater problem with serious crimes such as homicide and crime committed with firearms than England, the English appear to have more of a problem with property offenses and assaultive behaviors.

At the time of this writing, new statistics were released. The complete Crime in England and Wales 2005/2006 and preliminary reports from the Uniform Crime Reports, as well as reports from the 2005 National Crime Victimization Survey were available online. To briefly summarize, every crime report, both U.S. and British, noted that crime was continuing the downward trend from the mid 1990s, and had stabilized or remained unchanged at a level not seen in decades, as was noted earlier in the chapter. Unfortunately, this decline seems to have stalled, and in some cases, a slight resurgence upward of some crime rates on both sides of the Atlantic may be heralding a change, which is evidenced by a few of the latest crime statistics. Whether these variations are within the "margin of measurement error" in how agencies are utilizing their new measuring techniques for crime may be open to question. But because both English and U.S. agencies are finding the same trends in their crime statistics, it appears fair to conclude that perhaps a new era of increased criminality may occur, with both English and U.S. agencies measuring their crimes more accurately in the end.

■ Crimes Reported to Survey Interviewers

In the United States, there has been a gradual increase in the percentage of victimizations reported to the police. Whereas 42% of all violent and 34% of all property offenses covered by the victimization surveys were reported to the police in 1993, by 2004 these figures had risen to 50% and 39%, respectively (U.S. Department of Justice, 2005, p. 1). There are fluctuations in the year-to-year reporting rates, but most offenses, with the exception of rape since 2002, show overall increases in reporting to the police (U.S. Department of Justice, p. 11). In 2004, 61% of robberies, 64% of aggravated assaults, 36% of rapes/sexual assaults, and 45% of simple assaults were reported to the police (U.S. Department of Justice, p. 10). For property crimes, 32% of household thefts, 41% of personal thefts, 53% of burglaries, and 85% of all motor vehicle thefts were reported.

In England, the reporting rate of offenses has been fairly stable since 1997, and as was seen with the U.S. rates of reporting, varies considerably by offense. Whereas 31.1% of all British Crime Survey offenses were reported to the police in 1981, this figure rose to 34.1% in 1983, 36.1% in 1987, and 43.0% in 1991, and has settled at 42% in 2003/2004 for offenses comparable with those recorded by the police. In 2003/2004, 95% of victims reported the theft of a vehicle, 78% reported burglary with loss, and 53% reported robbery. For violent offenses, overall, 62% of violent offenses were recorded by police, 48% reported wounding, and 30% reported a common assault with no injury, the latter two of which were comparable across police-recorded and victim surveys (Simmons, 2005, p. 34).

These summary data on reporting rates give an indication of reporting trends in both England and the United States. Because the two surveys examine somewhat different criminal offenses, and because different offenses are included in categories such as "offenses of personal violence" and "household offenses," it is necessary to examine specific offenses in order to obtain a more accurate assessment of the comparative reporting trends. **Table 4-11** presents reporting rates for five reasonably comparable offenses. The three offenses of violence (robbery, aggravated assault, and simple assault) are all more likely to be reported to the police in the United States than in England. For the three violent offenses, the differences in reporting rates vary depending on the offense (ranging from 7.3%

Table 4-11	Percentage of Certain Selected Offenses Reported to the Police in England and the United States for the Years 1981, 1991, and 2004					
	1981		**1991**		**2004**	
Violent Offenses	**U.K.**	**U.S.**	**U.K.**	**U.S.**	**U.K.**	**U.S.**
Robbery	46.5	55.8	47.2	54.5	53	61
Aggravated Assault	40.2	52.2	47.7	58.4	48	64
Simple Assault	25.1	39	25.5	41.5	30	45
Property Offenses						
Burglary	66.2	51.1	73	49.9	78	53
Motor Vehicle Theft	94.9	87	98.8	92.4	95	85

to 9.3% for robbery, 10.7% to 16% for aggravated assault, and 13.9% to 16.0% for simple assault). The two property offenses (burglary and completed motor vehicle theft), however, are more likely to be reported to the police in England than they are in the United States. For burglary, the differential in reporting rates is considerable, ranging from 15.1% in 1981 to 25% in 2004. In sum, the general trend appears to be that property crimes are more likely to be reported to the police in England, whereas crimes of violence against the person are more likely to be reported to the police in the United States.

■ Conclusion

Those who generate and analyze crime data in the United States and England recognized there were inherent problems in how crime data were collected as well as limitations as to what the collected data could impart. The gaps between the data culled from the police-recorded statistics and crime surveys were also acknowledged, and both countries made adjustments as to definitions and how crime statistics are collected and recorded accordingly. Although crime surveys are not the end-all to crime statistics, the detail-oriented approach to collecting crime data in surveys provides a paradigm from which much may be learned about criminality, offenders, victims' issues, crime prevention, police accountability, and police effectiveness. Paired with the improved police-recorded statistics in both countries, hopefully a truer picture of crime may be revealed, particularly after the effects of the changes brought about by the U.S. National Incident-Based Reporting System and the English National Crime Recording Standard have been overcome or have been taken into account on a consistent basis.

As mentioned before, it is our opinion that the English will reap the benefits of these changes sooner than their U.S. counterparts for several reasons. First, because the English criminal justice system, particularly law enforcement, is nationalized, the implementation and "tinkering" needed to make effective changes in crime statistics will happen more quickly than in the United States, particularly because the implemented changes in the United States are on a more voluntary than mandatory basis. Second, because the victim surveys and police-recorded statistics are combined in the same report in England, meaningful comparisons and determinations can be made that can directly affect public policy, as opposed to those that would be made on a more academic basis in the United States. Finally, as we have seen, each type of crime measure has its advantages and disadvantages. Police-recorded crime does well at tracking homicides and those types of crimes warranting police reports for insurance purposes (i.e., commercial/business and motor vehicle thefts), whereas victim surveys are best for crimes not reported to the police (i.e., household crimes and crimes against the person). Thus, although reporting and recording are improving, crime statistics must still be viewed with some caution and are still very much variable, no matter how they are collected, from whom they are collected, and from which country they derive.

CHAPTER SPOTLIGHT

- Serious criminal offenses, such as murder, rape, robbery, and burglary, are defined in a similar fashion in both the United States and in England and Wales. The same kinds of wrongful acts against people and property are regarded as crimes in both jurisdictions.

- However, the jurisdictions differ in their approach to "victimless" crimes. The United States is more likely to use criminal law to regulate gambling and consensual sexual activities.

- Victim surveys can provide useful and reliable information on certain crimes, such as assault or theft. They provide information on crimes not reported to the police.

- The United States and England and Wales both use "uniform reporting" systems for collecting crime data. The United States uses the FBI Uniform Crime Reports (UCR). England and Wales has the Criminal Statistics: England and Wales (CSEW).

- Both the UCR and CSEW established a hierarchical classification of crimes, with only the most serious crime within a particular incident being counted towards the statistics.

- The systems do differ in that under the CSEW the definitions of crimes recorded correspond to the definitions of specific offenses. In the United States, where crimes are defined differently among the states, the UCR does not conform to the definitions of any one state.

- England and Wales is a unitary jurisdiction, in which the government has the authority to compel uniform recording of crime statistics. In the United States, the FBI does not have this authority, and recording is voluntary.

- The UCR underwent improvements in 1989 and the National Crime Victimization Survey (NCVS) was broadened. The intent was to improve the accuracy of criminal statistics in the United States. From 1995, attempts were made to record all crimes within an incident under the National Incident-Based Reporting System (NIBRS).

- Likewise, in England and Wales a new scheme was adopted in 2001: the National Crime Recording Standard (NCRS). Counting rules were altered and a "victim-centered approach" adopted. The intention was to have a better understanding of crime, and therefore a more effective response.

- The victim-centered approach has drawbacks because victims are less stringent in their recording and have less experience than the police in interpreting crimes. Therefore, there is a risk that the crime statistics could be artificially inflated.

- Violent crime rates are disproportionately affected by the errors in victim-centered reporting.

- In England and Wales the statistical errors were exacerbated by the mass multiple homicides of one man. Local physician Harold Shipman is estimated to have murdered at least 215 of his patients, causing an apparent inflation in the murder rate. The sheer size of the United States' population makes it unlikely that any one person could similarly affect the U.S. crime statistics.

- Criminal homicide rates in the United States are consistently higher than those in England and Wales. However, the gap has narrowed, which is true of other crimes as well. The exceptions to this are crimes of robbery and common assault.
- Crimes of violence in the United States are far more likely to be committed with a firearm than they are in England and Wales.
- Property crime is higher in England and Wales than in the United States. This includes burglaries and motor vehicle thefts. However, such crimes are more likely to be reported to the police in England and Wales than in the United States.

KEY TERMS

Buggery: Sodomy or bestiality.

British Crime Survey (BCS): A systematic study of victimization in Britain conducted by the Home Office, whereby victims are queried about victimization experiences in the previous year. Started in 1982, the study was conducted in 1982, 1984, 1988, 1992, 1994, 1996, 1998, 2000, 2001, and is now done yearly in conjunction with the Criminal Statistics: England and Wales (CSEW) report.

Criminal Statistics: England and Wales (CSEW): An annual report published by the Home Office concerning police-recorded crimes; it uses crime data collected by the British Crime Survey. The goal is to present a comprehensive description of crime trends and patterns from 43 police agencies in England and Wales.

Kerb crawling: Soliciting from a motor vehicle for the purposes of prostitution.

National Crime Recording Standard (NCRS): Started in April 2002, this is a method for recording crime in England and Wales whereby all incidents, whether they are criminal or not, are recorded in the police statistics. The source for the statistics may be from victims, witnesses, or any other involved party. The goal is to increase validity and reliability of crime statistics across Britain and to have a more victim-oriented approach to the collection of criminal statistics.

Self-report studies: Studies in which those sampled are asked whether they have engaged in particular types of criminal activity. Usually confidential questionnaires are the research tool and the data are compared with official crime statistics and victims' surveys to determine a "truer" level of crime and an idea of who the offenders may actually be.

Victimization surveys: Studies such as the United States' National Crime Victimization Survey and Britain's British Crime Survey, which query a sample of people about their victimization experiences over a fixed time period. The studies also collect information on whether and why people report their victimizations to the police. Such studies give information concerning the "dark figures" of crimes, or those crimes that are not reported to the police.

PUTTING IT ALL TOGETHER

1. What were the issues that prompted change in the collection of crime statistics in England and the United States?

2. How did the changes to crime statistics in the 1980s and 1990s improve the reliability and validity of crime statistics?

3. What effect did the new crime recording scheme have on crime statistics initially, and why was this the case?

4. What are the trends for English and American violent crime? What are the trends for property crimes?

5. Why do English crime statistics improve with age?

6. Why will the English see the effects of the changes in recording practices before the United States?

7. Why was the British Crime Survey combined with the police-recorded statistics?

8. How does the reporting of crime compare between nations as it pertains to violent and property crimes?

9. What are the advantages and disadvantages of the respective types of crime data collection (i.e., police-recorded, victim surveys, and self-reports)?

10. How do firearms affect the crime statistics comparatively in England and the United States?

11. With what crimes are the British becoming more comparable with the United States, and why might this be the case?

ENDNOTES

1. The hierarchy rule still applies, although modified slightly. In a sequence of crimes or if the crime contains more than one type of crime, the most serious crime will be counted. A sequence of incidents between the same offender and the same victim counts as one incident if reported to police all at once. In the case of multi-victimization, offenses are counted separately if the offender(s) acted independently. There is also one crime per victim. For personal crimes, the victim is the person to whom the crime happened. For property crimes, the victim is the person who owns the property (Home Office, 2006).

REFERENCES

Clegg, M., Finney, A., & Thorpe, K. (2005). *Crime in England and Wales: Quarterly update to December 2004.* Home Office Statistical Bulletin 07/05. Retrieved May 1, 2006, from http://www.homeoffice.gov.uk/rds/pdfs05/hosb0705.pdf.

Dodd, T., Nicholas, S., Povey, D., & Walker, A. (2005). *Crime in England and Wales, 2003/04 annual report.* Retrieved May 1, 2006, from http://www.homeoffice.gov.uk/rds/pdfs04/hosb1004.pdf.

Federal Bureau of Investigation. *NIBRS: General information online.* Retrieved June 1, 2006, from http://www.fbi.gov/ucr/faqs.htm.

Federal Bureau of Investigation. (1981). *Uniform crime report for the United States.* Washington, D.C.: U.S. Department of Justice.

Home Office. *Recorded crime statistics 1898–2002/03.* Excel file. Retrieved June 4, 2006, from http://www.homeoffice.gov.uk/rds/pdfs/100years.xls.

Home Office. *Recorded crime statistics, 1898–2002/03 Violent Crimes.* Retrieved June 10, 2006, from http://www.homeoffice.gov.uk/rds/recordedcrime1.html.

Home Office. (1989). *Criminal statistics: England and Wales 1988.* London: H.M.S.O.

Home Office. (1992). *Criminal statistics: Counting rules for recorded offences.* London.

Home Office. (2006). *Counting rules for recorded crime.* Retrieved July 6, 2006, from http://www.homeoffice.gov.uk/rds/pdfs06/countgeneral06.pdf.

Langan, P., & Farrington, D. *Crime and Justice in The United States and in England and Wales, 1981–96.* Bureau of Justice Statistics. October 1998. NCJ 169284. Retrieved May 30, 2007, from http://www.ojp.usdoj.gov/bjs/pub/pdf/ cjusew96.pdf.

National Atlas. *Summary of the Uniform Crime Reporting (UCR) Program 2003.* Retrieved June 6, 2006, from http://nationalatlas.gov/articles/people/a_crimereport. html.

Povey, D. (Ed.). *Crime in England and Wales 2003/2004: Supplementary Volume 1: Homicide and gun crime.* Retrieved June 6, 2006, from http://www.homeoffice .gov.uk/rds/pdfs05/hosb0205.pdf.

Rantala, R. R., & Edwards, T. J. (2000). *The effects of the NIBRS on crime statistics.* Retrieved June 6, 2006, from http://www.search.org/files/pdf/BJSeffects.pdf.

Simmons, J., et al. (2002). *Crime in England and Wales 2001/02.* Retrieved June 23, 2006, from http://www.homeoffice.gov.uk/rds/pdfs2/hosb702.pdf.

Simmons, J., Legg, C., & Hosking, R. (2003). *National Crime Recording Standard: An analysis of the impact of recorded crime, 2002/03, Part one.* Retrieved June 23, 2006, from http://image.guardian.co.uk/sys-files/Guardian/documents/2003/07/17/NCRS1.pdf.

Smith, J. (2003). *The nature of personal robbery.* Retrieved June 6, 2006, from http://www.homeoffice.gov.uk/rds/pdfs2/hors254.pdf.

Thorpe, K. (2004). *Comparing BCS estimates and police counts of crime 2003/2004.* Technical paper. London: Home Office.

U.S. Department of Justice. (1995). *National Crime Victimization Survey redesign.* Retrieved June 6, 2006, from http://www.ojp.usdoj.gov/bjs/pub/pdf/redesfs.pdf.

U.S. Department of Justice. (2003a). *Criminal victimization in the United States, 2002 statistical tables.* Retrieved June 23, 2006, from http://www.ojp.usdoj.gov/bjs/pub/pdf/cvus02.pdf.

U.S. Department of Justice. (2003b). *Press release: Uniform Crime Reporting Program releases crime statistics for 2002.* Retrieved June 6, 2006, from http://www.fbi.gov/pressrel/pressrel03/ucr2002.htm.

U.S. Department of Justice. (2005). *National Crime Victimization Survey: Criminal victimization, 2004.* Retrieved June 6, 2006, from http://www.csdp.org/research/cv04.pdf.

U.S. Department of Justice. (2006a). *National Crime Victimization Survey: Criminal victimization, 2005.* Retrieved June 23, 2006, from http://www.ojp.usdoj.gov/bjs/pub/pdf/cv05.pdf.

U.S. Department of Justice. (2006b). *National Crime Victimization Survey: Property crime trends, 1984–2003 per 1,000 inhabitants.* Retrieved June 23, 2006, from http://www.ojp.usdoj.gov/bjs/glance/tables/proptrdtab.htm.

U.S. Department of Justice. (2006c). *National Crime Victimization Survey: Violent crime trends, 1973–2005.* Retrieved June 23, 2006, from http://www.ojp .usdoj.gov/bjs/glance/tables/viortrdtab.htm.

U.S. Department of Justice, Federal Bureau of Investigation. *Crime in the United States by Volume and Rate 1984–2003.* Retrieved June 23, 2006, from http:// www.fbi.gov/ucr/cius_03/xl/03tbl01.xls.

U.S. Department of Justice, Federal Bureau of Investigation. (2004). *Crime in the United States 2003 (UCR violent crimes).* Retrieved June 23, 2006, from http:// www.fbi.gov/ucr/cius_03/pdf/03sec2.pdf.

U.S. Department of Justice, Federal Bureau of Investigation (2006). *Crime in the United States 2005, Table 1A.* Retrieved June 23, 2006, from http://www.fbi.gov/ ucr/05cius/data/table_01a.html.

Walker, A., Kershaw, C., & Nicholas, S. (2006). *Crime in England and Wales 2005/06.* Retrieved June 23, 2006, from http://www.homeoffice.gov.uk/rds/pdfs06/ hosb1206.pdf.

Law Enforcement in England

II

The History and Organization of the Police

5

Chapter Objectives

After completing this chapter, you will be able to:

- Explain the historical beginnings of the English police force.
- Identify the early reformers and their role in the development of the English police force.
- Discuss the more recent history of policing in England and its comparison to policing in the United States.

■ Introduction

In this chapter we will examine the historical development of the English policing system as it rose to become one of the preeminent models for policing throughout the Western world. Because of England's central role in the development of modern law enforcement, particular attention will be paid to the historical precedents of modern policing systems. Primary focus will be on the forerunners of law enforcement structure and practice in the United States. In addition, we will discuss the similarities and differences in the development of the police in England and the United States and present a brief description of the organization and structure of the English police.

Perhaps the most frequent comment heard from American tourists and students visiting England concerning the police is about the polite, caring, and patient attitude they display toward travelers, an attitude that has always seemed to characterize the English police. This image has not been achieved as easily as one might imagine. In fact, much of this image is attributable to the efforts of the police <u>reformers</u> circa 1829. In addition, the English police have recently undergone critical scrutiny and questioning from the public concerning their traditional methods of enforcing the laws and keeping the "King's (or Queen's) peace."[1] How the English police have responded to this challenge, what direction they will take in the future, and how the public will respond are important questions that need to be addressed. However, before a discussion of that nature can take place, attention must be directed toward the history of policing in England, which, of course, is also a study of the historical roots of the U.S. systems. This Anglo-American system of policing, which has evolved from Germanic roots, differs from the system that eventually became the basis for the continental European style of policing, which is based on the practices of the ancient Romans.[2] Although not as decentralized as in the United States, the English system gave impetus for the emergence of the policing systems in the United States, and a detailed examination of its organization, structure, and functions is required for a true understanding of how modern policing has developed both in England and in the United States.

■ A History of English Policing

Why does it seem that the English police have set the standard for effectiveness throughout the modern world? Melville Lee may have articulated it best. He wrote: "The members of the English police are public servants in the sense of the term; not servants of any individual, of any particular class, but servants of *the* whole community—excepting only that part of it in setting the law in defiance has thereby become a public enemy" (1971, p. xv). Lee recognized that the essence of good police work depends upon good public relations, and that each police officer must operate under the assumption that "he has the sanction and approval of the great majority of his citizens" (p. xv). Furthermore, the police cannot be expected to perform their duties effectively if their every act is questioned or if every action is placed in the context of surmounting obstacles to ensure that justice is served (p. xv). Modern police systems throughout the world would probably agree with the following statement, which has been a cornerstone of the English policing system since the reforms of the 1800s: "The strength of a democracy and the quality of life enjoyed by its citizens are determined in large measure by the ability of the police to discharge their duties" (Goldstein, 1977, p. xx).[3]

Early Beginnings

Although there were many milestones in the development of the English policing system, the early beginnings (prior to the Norman conquest in 1066) have several important "markers." The <u>frankpledge system</u>, <u>tithings</u>, <u>hundreds</u>, <u>constable</u>, <u>shire</u>, <u>hue and cry</u>, <u>posse comitatus</u>, and <u>justices</u> are just a few of the important concepts that may be recognizable as precursors of the systems in the United States. The origins of these markers can be found in the tribal customs and laws of the early Danish and Saxon invaders (Critchley, 1967; Lee, 1971; Reith, 1952a; Smith, 1985). The earliest references to some form of a policing official are to a <u>tithingman</u>, who was responsible for collecting fines levied on people in his tithing. All males by the age of 12 had to be brought into a tithing, a group of 10 men. These groups were then organized into a collection 10 tithings (called a hundred). The individual assigned the responsibility for each hundred was called a **hundredman**. Because the Anglo-Saxons had divided the realm into shires, the hundredman was obliged to report to the **shire reeve**, who was the overseer for the ruler. While in charge of order for the shire, the sheriff (as he came to be known) had responsibility not only for punishment for failure to pay fines, but also for other "breaches of the peace."

The men in a tithing were mutually responsible for each other's behavior. If any one man committed a wrong, the rest were held liable. This laid one of the foundations of a kinship-style policing system. After the Norman conquest of 1066, the tithings evolved into the frankpledge system, which was called the "most important police institution in the Middle Ages" by Critchley (1967, p. 3).

Of particular interest to us is the development of the tithingman. His title ceased having any connection to its original meaning and he later became known as the "**chief pledge**" or "**capital pledge**" (Reith, 1952a, p. 28). From leader of the tithing he developed into a leader of men and became responsible for apprehending people who broke the law. Between the end of the twelfth century and the fourteenth century his duties and prestige increased and he became known as the **parish constable** or **village constable**, so called because of the ecclesiastical parish system of local government in England during the reign of Queen Elizabeth I.[4] As was the custom in those times, each parish had one elected or appointed constable. His tenure was for one year, he was not paid, and all able-bodied men were expected to serve in this capacity. Eventually this gave way to a new system in London and towns called the **watch system**. This watch system normally functioned only at night and, depending upon the size of the town, could consist of as many as 16 men whose duty was to supplement the responsibility of the constable. Normally, they would be stationed by the constable at all the gates of a walled town or city and were given arrest powers. If a stranger or villain was noticed (and refused to stop when asked), they were allowed to impose the ancient Saxon practice of "hue and cry," which literally meant that all persons (a "posse comitatus") in the town or in nearby towns were required to pursue the fugitive. The Statute of Winchester of 1285, which required work to be stopped so that all men could take part when a posse was summoned, backed this provision. This law also assigned penalties for those who did not join in the pursuit of the villain (Critchley, 1967; Lee, 1971; Reith, 1952a). As Critchley notes:

> *The Statute of Winchester, embodying a fusion of Saxon and Norman ideas, may thus be conveniently regarded as marking the end of the first police "system" in England, which can be seen to pivot largely round the part-time constable, a local man with a touch of regal authority about him, enshrining the ancient Saxon principle of personal service to the community and exercising powers of arrest under the common law. (p. 7)*

The Decline of the Police Constable

The next 500 years, from the late thirteenth century to the nineteenth century, saw a decline in the office of constable, who became known as the parish constable.

With the passage of the Justices of the Peace Act of 1361, a new post of justice was created, with more power and prestige than that of the parish constable. As the social and political importance of the justices increased, the corruption and eventual decline in standards of integrity of the justices also grew. The practice of "<u>**trading justice**</u>" by exacting a fee for every act performed by a justice amplified the potential for evil in this system (Critchley, 1967, p. 19). The justice was paid in proportion to the number of persons he convicted, so he had little motivation to resist and eventually gave in to temptation. Enterprising villains could match or exceed the standard fee, thereby becoming immune from prosecution. This contributed to the decline and eventual disappearance of the parish constable.

Although the office of parish constable survived the Civil War and the chaos of the Restoration, it failed to survive the Industrial Revolution, but not before it had been transplanted to the United States. This office became the foundation for both simple rural and complex urban police organizations in the United States. It survives to this day, and vestiges of its history and development are still apparent in both U.S. and English policing.

Many factors contributed to the downfall of the parish constable in England. During the disorder of the reign of Charles II, the emergence of highwaymen, bandits who preyed on travelers, brought about attempts to control them through policing. A thousand nightwatchmen (called **bellmen** or "**Charlies**" after Charles II) were appointed to assist the constables. At the same time (1692), the government set up a system of monetary rewards and pardons for information or assistance that led to the arrest of highwaymen. Unfortunately, this system led to increased corruption in the cities and towns, and soon included all the thieves and robbers in London. As a consequence, local parishioners demanded that they be allowed to hire their own private deputies. These deputies quickly became professional criminals and spawned a sophisticated network of crime in the city. The parish constables were no match for these criminals. By the end of the eighteenth century, this ineffectual system in London began to spread to the towns and provinces of the countryside and led to uncontrolled crime throughout England.[5]

Inevitably, the City of London and the rest of England were in great turmoil. Crime was rampant. The population in England had doubled over the course of the eighteenth century to 12 million people. Whole districts of London were regarded as sanctuaries in which thieves and villains enjoyed almost complete immunity from prosecution.[6]

It was against this background of lawlessness in England that the first anguished cries for police reform came and the first reformers emerged. The early reformers (e.g., Henry and John Fielding, Patrick Colquhoun) were more focused on police reform in London and its surrounding areas than in the rural areas. But before proceeding to a discussion of police reform in London, it should be noted that some reform did take place in the rural areas; the concept of paid constables in place of part-time amateurs was gaining acceptance. The full-time village constables, decrepit and inefficient as many were, were linked to the reform occurring in London (Critchley, 1967, p. 27).

The City of London Police

Although the strife, crime, and turmoil of the surrounding metropolitan areas continued during most of the eighteenth century, one small area within London was more successful in governing and policing its inhabitants. As a result of some ancient charters and enactments, the old walled and gated section of London was, and still is, a distinct unit known as "the City." By hiring more watchmen and constables the authorities were better able to police the City, and with the assistance of some stringent laws and rules they enforced compliance. Although these "Charlies" were inefficient, derisive, and given to drinking (Critchley, 1967, p. 30), the City was probably the best policed part of London. It is important to note that the City of London force continues to be independent from the **Metropolitan Police Force**, which has always had the responsibility for policing only the areas surrounding the one square mile of the City. The history of the City Police is rich in controversy and conflict, and the various attempts to abolish the force and the City's determination to maintain a separate force have been chronicled by many authors (e.g., Rumbelow, 1971).

The beginning of English policing until approximately the first quarter of the eighteenth century is primarily the story of the tumultuous development of the tithingman, the emergence and then the downfall of the parish constable, and the chaotic conditions in which this occurred. However, the next century of English policing saw the emergence of the police as a coherent force with the collection of watchmen and constables developing a unity of purpose and mission and becoming concerned with professional conduct, integrity, and the means to defeat the lawlessness that characterized London. A succession of police reformers effected changes in the way London was policed. It is widely agreed that the reforms came about due primarily to the single-minded vision of several individuals who eventually shaped the police to become the force it is today. Beginning with Henry Fielding in 1748, and ending with Robert Peel in 1829, the changes wrought by these reformers left an indelible influence upon all policing.

The Reformers

Henry Fielding was a novelist in London in 1748 when he was appointed to the office of Chief Magistrate of Bow Street. He had been widely characterized as a bohemian for his lifestyle and his affinity for alcohol (Reith, 1952a, p. 131); however, his writings were acclaimed as being full of human compassion and understanding (Critchley, 1967, p. 32). Fielding was a pioneer in critical thinking about crime and penal policy. Through a series of pamphlets, most notably "An Inquiry into the Causes of the Late Increase of Robbers" (1751), he developed his thesis that policing did not necessarily need to be reactive in nature, but could be proactive. He believed that citizens could band together and help to *prevent* crime. During his time as a magistrate at the Bow Street office, he established a group of men (later to be called the "**Bow Street Runners**") who were paid to seek out and prevent crime in the area surrounding Bow Street. These individuals were not distinctive in dress or style, but carried a staff to set them apart from the citizenry. Together with the citizens and a small grant from the government, they were successful in ridding the area around Bow Street of crime.

Henry Fielding's blind brother, John, succeeded Henry on the bench upon Henry's death in 1754 and carried on the work of his brother. The importance of their work, however, was not recognized for some time despite the practical success they had in the Bow Street area. Henry's idea of a paid magistrate with a preventative force at his disposal was not immediately adopted by the government. In fact, most of the people in London and the rest of England were very suspicious of a police system as suggested by the Fieldings. The people felt that the word *police* had a sinister tone to it, and reminded them too much of the French *gendarmerie* (Critchley, 1967, p. 35), whose centralized power they feared and mistrusted. In 1785 a bill was introduced into Parliament to establish a paid magistrate and a salaried metropolitan police force. It was greeted with widespread resistance, primarily from the City of London, and was eventually withdrawn. However, this bill anticipated Robert Peel's 1829 bill and can be considered the precursor to the first Metropolitan Police Force of London.

One interesting point to note is that Ireland adopted most of the bill in 1786 and began to lay the foundations for reform in that country, thus providing the groundwork for the early beginnings of the **Royal Irish Constabulary**. In fact, Robert Peel, whose contribution to police reform is described later, was Chief Secretary of Ireland between 1812 and 1818 and worked on police reform in Ireland. However, Ireland did not establish a metropolitan force in Dublin until 1838, after the success of Peel's force in London. Nevertheless, many historians believe that Dublin should be credited with the establishment of the first organized metropolitan police force in 1786 (Critchley, 1967, p. 38; Palmer, 1988).

Mention should also be made of another reformer who predates Robert Peel. Patrick Colquhoun, a Scotsman who originally came to London as a tradesman, was appointed a magistrate in 1801 and endeavored to carry on the ideas and work of the Fielding brothers. A prolific writer, Colquhoun proposed the "new science of preventative police" (Reith, 1952a, p. 137), focusing on the prevention and detection of crimes. He is credited with providing impetus for the establishment of the private Marine Police in 1798, which was set up to patrol the river Thames for thieves and villains. The success of this force led the government to establish the Thames River Police in 1800. Based on these facts, many support the notion that the Thames River Police was actually the first regular professional police force in London (Critchley, 1967, p. 43). Although Colquhoun's ideas were successful in practice, his plans to extend this type of force to all ports and towns in the country were not acted upon for some 20 years, in the time of Robert Peel.

Between 1800 and 1829, there was a rapid expansion of what had become the Bow Street horse patrol. This patrol became the first uniformed police force in the country in approximately 1805 with its distinctive dress of blue coats with yellow buttons, scarlet waistcoats, blue trousers, Wellington boots, and black hats. This combination of blue and red eventually led to the nickname "**Robin Redbreasts**." Along with the horse patrol, the foot patrols from Bow Street were expanding and given more local discretion. In 1822, when Robert Peel took the office of Home Secretary, a new uniformed day patrol was organized of mostly old soldiers. Thus, by 1828, on the eve of the formation of the Metropolitan Police, there were already a large number of professional full-time police officers in London. In addition, there were also a substantial number of watchmen employed by the City of London and by the parishes, assigned to patrol the more than 1.5 million inhabitants of London. At best, however, these individuals could do little more than catch a few criminals and attempt to bring them to justice.

Sir Robert Peel and the Metropolitan Police Service

On April 15, 1829, Robert Peel introduced a "Bill for Improving the Police in and near the Metropolis." The bill proposed a police office in charge of justices, the creation of paid constables for what was to be called the Metropolitan Police district (excluding the City of London), the replacement of the watch rate by a new police rate, and the appointment of an official to be called the Receiver, whose job was to handle all the money resulting from the act (Critchley, 1967, pp. 49–52). The Metropolitan Police Act became law on July 19, 1829, and represented the culmination of the hard work and vision of its predecessors, Henry and John Fielding, Patrick Colquhoun, and even Jeremy Bentham.

Although Peel generally receives most of the credit for being the "father of English policing," he should also be noted for his knowledge and experience—as a politician and for his ability to get the bill passed with virtually no opposition, a feat all the more significant in light of the 75 years of wrangling that preceded it. In addition, he appointed the first two justices, or commissioners, who were assigned the monumental task of organizing and implementing the institution set up by the act. He chose Charles Rowan, a retired military officer, and Richard Mayne, a barrister, who formed a now famous partnership to lead the Metropolitan Police into its place in the history of law enforcement. They took as their headquarters an office on Whitehall Place with a rear entrance onto a narrow lane to the east known as Scotland Yard. Today, after many relocations, the administrative headquarters for the Metropolitan Police is located a short distance away in a multistory office building on Victoria Street, and is called New Scotland Yard. Although always referred to as New Scotland Yard, tentative plans are currently being discussed to move many of the administrative functions of New Scotland Yard to a 27-story high-rise building near Earl's Court in west London.

The "New Police"

The task of implementing a new police organization in London was a formidable one for Rowan and Mayne. In addition to recruitment, they had to restore dignity and credibility to the police in light of rising opposition from the citizens. This was true not only in London, but also throughout the provinces and counties of England. The two men immediately set about structuring the Metropolitan Police by dividing the district (roughly a 7-mile radius emanating from Bow Street and Charing Cross) into 17 police divisions, each with 165 men for a total of about 2,800 police constables. For each division, there was a superintendent in charge, with 4 inspectors and 16 sergeants. Each sergeant had control of 9 constables. In addition, the commissioners decided to outfit the men in a nonmilitary uniform of a blue tailed coat, blue pants, and a glazed black top hat with a thick leather crown. They were required to carry a rattle and a short truncheon concealed under the coat (Critchley, 1967, p. 51). In August 1829 the recruitment of the 2,800 men for the force was begun, and at 6:00 p.m. September 29, 1829, the first groups of New Police marched out of the station house at Scotland Yard to begin their first patrol (Lee, 1971, pp. 236–239). Regarded suspiciously by Londoners, these New Police were often referred to as "**peelers**" or "**bobbies**" due to their relationship to Sir Robert Peel. The nickname "bobby" has stuck and is associated throughout the world with the London police officer.

The New Police encountered strong opposition from the citizenry, who generated serious organized attempts to abolish them. The first decade of operation was fraught with difficulty and adversity; however, they did persevere to forge the reputation that would eventually become world famous in law enforcement. As Critchley portrays it: "Their imperturbability, courage, good humor, and sense of fair-play won first the admiration of Londoners and then their affection" (Critchley, 1967, p. 55).

The ensuing decades until the twentieth century were colorful and exciting times for the police in England, but far beyond the scope of this book; however, they are important years because they set the foundation for the current structure of English policing and are paralleled by the development of policing across the Atlantic Ocean.[7]

The development of the Metropolitan Police in London served as a prototype for the establishment of similar forces elsewhere. The Municipal Corporations Act of 1835 and the County Police Act of 1839 provided for the establishment of police forces outside London. The County and Borough Police Act of 1856 required the establishment and maintenance of police forces. A second important accomplishment was the assignment of powers to the Home Secretary for control of the police throughout the country, thus allowing for a form of nationwide policing. This served to reconcile the concept of central supervision with local management outside London and struck a balance that remained virtually unchanged until the middle of the twentieth century (Critchley, 1967, p. 101). In fact, during the second half of the nineteenth century, change was quite slow in coming to the police forces throughout England, while significant social and governmental change was rapidly occurring throughout the country (Home Office, 1960, para. 61).

■ Twentieth-Century Policing in England

The dawn of the twentieth century found policing in England to be mainly a collection of various remnants of the preceding century. The rural forces remained under local control, and relatively sparse information is recorded concerning their activities (Critchley, 1967). Surviving police histories consist mainly of accounts of the activities of the two London forces: the Metropolitan and City police.[8] For the most part, the constable was represented as a working-class figure who was often the focus of comedy and stage depictions as an individual with singularly low socioeconomic roots and an uneducated background. As long as this image was supported, the English citizenry was not concerned about the police and policing; however, as many historians have noted, the introduction of the automobile changed the direction of modern policing in England (Critchley, 1967, pp. 176–177; Palmer, 1988, pp. 8–10).

At first hostile to the new automobile, the police eventually accepted it and the problems it brought. The relationship between the police and the motorist reflected other types of conflicts that can arise between the police and the people they serve. Complaints about unnecessary suspicion, heavy-handedness, and inflexibility, among others, can be the result of what is perceived to be improper handling of simple motor-related situations (Critchley, 1967, pp. 176–179). Today in England, the United States, and most industrialized countries, the relationship between the citizen and the police on the crowded motorways is still a delicate one, and often a source of emotional conflict.

The recent history of the English police from the beginning of the twentieth century is marked by many significant events. Advances in science changed the nature of detective work, and social changes included the introduction of female and minority police officers. The two world wars had an impact on police work. The establishment of the Hendon Police College and Bramshill Police College signaled the new status of police officers as professionals. In addition, riots and urban unrest and the resulting modifications of the law brought about significant changes in policing. All of this will be discussed in more detail in the following chapter.

■ Conclusion

From its earliest Saxon beginnings, the police system in England was very different from the French *gendarmerie*. Indeed, as Stead (1975) put it: "France's powerful police system was ancient long before they [the English] were born" (p. x). Although the French gendarmerie had long been a powerful centralized organization, the development of the police in England was slow and met with suspicion by the citizens, always protective of their freedom. As some saw it, the "British police institution was made not by, but in spite of, the British people" (Reith, 1952a, p. 131).

If the English were reluctant to empower the police, the Americans were even more extreme in their attitudes. Because they had renounced the monarchy, anything smacking of royalty, aristocracy, or state corruption was repugnant to them. Because they had rejected tyranny from afar, Americans favored local control and opposed a national or federal police system. Theirs was a country founded on liberty and order, which they felt should most properly flow from local control. This is evidenced by the fact that the Federal Bureau of Investigation and the Central Intelligence Agency are of somewhat recent origin (1908 and 1947, respectively), and are even now regarded suspiciously by many Americans. Furthermore, there are nearly 19,000 separate organized police forces in the United States, characterized by a highly fragmented organizational structure (U.S. Department of Justice, 2000). Compare that number to the 43 forces in England. Although it is unlikely that a single U.S. police system will ever be established, the enforcement systems in the United States have at least become more closely integrated as a result of the sophisticated communication systems that now exist.

In many ways the systems in the United States have paralleled the development of the English police since 1829. There was a similar resistance from the public and then a gradual acceptance of the experiment with both the New Police in England and the development of the policing bodies in the United States. There were obvious differences between them regarding firearms and uniforms, but these were relatively minor. The Americans felt that uniforms were too similar to royal garb and did not wear them until several years after the English adopted them. American police began carrying firearms in 1850, after increases in urban immigration, the rates of violent crimes, and the use of guns by criminals (Palmer, 1988, pp. 19–20). Finally, whereas their English counterparts were responsible to unelected county magistrates and, in London, to the government represented by the Home Secretary, many of the police forces in the United States reported to locally elected officials. In fact, the period following the American Revolution was characterized by an interwoven relationship between police and politics.

As in England, the first police forces established in the United States were in the cities, with the notable exception of the frontier border patrol, the Texas Rangers (1835). The earliest urban police forces in the United States included New York (1845), New Orleans and Cincinnati (1852), and Boston, Philadelphia, and Chicago (1855).

There is also a parallel between the histories of the police in both countries vis-à-vis the development of their largest cities: New York and London. Just as the New Police were beginning in London, a similar system of night watchmen and civilians was employed in New York. It is reported that in 1825, 100 night watchmen looked after the well-being of the city of New York (population 125,000) preceding the establishment of its first organized police force (Richardson, 1970, p. 9).

In contrast to England, however, calls for reform of the New York City police did not come from the working classes but from city council members and other politicians. This agitation continued until the first force was established in 1845. The officers were uniformed (though not until 1856 because of objections to the similarity between the blue-coated uniforms and British uniforms) and armed with two-foot-long batons. The Police Act of 1857 transferred control of the New York City Police Force to the governor of New York State. This placed the governor of New York in a role similar to the Home Secretary of England, both having control over their respective police forces (Richardson, 1970, pp. 96–108). Another similarity between New York and London was the fact that the mayor of New York City would not recognize the new Albany-controlled police force in his city. For 3 months there existed two police forces in New York City, suggestive of the division between the City of London police and the Metropolitan Police of London. Unlike the City of London police, however, the municipal police of New York City succumbed to pressure and accepted a court order to disband. In 1870 control of the police was returned to New York City, and as is true for the vast majority of police forces in the United States, it is now subject to local municipal control.

Although today police in the United States routinely carry firearms and ride in automobiles, the English still rely primarily on foot patrols and limited use of firearms. However, as we will see in Chapter 6, these and other current practices are being challenged, and change appears to be forthcoming for the traditional English "bobby." Just as American law enforcement has been significantly affected by the sociopolitical conditions in a changing United States, the English are seeing vast changes in their system of policing. The nature and direction of these changes are subject to speculation; however, they are inevitable given the forces of change currently at work in England as the police move forward in the twenty-first century.

CHAPTER SPOTLIGHT

- Important markers to the development of the English police system include the frankpledge system, tithings, hundreds, constable shire, hue and cry, posse comitatus, and justices. All of these may be recognizable as precursors to systems in the United States.

- After the Norman conquest of 1066, the tithingman developed into a leader of men, who became responsible for apprehending people who broke the law. As his duties and prestige grew over the next 200 years, he became known as the parish constable or village constable. This elected or appointed office would require the newly appointed constable to serve for one year, unpaid. All able-bodied men were expected to serve. This gave way to a new system known as the watch system.

- The Statute of Winchester may be regarded as marking the end of the first police "system" in England, which pivoted largely around the part-time constable: a local man with regal authority to arrest people under the common law.

- The Justices of the Peace Act of 1361 created a new office with more power and prestige than that of the parish constable. However, even with the increase in social and political importance for the justices, corruption and decline in standards of integrity of the justices also grew.

- The English government set up a system of monetary rewards and pardons for information or assistance that led to the arrest of highwaymen: bandits who preyed on travelers. Unfortunately this led to increased corruption within cities and towns. Parishioners demanded to be allowed to hire their own private deputies, but this system proved to be ineffectual.

- The walled and gated section of London was, and still is, a district known as "the City." The City, by hiring more watchmen and constables, and having stringent laws and rules, was better able to police the area and enforce compliance. The City of London Police continues to be independent from London's Metropolitan Police Force. The City police have the responsibility for policing only the areas within the one square mile of the City.

- A number of reformers influenced policing including Henry Fielding and Robert Peel. Henry Fielding, a novelist, became Chief Magistrate of Bow Street in London in 1748. It was here that he established a group of men, the "Bow Street Runners," who were paid to seek out and prevent crime in the area surrounding Bow Street.

- Henry Fielding believed that policing could be proactive rather than reactive and that citizens could band together and help to prevent crime. He also believed that paid magistrates should have a preventative force at their disposal, a suggestion that was not immediately adopted by government. Henry was succeeded by his brother, John, in 1754.

- In 1785 Parliament passed a bill that established a paid magistrate and a salaried metropolitan police force. It was withdrawn due to widespread resistance. However, it can be considered a precursor to the first Metropolitan Police Force of London.

- Ireland, then part of the United Kingdom, adopted most of the 1786 bill and began laying the groundwork for the Royal Irish Constabulary. However, the first metropolitan force was not established in Dublin until 1838.

- Patrick Colquhoun, a Scotsman, was a tradesman who was appointed a magistrate in 1801. He proposed the "new science of preventative police," focusing on the prevention and detection of crimes. He is credited with providing impetus for the establishment of the private Marine Police in 1798, which patrolled the river Thames. Consequently the government established the Thames River Police in 1800.

- The Bow Street horse patrol became the first uniformed police force in England, in approximately 1805. The distinctive dress included blue coats with yellow buttons, scarlet waistcoats, blue trousers, Wellington boots, and black hats, which led to the nickname "Robin Redbreasts."

- In 1829 Robert Peel introduced a "Bill for Improving the Police in and near the Metropolis" that proposed a police office in charge of justices, the creation of paid constables for what was to be called the Metropolitan Police district (excluding the City), the replacement of the watch rate by a new police rate, and the appointment of an official to be called the Receiver, responsible for handling all monies resulting from the act. The Metropolis Police Act became law in 1829.

- Charles Rowan and Richard Mayne, the first two appointed justices or commissioners, along with Robert Peel, set up headquarters for the Metropolitan Police (The Met) in an office on Whitehall Place with a rear entrance onto a narrow lane to the east known as Scotland Yard. Today the administrative headquarters for the Met has moved to a new location and is called "New Scotland Yard."

- The Met was set up by dividing the district into 17 police divisions, each with 165 men. Within each division there was a superintendent in charge with 4 inspectors and 16 sergeants. Each sergeant had control of 9 constables.

- The New Police were often referred to as "peelers" or "bobbies" due to their relationship to Sir Robert Peel. Today officers in London are still referred to as "bobbies."

- The Municipal Corporations Act of 1835 and the County Police Act of 1839 provided the establishment of police forces outside London.

- The County and Borough Police Act of 1856 required the establishment and maintenance of police forces, and allowed for a form of nationwide policing by assigning powers to the Home Secretary for control of the police throughout the country, a concept of central supervision with local management outside of London.

- With the advent of automobiles, the relationship between the police and motorists was often fraught with conflict as complaints about unnecessary suspicion, heavy-handedness, and inflexibility arose due to what was perceived to be improper handling of simple motor-related situations. This is still the case today as the relationship between motorists and the police is a delicate one.

- The systems in the United States have paralleled the development of the English police since 1829. In the United States the first police forces were also established in the cities including New York, the notable exception being the Texas Rangers, a frontier border patrol.

- The Police Act of 1857 transferred control of the New York City Police Force to the governor of New York State. This placed the governor of New York in a role similar to the Home Secretary of England, both having control over their respective police forces. For 3 months there existed two forces in New York City. In 1870 control of the police was returned to New York City.

KEY TERMS

Bellmen/Charlies: The 1,000 nightwatchmen who were appointed to assist the constables in their attempt to control the emergence of highwaymen.

Bow Street Runners: A group established by Henry Fielding who were paid to seek out and prevent crime in the area surrounding Bow Street.

Chief pledge/capital pledge: Formerly known as the "tithingman" and developed into a leader of men who became responsible for apprehending people who broke the law. Eventually became known as the parish constable or village constable.

Constable: An English police officer.

Frankpledge system: The new name for the tithings following the Norman conquest of 1066; it was considered the "most important police institution in the Middle Ages."

Hue and cry: An ancient Saxon practice that was often imposed by the watch system when a stranger or villain was noticed. It called for all persons in the town or in nearby towns to pursue the fugitive.

Hundreds: A collection of 10 tithings.

Hundredman: The individual assigned the responsibility for the hundred, who was obliged to report to the shire reeve.

Justices: Also known as commissioners; a position created with the passage of the Justice of the Peace Act of 1361 that held more power than the constable and was assigned the monumental task of organizing and implementing the institution set up by the act.

Kin police: Resulting from the tithing where men were mutually responsible for each other's behavior; the current day phenomenon where police are under local control, as opposed to government control.

Metropolitan Police Force: The police force that is responsible for policing the area of London that surrounds the one square mile of the City of London.

Parish constable/village constable: Formerly known as the "chief pledge" or "capital pledge," he was elected or appointed to his position for a tenure of one year. This position was not paid, and all able-bodied men were expected to serve in this capacity.

Peeler/bobby: The nickname given to police constables in London (circa 1829) because of their relationship to Sir Robert Peel.

Posse comitatus: The group of persons from the town or the nearby towns who were required to pursue the fugitive when the watch system practiced the "hue and cry."

Reformers: The individuals who are credited with shaping the English police force into what it is today (e.g., Robert Peel and Henry Fielding).

Robin Redbreasts: The Bow Street horse patrol, which became the first uniformed police force in the country. The combination of blue and red in their uniform led to the nickname.

Royal Irish Constabulary: What some people argue to be the first organized metropolitan police force in Ireland, although England is most often credited with housing the first organized force.

Shire: The division of land determined by the Anglo-Saxons.

Shire reeve: The overseer for the ruler who became known as the sheriff and had the responsibility for "breaches of the peace."

Tithings: A group of 10 men who were mutually responsible for each other's behavior.

Tithingman: Considered the earliest reference to some form of policing official; responsible for collecting fines on people in his tithing. Later became known as the "chief pledge" or "capital pledge."

Trading justice: A practice in which a justice was paid a fee for each act he performed. This practice often amplified the potential for evil in the criminal justice system.

Watch system: A group of as many as 16 men who normally functioned only at night and served to supplement the responsibility of the constable by being stationed at all of the gates of a walled town or city. These men were given arrest powers and were able to impose the "hue and cry" if a stranger or villain was noticed.

PUTTING IT ALL TOGETHER

1. What historical events led to the development of the police constable and the English police force?

2. What was the role of the early reformers in the development of the English police force?

3. How does the English style of policing compare to the style of policing in the United States?

ENDNOTES

1. "King's peace," so called because the King guaranteed, or at least promised, his subjects a state of peace and security in return for their allegiance. First noted during the Anglo-Saxon period, it has continued to be cited until modern times (Lee, 1971, p. I).

2. From Saxon times to the early eighteenth century, American and English police histories have a commonality founded upon the principle of **kin police** (police under local control, as opposed to government control). This principle has been more prevalent in the United States than in England,

as can be seen by the multiplicity of city, county, state, and federal forces in the United States (Stead, 1975, p. x).

3. Although stated by an American author, this would appear to be appropriate in England as well as the United States.

4. The term *constable* first appears in a statute in 1252 and comes from the words *comes stabuli*, meaning "master of the stable" (Critchley, 1967, p. I).

5. For a complete and detailed account of this breakdown of the parish constables, see Reith (1938, 1943).

6. For a colorful and interesting description of that period, see the "Story of Jonathan Wild" (Reith, 1952b). It is an amusing account of the life of one of the first organized crime bosses in England and how he made his system work with impunity for several years in London.

7. For further descriptions of this period, see the excellent works of Miller (1977), Emsley (1984), Mather (1959), Best (1972), Richter (1981), Smith (1985), Rumbelow (1971), and Palmer (1988).

8. For a discussion of the paucity of literature on this period of English policing, see Palmer (1988).

REFERENCES

Best, G. (1972). *Mid-Victorian Britain, 1851–1875*. New York: Schocken.

Clarkson, C., & Richardson, J. (1889). *Police! History of the Metropolitan Police*. London: Leadenhall.

Critchley, T. A. (1967). *A history of police in England and Wales 1900–1966*. Letchworth, Hertfordshire, England: Garden City.

Emsley, C. (1984). *Policing and its context, 1750–1870*. Philadelphia: Open University Press.

Goldstein, H. (1977). *Policing in a free society*. Cambridge: Balinger.

Home Office. (1960). *Interim report of the Royal Commission*. London: H.M.S.O.

Lee, M. (1971). *A history of police in England* (reprint). Montclair, NJ: Patterson Smith.

Mather, F. (1959). *Public order in the age of the Chartists*. Manchester, England: Manchester University Press.

Miller, W. (1977). *Cops and bobbies: Police authority in New York and London, 1830–1870*. Chicago: University of Chicago Press.

Palmer, S. (1988). *Police and protest in England and Ireland: 1780–1850*. Cambridge: Cambridge University Press.

Reith, C. (1938). *The police idea*. Montclair, NJ: Patterson Smith.

Reith, C. (1943). *British police and the democratic ideal*. Montclair, NJ: Patterson Smith.

Reith, C. (1952a). *The blind eye of history: A study of the origins of the present police era*. Montclair, NJ: Patterson Smith.

Reith, C. (1952b). Story of Jonathan Wild. In Reith, C. (Ed.), *The blind eye of history: A study of the origins of the present police era* (pp. 34–53). Montclair, NJ: Patterson Smith.

Richardson, J. (1970). *The New York police: Colonial times to 1901*. New York: Oxford University Press.

Richter, D. (1981). *Riotous Victorians*. Athens, OH: Ohio University Press.

Rumbelow, D. (1971). *I spy blue: The police and crime in the city of London from Elizabeth I to Victoria*. London: Macmillan.

Smith, P. (1985). *Policing Victorian London*. Westport, CT: Greenwood.

Stead, P. J. (1975). Introduction. In C. Reith (Ed.), *The blind eye of history: A study of the origins of the present police era* (pp. 1–6). Montclair, NJ: Patterson Smith.

U.S. Department of Justice, Bureau of Justice Statistics. (2000). *Law enforcement statistics*. Washington, D.C.: U.S. Department of Justice.

The Modern English Police

6

Chapter Objectives

After completing this chapter, you will be able to:

- Describe the past and present structure and organization of the English police.
- Explain the path one must take to become an English police constable.
- Describe the duties and roles of police constables and the public's response to how they perform them.

■ Introduction

As you saw in Chapter 5, the history of the English police is fraught with colorful and exciting periods. There have always been conflicts between the advocates of a national centralized police force and those who, fearing it, supported local control of the police. Bitter opposition has been aroused among the citizens whenever there has been an attempt to strengthen the efficiency and powers of the police.

In this chapter we will examine the current structure and organization of the English police, police recruitment and training, and the issue of women and minorities in policing. We will then turn our attention to an investigation of police powers and procedures, the image of the police in English society, the police response to the increasing violence in England, and new directions and challenges to the traditional structure of the police.

■ Current Structure and Organization

As of World War II, there were 183 police forces in England and Wales. In the mid-1960s, after the 1962 Willink Commission[1] recommended a reduction, and economic conditions necessitated it, the number of forces was reduced to 43 in England and Wales, which remains the number of forces in service. Among the 43 English forces are the two police forces for the London area: the Metropolitan Police and the City of London police. In the 1960s and 1970s several changes were made in the boundaries of the English police forces, primarily due to local wrangling between political figures in neighboring areas. More recently there has been talk of reducing the number of police forces by combining some of the smaller forces (Her Majesty's Inspectorate of Constabulary [HMIC], 2006). However, this appears to have been effectively abandoned after only two forces agreed to merge and then discovered that the costs were prohibitive. The BBC reports that the entire review process for mergers cost police forces over 6 million pounds (BBC News, 2006).

The boundaries of the 41 police forces outside of London coincide roughly with the boundaries of the local governments. These forces are headed by chief constables who report to a **local police authority**. In some cases, particularly in large urban areas, two or more local governments have combined their forces to make one larger police force. In the 41 forces outside London, the administrative structure is a tripartite system composed of the chief constable, the local police authority, and the central government (Home Office), which provides half of the budgetary appropriation for the force. The local police authority, which provides the other half of operating funds, is composed of a maximum of 17 members, 9 of whom are locally elected councilors, 3 of whom are magistrates, and 5 of whom are independent members. In addition, there is an Inspectorate from **Her Majesty's Chief Inspector of Constabulary Office**. Originally created in 1856, this office is responsible for the formal inspection and assessment of all 43 police forces and acts in a key advisory role to the Home Office, the chief constables, and the local police authorities (HMIC, 2006). Accountability is sought at the local level by having a member of the local police authority available to answer "questions on the discharge of the police authority's functions" at council meetings (Police and Magistrates' Courts Act 1994, Section 12), and at the national level by having the Home Secretary answerable to Parliament.

The basic organization of an English provincial police force is depicted in **Figure 6-1.** The structure is similar throughout England and Wales. The London Metropolitan Police Force is headed by a commissioner who is appointed by the Crown on the advice of the Home Secretary. Prior to passage of the Greater London Authority Act of 1999, the commissioner had reported directly to the Home Secretary. Now the commissioner reports to the Metropolitan Police Authority, which was established by that act. The City of London Police commissioner reports to the Common Council of the City of London.

The local police authority is responsible, subject to the approval of the Home Secretary, for selecting the chief constable and the assistant chief constables. Again subject to the approval of the Home Secretary, the local police authority may remove a chief constable or assistant chief constable (Police Act 1996,

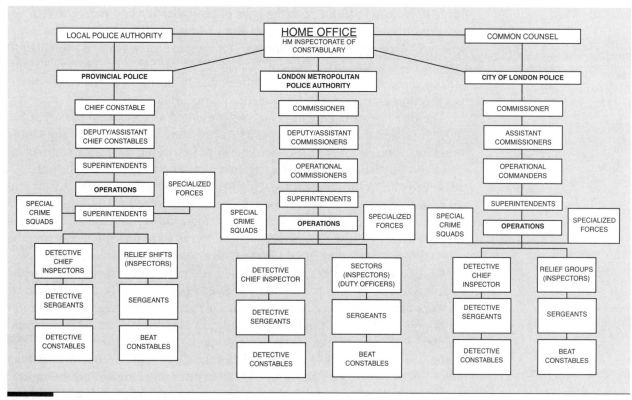

Figure 6-1 Organizational Chart for English Law Enforcement

Section 11), though in this case the Home Secretary may also require a police authority to "call upon a chief constable (or assistant chief constable) to retire in the interests of efficiency or effectiveness" (Section 42).

Although there is, as stated above, a tripartite system composed of the chief constable, the local police authority, and the central government, the central government represented by the Home Secretary has long been able to exact compliance in matters of policy, budget, and personnel. This has tended to encourage conformity, cooperation, and efficiency throughout the country. Although the powers of the Home Secretary are vast, there have been no significant complaints of abuse of this power. However, the 1980s signaled an increase in "**guidance circulars**" issued to the local forces from the Home Secretary's office, indicating a possible shift of the center of power from the local authorities to that office (Rowe, 1986a, p. 11). The 1990s and early twenty-first century saw the development of a national policing plan and national objectives, with local plans and objectives tied into the national plan and objectives. The Police and Magistrates' Courts Act of 1994 enabled the Home Secretary to "determine objectives for the policing of the areas of all police authorities" and to set performance targets for achieving those objectives (Section 15). These provisions were set out again in the Police Act of 1996 (Sections 37–38), which more clearly directed than before that the local police authorities, in carrying out their functions and establishing local policing objectives, should have regard to objectives set by the Home Secretary and comply with any directives issued by him or her (Section 6). The

Police Reform Act of 2002 took the issue of centralization one step further by establishing a National Policing Plan (Section 1). This plan, which sets out the Home Secretary's "strategic policing priorities" for the next three years, is presented to Parliament each year. In developing the plan, the Home Secretary is to consult with persons representing the interests of the police authorities and the chief constables (Section 1), thus maintaining the traditional tripartite approach to policing.

The first **National Policing Plan** was presented to Parliament in November 2002. It covers the years 2003 through 2006 and, in the words of the Home Secretary, "sets out a clear national framework for raising the performance of all forces and publishes the indicators against which performance will be judged." It also is to "form the basis for local plans to be drawn up by Chief Officers and police authorities to improve standards and ensure forces are responsive to the needs and priorities of their communities" (Home Office, 2002, p. 2). Following the presentation of the first plan, two more plans were presented covering the years up to and including 2008 (Home Office, 2003; Home Office, 2004).

On December 31, 1991, there were 127,100 police officers employed by the 43 police forces. By March 31, 2005, the number of police officers employed by the 43 police forces had risen to 142,795. This was a 2% increase over the number of police officers in England and Wales in March 2004. In 2005 the mean number of officers per 100,000 population was 270, ranging from a low of 184 for the Surrey police force to a high of 433 for the City of London and Metropolitan police forces. There are also about 83,000 civilians working within the various police forces. These include 71,496 civilian police staff who provide technical and clerical services; 2,234 police traffic wardens; 6,702 community support officers, civilians entrusted with limited enforcement powers to deal with lower-level crime, disorder, and anti-social behavior; and 2,234 designated officers who are skilled civilian staff assigned to specific police roles (Bibi, Clegg, & Pinto, 2005).

In addition to the officers on the 43 forces, there are several special branches of law enforcement, including regional crime squads designed to combat specific types of crime, such as child pornography, and squads dealing with drugs and soccer violence. In 2006 a new law enforcement agency, the Serious Organized Crime Agency (SOCA), was created "to reduce the harm caused to people and communities in the UK by serious organized crime" (SOCA, 2006). Finally, there are a number of specialized forces, such as the British Transport Police, the Ministry of Defense Police and Guarding Agency (formerly the Ministry of Defense Police), and the Civil Nuclear Constabulary that have narrowly defined areas of authority (HMIC, 2006, p. 44).

Given this brief overview of the current organization and structure of the English police, it is appropriate to turn now to questions more germane to the functions of modern policing in England, such as: How does one become a police officer in England? And, what can one expect in terms of a career in policing?

■ Recruitment and Training

Due to the unitary nature of the English system of policing, the recruitment qualifications are uniform throughout the country. Anyone over 18½ years of age can apply to become an officer in any force in England. There is no upper age limit for appointment as a police officer in England and Wales; however, the typical retirement age for a police officer is 55. Applicants are required to be British citizens, EU/EEA nationals, or Commonwealth citizens. Foreign nationals are permitted to apply as long as no restrictions have been placed on their stay in the United Kingdom (Police—Could You?, 2007a).

Cadet programs continue to exist in 10 U.K. forces (Police—Could You?, 2007b), though the Metropolitan Police had to suspend its cadet school at the Peel Centre Police Training Facility, Hendon, in 1993 due to lack of funding. An individual can apply to the cadet program as early as 14 years of age and be subject to much of the preparatory training that recruits undergo (although no formal police training is given to cadets until they have been accepted into the police force). At the age of 18½ they can apply for formal police recruit status.

There are no formal educational requirements for police officer applicants. Applicants are required to take two written exams to ensure that they possess an appropriate understanding of the English language. Applicants are also required to take a numeracy exam (Metropolitan Police Service, 2007). In past years, applicants to the English police were required to take a written examination or have passed at least four O-level (Ordinary) examinations; however, the actual requirement of having passed four O-Level examinations was replaced by the General Certificate in Secondary Education (GCSE) in the late 1980s. (A GCSE is roughly the equivalent of a General Education Degree (GED) in the United States; it is taken at the age of 16, when students in both England and the United States can leave secondary school.) The written examination was designed to evaluate the applicant's aptitude for police work before moving to the next stage of the recruitment process. Recently, as part of a large-scale recruitment strategy, this examination was eliminated. In addition to age and requirements, police recruits must meet certain physical, intelligence, and moral standards before they are accepted into a force and assigned to a probationer.

After acceptance as police officers, all recruits are given a two-year probationary period. The initial stage of training is a 14–17-week course of instruction at one of the six district training centers in the country. In addition to the practical aspects of police training, the recruits receive training in cultural diversity, gender and racial diversity, integrity, elimination of discriminatory behavior, and sound decision making (Metropolitan Police Report, 2006). The focus on these topics had been prompted by the deficiencies in training noted in a 1999 Home Affairs report (McPherson, 1999). After this initial course of instruction, new recruits spend the remainder of their two-year probationary period training in the community in which they will serve (House of Commons, 1999).

The **High Potential Development Scheme (HPD)** is designed to provide guidance and support to those police officers who possess desired leadership skills (Police—Could You?, 2007b). All new and existing officers (up to chief inspector) are eligible to apply for this program. In fact, from the moment a person applies to the police service, he or she is able to show an interest in the HPD program. The application for entry into the police service provides new applicants an opportunity to apply for the HPD program. If the new applicant is successful in gaining entry into the police service, an additional application for the HPD program is sent. Persons who were not originally interested in the HPD program, but were later deemed to be eligible, are also sent applications for the HPD program. Officers already serving in the police service may ask their line manager for an application. After the initial HPD application process, successful candidates complete two written examinations. After successful completion of this stage, candidates participate in a two-day interactive assessment, which includes an interview.

HPD participants have the opportunity to participate in a variety of modular courses. The progress of participants is assessed every six months by line managers. Participants are promoted through the HPD promotion process, not their local force promotion process. Accelerated promotion is a main characteristic and incentive in this program. The Metropolitan Police Service suggests that, through this program, it is possible to achieve the rank of Chief Inspector in as little as five years (Metropolitan Police Service, 2004b). Pay, however, does not differ. Participants in the HPD program receive the same pay as nonparticipants of the same rank.

Another recruitment program allows a limited number of university graduates to enter one of the forces and advance more rapidly through the ranks. After a series of interviews, those selected are appointed as probationer constables. After successfully completing the probationary period of training and passing the examinations, they are posted to the accelerated promotion course at Bramshill Police Staff College (now referred to as the Leadership Academy for Policing at Bramshill), which is described in more detail later in this chapter.

Due to its size, the London Metropolitan Police has its own training center, the Peel Centre, at Hendon. It is interesting to compare this facility and the length of probationary training with that of the old "peelers" or "bobbies" in 1829. In those days new police officers were older (usually ex-servicemen) and were turned out onto the streets for on-the-job training with experienced officers. Today's recruits, though younger, are better trained and more prepared for the job of police constable. As an example of police training in England, we will focus on the metamorphosis of the Metropolitan Police training center at Hendon.

Training at the Peel Centre

At the turn of the twentieth century, Metropolitan Police officers received four weeks of basic training at the Peel House in central London. In 1934, a five-year experiment with accelerated police training was established at Trenchard Police College at Hendon, which later became known as Hendon Police College.

Situated in buildings originally designed for a country club and later used as a factory, the advanced training was designed for senior officers. The fact that some of these officers were drawn from outside the ranks met with great resistance from the rank and file members of the police forces, because for the past hundred years senior ranks had been filled by officers from the lower ranks. Consequently, the term of the original police college was short-lived. It was officially closed in 1939, but its connection to the Metropolitan Police continued with some training functions (e.g., driving school) still carried out there until all training was moved to the Peel Centre, which opened in 1974 (Stead, 1985, p. 82).

The building of this new complex brought all training for the Metropolitan Police under one roof. As a purpose-built training center, the Peel Centre has some of the finest facilities for any police force in the Western world. Composed of technical, academic, sports, and residential facilities, it is a state of the art complex dedicated to training police officers. A May 2006 report from the Metropolitan Police Authority suggested that the Hendon Police College might close, however, and all the land and existing structures might be sold. The desire to decentralize training and conduct training at 15 local centers was given as a rationale for this proposed course of action. This, the MPA believed, would enhance recruitment efforts as well as result in cost savings (Marzouk, 2006; Metropolitan Police Authority, 2004a).

Traditionally, the curriculum has consisted of subjects such as law, human awareness, community relations, physical fitness, first aid, and driver training, and has employed simulation techniques to supplement the conventional lecture-study method. In recent years, however, the nature of the training has shifted to a greater emphasis on human awareness and human relations as opposed to the more rote technical skills training of the past. The underlying assumption seems to be that police officers need to become active participants in the community and not merely enforcers of the law. In the mid-1980s the entire curriculum was dramatically overhauled to reflect this theme.

There is now an increased reliance on simulations and videotaped instructional activities with a move to more colloquial instruction ("street talk"). Military-style drilling and self-defense are included, but firearms training is still not part of the program (Rowe, 1986b; Metropolitan Police, 2006). Although it has always appeared odd to people in the United States, the routine carrying of firearms is still resisted by the majority of the English police. Although there is increasing demand from the younger officers and some factions in society to include weapons training in the training program, the majority of police constables still do not carry firearms. Only 6.9% of the 30,035 Metropolitan Police officers and 4.2% of the police officers in all of England and Wales are Authorized Firearms Officers (AFO) (Home Office, 2004a). The constables who currently carry firearms are usually members of a special branch (e.g., Diplomatic Protection Group, forces at Heathrow and Gatwick International Airports, and certain other highly specialized units that may require the carrying of firearms; Gould & Waldron, 1986, p. 205). However, in May 1994, the Metropolitan Police increased the number of armed response vehicles from 5 to 12 and, in contrast to the previous practice of keeping the firearms in locked boxes in the vehicles, allowed the officers to carry them openly (Campbell, 1994, p. 1). Furthermore, the number of armed response vehicle incidents in England and Wales increased from 8,179 in 2001 to 13,218 in 2004. Approximately 18% of these incidents involved the Metropolitan Police (Home Office, 2004a). Additionally, the number and deployment of armed response vehicles increased again in light of the London Tube bombings of July 7, 2005, specifically to combat the threat of suicide bombers (BBC News, 2005).

After successfully completing the initial 14–17-week program and passing the written examinations, the recruit is assigned to a force, usually in his or her home area, for instruction in the operational practices of that particular force. However, this is not the end of formal education; the recruit may return routinely to the district training center for further instruction during the rest of the probation period, or return to Hendon in the case of the Metropolitan Police.

Throughout their careers, all officers continue in-service training to retain needed skills and to learn new ones.

Learning the proper use of discretion is one of the major goals of training. Great emphasis was placed on this after the Brixton riots of 1981 when police-community relations hit an all-time low in minority communities. The unrest resulted in a full-scale riot, causing great damage and injury to both the police and the community. The Brixton riots, however, did have some positive consequences. They gave authorities the impetus to examine the nature of the training of police officers and to ensure that heightened sensitivity to human relations would become the cornerstone of the recruit training not only in London, but also throughout the country.

As is true of all the district training centers, the instructors at the Peel Centre are active officers with many years of experience. They are specially selected for their knowledge and their ability to transmit it. During the 14–17-week training program, the 160 to 200 recruits in each class live in residence dorms during the week and are allowed to visit their homes on the weekends; however, study time is at a premium and weekends are usually reserved for catching up. There are eight sessions per year. A typical morning for a recruit might consist of classes in law, theft, and civil disputes and participation in police-citizen simulations; in the afternoon there might be physical fitness training and instruction in interpersonal skills and nonverbal communication. During the course, recruits are continually evaluated both in the classroom and in on-the-street simulations. During the fourth week of training the recruits are sent to a police station for one to two weeks of more practical training. In most cases the recruits select the station to which they are sent, and they will usually be assigned to that station after completion of their training course. After the recruits have passed the two-year probation period, they are confirmed as Metropolitan Police constables if they have been training at Hendon, or as police constables in their assigned constabulary.

The new police constable is usually assigned responsibility for a given beat. In the rural areas, the constable may be assigned for months or even years to a particular beat; such a lengthy term is unusual in urban areas, because there is great demand for constables throughout the various specialized positions in the forces (e.g., drug investigators, dog handlers, armed officers) and many officers apply to leave the beat patrol when vacancies occur. Because of national standards of pay, conditions of service, and pension rights, transfers are common among forces in England, and officers usually can maintain their seniority when transferring between forces. This is in stark contrast to the situation in the United States.

Promotion is from constable to sergeant, inspector, chief inspector, superintendent, chief superintendent (which was eliminated as of April 1, 1995), assistant chief constable, deputy chief constable, and chief constable. The London Metropolitan Police has different ranks at the upper levels: commander, deputy assistant commissioners, assistant commissioners, a deputy commissioner, and a commissioner (Metropolitan Police Service, 2004c). Upward mobility through the ranks is similar to that in the United States in that it is through examination, interview, and time in service. In addition, in England upper ranks are often filled by officers who have successfully completed additional courses at the Leadership

Academy for Policing at Bramshill. This is a unique institution and deserves closer examination.

Leadership Academy for Policing (Formerly Known as the Bramshill Police Staff College)

In light of the previously mentioned controversy surrounding the Trenchard College (later called the Hendon Police College) in pre-World War II days, the Post War Committee on Police, appointed in 1944, cautiously suggested the establishment of a National Police Staff College for the training and education of senior ranks within the English police forces (Critchley, 1967, p. 246). The committee felt that positions should be filled from within the ranks, but because few constables had college training, prospects for finding suitable candidates were not very good. It was believed that men (at that time, only men of senior ranks were eligible) destined for promotion and leadership roles needed the opportunity to study and pursue further command-level positions. The government responded to this need and in June 1948 opened its first national police college in Warwickshire. It remained there until 1960; then the college moved to its present home at Bramshill House, a stately mansion with an estate of some 300 acres in Hampshire (Rowe, 1986b, p. 12).

The early curriculum included English literature and other liberal arts courses as well as courses of police study and a full program of physical fitness training. Thus, Bramshill College was able to provide an upper-level learning experience for senior police officers. Over the years, the college has developed a national and international reputation for excellence in the education of police officers. Both police officers and academics throughout the world have attended the various courses of study at Bramshill. These courses vary in content and focus, and are designed for different types of individuals.

The "special course" is usually considered early in an officer's career because it is designed for bright young men and women of the rank of sergeant. In addition, there are usually a few slots available for university graduates seeking a career in the police with accelerated promotion. Initially lasting for one year, the structure of the course was changed to a "sandwich"-style format with a combination of a few months in residence at the college and time back in the home force. Upon successful completion of the inspector's examination, promotion to that rank is common.

Other courses include a "junior command" course for officers of chief inspector or inspector rank; an "intermediate command" course for officers likely to ascend to division command–level rank; and a "senior command" course for those at the ranks of assistant chief constable and above. In addition, several short courses are offered throughout the year that provide specialized training for officers throughout the forces. Finally, an interesting development was the institution of an "overseas command" course for senior-level command officers from other countries. This course contributed greatly to the international reputation of Bramshill.

As always, there are pressures on every aspect of policing in England, and Bramshill is not immune. There are those who believe that Bramshill should play a greater role in the training of officers in England. This would, of course, depend upon time, resources, and budget considerations. In any case, Bramshill is a leading institution in the field of law enforcement education throughout the world, and its reputation as a leader in the development of senior-level command officers is well deserved.

Following a Home Affairs report in 1999 detailing deficiencies in police training, including the issue of cultural diversity training discussed earlier, the Home Office created the **Central Police Training and Development Authority (Centrex)**. Part 4 of the Criminal Justice and Police Act of 2001 authorized Centrex to provide police training and police training facilities. Bramshill, no longer called the Police Staff College, is one of Centrex's police training facilities and houses the Leadership Academy for Policing at Bramshill and the National Police Library. The Leadership Academy for Policing is responsible for providing leadership training for all police officers to promote professionalism within the police forces of England and Wales; the National Police Library has one the world's largest collections of policing literature (Centrex, n.d.). Ironically, in May 2006, it appeared that the future of Centrex was uncertain due to an increasing lack of funding and other resources. Among the rank and file police officers there was a "sense" that there would be a reversion to the pre-Centrex system with more local funding and control (private conversations by author with several command-level Metropolitan Police officers, May 2006). During 2006, the Police and Justice Bill was introduced which, if passed, would provide for the replacement of Centrex with the **National Policing Improvement Agency (NPIA)** (Home Office, 2006). In April 2007, the Home Office replaced Centrex with the NPIA (NPIA, 2007).

■ Recruitment of Women and Minority Police Constables

The development of the role of women and minorities in the ranks of the English police has been slow and troubled. In 1919 there were approximately 150 policewomen, increasing to about 230 by the time of World War II (Critchley, 1967, p. 215). The government acknowledged the value of female officers and three different commissions—the 1920 Baird Committee, the 1924 Bridgman Committee, and the 1929 Royal Commission on Police Powers and Procedures—advocated increased recruitment of women (Critchley, p. 217). However, the rank and file of the police forces still regarded women as unnecessary and a luxury that the economy could not support. Women were considered more properly suited for unpaid social service work.

In fact, it was the work of voluntary organizations during World War I that laid the groundwork for the increased emphasis on female police officers and their patrol work throughout the country. By 1939, on the eve of World War II, 45 of 183 forces were employing policewomen. Although the Home Secretary had standardized pay and working conditions for the female officers throughout the country, their duties were still very traditional and included working with women and children, locating missing persons, handling female prisoners, and only limited patrol duties (Critchley, 1967, p. 218). This resistance to female police officers, although diminishing, has carried on through the years. The late 1980s and early 1990s saw a slight increase in the percentage of female police officers on the 43 forces, rising from 9.4% (11,304 of 120,707 officers) in 1985 to 11.8% (15,061 of 127,127 officers) in 1991 (Home Office, EO Monitoring, 1992). Although the overall percentage remains relatively low, some progress has been made. For example, as of March 2005, women accounted for approximately 21% of the police officers in England and Wales (Bibi et al., 2005).

The history of minorities in the English police shows a similar pattern, although it was not until the 1970s that much concern about minorities surfaced. In 1999, an inquiry into the murder and investigation of a young, minority male charged the Metropolitan Police Service with institutional racism. This report, the Stephen Lawrence Inquiry, suggested that the Metropolitan Police Service's investigation into this murder was plagued with incompetence, racism, and poor leadership (McPherson, 1999). Since this inquiry, it has been the Home Office's goal to increase and improve the recruitment of minority and female police officers (Home Office, 2005). In fact, the McPherson Report led directly to passage of the Race Relations (Amendment) Act of 2000, which required that chief officers be held responsible for acts of discrimination by individual officers and required that public authorities promote racial equality (Home Office, 2001). The **Cultural and Communities Resource Unit** was created by the MPS to better equip police officers with specialized knowledge of different neighborhoods and cultures, and the **Met Careers Team** (also called the Positive Action Team) was developed in response to the Race Relations Act of 2000 to promote careers in the police service to currently underrepresented groups (Metropolitan Police Service, 2004c).

The emphasis on minority recruiting in the late 1980s and early 1990s resulted in a doubling of minority police officers on the forces, from 761 in 1985 to 1,592 in 1991 (Home Office, Equal Opportunity Monitoring, 1992). However, these figures represent only 0.7% and 1.3% of the officers on the police forces, respectively, as compared to the approximately 5% minority representation in the total population. By March 2005, there were 5,017 ethnic minority police officers in the 43 police forces throughout England. This was an 8% increase over the previous year; however, this number still represents just 3.5% of all of the police officers in England. Approximately 55% of this increase was the result of an increase in the number of minority officers in the Metropolitan Police Service, where approximately 7% of the police officers are minorities. Throughout England, minority officers continue to be underrepresented in the upper ranks: minorities account for only 2% of officers with the rank of chief inspector and above (Bibi et al., 2005). This is an ongoing problem, and one that the English continue to face in the twenty-first century.

In the United States, police forces have been fairly successful in their efforts to recruit, retain, and promote minorities, but not without the strife and strain that accompany such efforts. Although illegal in the United Kingdom, affirmative action plans and fair employment practices have generally resulted in a much greater percentage of minority officers on police forces (especially urban forces) in the United States than in England. Because of these federal policies and the threat of financial liability in the absence of compliance, the law enforcement systems in the United States have obtained positive results. In contrast, in England it has been left to each of the individual forces to develop its own recruitment strategies with the government playing only a supportive role (McKenzie, 1993, p. 5). Without any government initiative, monitoring of these plans is left to each chief officer, resulting in the current situation in which less than 4% of the forces are composed of minorities and only 20% are women. With regard to the composition of the senior ranks, in 2005, women accounted for just 10% (Christophersen & Cotton, 2004) and minority officers accounted for only 2% (Bibi et al., 2005) of officers with a rank higher than chief inspector. This situation is in contrast to that in the United States, where women and minorities are represented across all ranks, including police chiefs (McKenzie, p. 9; Sabalow, 2005).[2]

Examples of the negative attitudes and limited opportunities for women and minorities in English police forces have been played out in both real life and fiction. In 1992 a female officer, Alison Halford, claimed that she had been denied promotion because of her sex. The case received sensational coverage in the media and was ultimately settled out of court (Grant, 1992, p. 20). At about the same time, the British television company Anglia Television began airing a series of stories about the Metropolitan Police entitled *Prime Suspect*. This program dealt with the problems of a female detective in a male-dominated force in London and provided an intense examination of the interpersonal problems faced by females and ethnic minority police officers. When it was shown in the United States, it received critical acclaim for its quality and insight into the inner workings of the English police. However, at that time it may not have been an accurate representation of the English police.

In England, as in the United States, many police officers and others believed that early U.S. television programs such as *Starsky and Hutch*, and *N.Y.P.D. Blue* did not provide a realistic representation of police work. However, more recent television shows in the United States such as the *Law and Order* series, *Crime Scene Investigation (C.S.I.)*, *C.O.P.S.*, and *Numb3rs* have attempted to be more realistic in their portrayal of police work with an increased number of women and minorities in pivotal police positions in the twenty-first century—albeit, with a dash of Hollywood included!

Finally, the situation facing women and minorities in English policing could be considered positive compared to the resistance facing homosexual police officers. Although perhaps more common in large urban police departments in the United States, the presence of declared homosexual officers (male or female) presents the English with a major challenge. In the United States, the issue of homosexuality is dealt with through legislation and the courts, as is the case with women and minorities. In England, however, the problem has, historically, been ignored, and in the words of one English observer there is perhaps "a covert belief that inaction in the hope of growing invisibility is the solution" (McKenzie, 1993, p. 11). Although moving cautiously, the Home Office has begun actively recruiting homosexual police officers by advertising in newspapers and magazines aimed at the gay and lesbian community, and has included this group in its definition of a diverse police force (Jones & Newburn, 2001). Additionally, the Gay Police Association, an organization partially funded by the Home Office, provides training and educational services regarding sexual orientation and policing issues (Gay Police Association).

■ Police Powers and Procedures

The 1950s and 1960s saw the English police enjoy an acceptance by the public perhaps unprecedented in their history. As Reiner states: "By the 1950s 'policing by consent' was achieved in Britain to the maximal degree it is ever attainable" (emphasis in original; 1985, p. 51). A public opinion survey carried out by the Willink Commission in the early 1960s disclosed an "overwhelming vote of confidence in the police," with 83% of those interviewed professing "great respect for the police" (Willink, 1962, pp. 102–103).

What led to this success? Much has been written about the factors that propelled English policing to the high level of acceptance it enjoyed in the 1950s, 1960s, and the early 1970s (see, e.g., Banton, 1964; Benyon, 1986; Gorer, 1955). Certain items seem to reappear in writings about this era of policing. There was a lack of overt conflict between the public and the police. Citizens identified with the police; accorded them respect, support, and legitimacy; and perceived that they themselves were participating in police matters of policy and procedure. Finally, the police were perceived to be both just and effective (Benyon, pp. 8–17). However, with passage of the Police Act of 1964, the seeds of distance between the police and the public had already begun to take root.

The Police Act of 1964

Following the report of the Willink Commission of 1962, the Police Act of 1964 "comprehensively revised and re-enacted the whole general law on provincial police administration, and provided for the complete or partial repeal of sixty-one Acts of Parliament dating back to 1801" (Critchley, 1967, p. 293). The act, which was passed in the face of some resistance from local authorities, provided inter alia for the amalgamation of police forces (Sections 21–24) and increased cooperation between forces, with chief constables authorized to enter into collaborative agreements (Sections 13–14); defined the functions of the Home Secretary (Sections 26–36), increasing the central influence of the Home Office and further enhancing the trend away from local participation and accountability; gave police constables powers throughout England and Wales (Section 19); and enlarged the scope of duties of the inspectors of constabulary (Sections 38–40).[3] After its full realization in June 1965, the effects of the Police Act of 1964 on the police and society began to emerge. As Benyon suggested in 1986, the act provided the basis for "policing developments for the last two decades—and the future" (p. 19).

With the passage of time, the major impact of the Police Act of 1964 can be fully appreciated. The high turnover of police personnel brought about by the larger police forces in the 1960s and 1970s resulted in deterioration in police-community relations. Outside of the larger urban areas there still were remnants of the traditional police constable at work. But it was clear that the face of English policing was changing and would never be the same. It was as if the police were now beginning to take on the look of a "national police service" (Critchley, 1967, p. 296) with many local units attempting to preserve their own loyalties and affiliations, yet they were always aware of an encroaching sense of "national purpose and unity" in policing on the horizon.

Twenty years after the Police Act of 1964, two other important pieces of legislation were enacted to greatly change the direction of police powers and the framework of safeguards for the public against abuse of these powers by the police. These were the Police and Criminal Evidence Act of 1984 and the Prosecution of Offences Act of 1985.

Background to the Legislation of the 1980s

The background to the enactment of these two acts originates with the Confait case of 1972. Maxwell Confait was found dead in his burning house; ultimately three teenage boys were convicted of the crime. After investigation and appeal, the verdicts were overturned due to scientific evidence showing that the boys could not have committed the offense. An official inquiry was launched and the police were found to have violated the boys' civil rights. The chairman of the inquiry, Sir Henry Fisher, suggested that a royal commission be appointed to look into the whole matter of police investigations (Fisher, 1977). The outcome was the appointment of the Royal Commission on Criminal Procedure of 1981, headed by Sir Cyril Phillips. The culmination of their efforts was the report of the Phillips Commission in 1981, which provided the impetus for the two acts (Phillips, 1981).

The Police and Criminal Evidence Act of 1984

The provisions of the two acts were wide-sweeping and intended to lead to increased effectiveness, consistency, accountability, and efficiency. The Police and Criminal Evidence Act of 1984 (PACE) and the accompanying **Codes of Practice** (which were modified most recently in 2003 and 2005) comprehensively changed and codified the law relating to police powers to stop and search, conduct road-checks, effect warrantless arrests, obtain search warrants, and detain suspects. In addition, PACE established a new framework for the investigation of police misconduct and provided for the tape recording of interviews conducted at police stations.[4]

Stop and Search

There had been concern about police stopping suspects on the basis of hunches or stereotypes. This concern appeared to be justified in the light of research showing that only a small percentage of stops and searches resulted in arrest. A study of stops and searches for controlled drugs, for example, revealed that from 1972 through 1978 less than one third of those searches resulted in the discovery of illegal drugs (Phillips, 1981, app. 2). And this proportion was considerably higher than the 9% of stops that Londoners reported resulted in an arrest or summons (Smith & Gray, 1985, p. 105). Whether the reality that "being young, male, (and) Afro-Caribbean . . . are all associated with a greater probability of being stopped and searched" is justifiable on legal criteria is, as in the United States, open to debate (Reiner, 1992a, p. 478).

Section 1 of PACE requires that police officers possess "reasonable grounds to suspect" that they would find "stolen or prohibited articles."[5] The Codes of Practice specify that the stop must be based on objective evaluation (e.g., a specific observation or reliable information) that the officer can use (code A.1). Although articulated facts have long been required under U.S. law, this represented a departure from past practices for the English police in that only "some grounds for suspicion" on the part of the constable had been required (Stone, 1986, p. 55). Now a "reasonable suspicion" has to be more than a negative perception of style or manner of clothes, hairstyle, or skin color. The actual codes of practice specifically address even minute aspects of the stop and search procedures (conduct of search, record of search, etc.). As a result of this act, the police constable was faced with a radical departure from past practice.

Roadchecks

Under Section 4 of PACE, police officers can, with appropriate authorization, conduct roadchecks in a locality if they have reasonable suspicion that there is someone in a car who is "unlawfully at large" or who has committed or is intending to commit a serious arrestable offense. Suspicion now has to be based on more than type of car or color and age of occupants, and must be carefully recorded and reported as an additional safeguard for the public. Such roadchecks can be conducted only to search for people, not for evidence. If the police want to search a vehicle, they have to rely on some statutory power other than the roadblock provisions. However, it has been suggested that the provision is still vague enough to allow for a wide latitude of interpretation and possible misuse of power (Hansen, 1986, p. 108). By the early part of the twenty-first century, both stop and search and roadcheck powers were increased under the Criminal Justice and Public Order Act of 1994 and post-2001 terrorism legislation.

Warrantless Arrest

The powers of warrantless arrest under PACE vary according to whether the officer is dealing with an arrestable offense (generally any offense that carries the possibility of a five-year sentence of imprisonment) or a nonarrestable offense. A constable may arrest without a warrant any person whom the officer has "reasonable ground for suspecting" has committed, is committing, or is about to commit an arrestable offense (PACE 1984, Section 24). A constable may effect the warrantless arrest of any person whom the constable has "reasonable grounds for suspecting" has committed or attempted to commit, or is committing or attempting to commit, any nonarrestable offense, provided that "service of a summons is impracticable or inappropriate" because of the presence of any of certain specified conditions (PACE 1984, Section 25(1)). These conditions include problems ascertaining the suspect's name or address and reasonable belief on the part of the constable that warrantless arrest is necessary to prevent physical harm or damage to property (PACE 1984, Section 25(3)).

After a person has been arrested, he or she must immediately be informed of his or her arrest and the grounds for it, even if the circumstances are obvious (PACE 1984, Section 28). Thus, an individual standing over a body with a smoking gun in his hand must still be immediately informed of the circumstances of his arrest.

The powers of warrantless arrest under PACE are broad and provide officers in the field with a great deal of discretion. Even the standard of evidence necessary for arrest, "reasonable grounds to suspect," is not as stringent an evidentiary standard as the standard of "probable cause" required in the United States.[6] Thus, police officers in the United States may well envy the broad arrest powers enjoyed by their counterparts in England.

Search Warrants

The provisions in PACE relating to search warrants were designed to supplement the existing laws and to fill some surprising gaps in the law. Prior to PACE, the police could only obtain a search warrant to search the scene of a murder or kidnapping after the suspect had been arrested (Zander, 1990, p. 20). Under Section 8 of PACE the police can obtain from a magistrate a warrant to enter and search premises for evidence of a serious arrestable offense. Although there had been considerable concern on the part of the medical profession that even confidential medical records would be subject to searches, such materials (along with counseling records and "items subject to legal privilege") were expressly designated to be outside the purview of the act (PACE 1984, Sections 8, 10, 12).

Detention

Perhaps some of the most vociferous opposition to the act came in response to its treatment of the sensitive issue of the detention of suspects. Prior to this time, the limits on the police powers to detain an individual suspected of a crime were somewhat unclear. The new provision that an individual must either be charged or appear in court within 36 hours after arrest (PACE 1984, Sections 41–42) represented a radical departure from past practice. Most of the concern revolved around the ability of the police to gather the type of evidence required for a successful charge to "stick" within that time frame. Some observers have noted that research indicates that only 2% of cases result in detention for more than 24 hours and that the vast majority of suspects can be released or charged within the time limit set by the act (Oxford, 1986, p. 70). The actual Codes of Practice on this point not only dictate the time limit, but also provide guidelines for when a review of detention should take place (within six hours after an arrest and then in intervals of nine hours). They also provide for regularly scheduled rest periods and meal and short rest breaks; at the completion of the 36 hours, the evidence of the case must be revealed to the detainee. As discussed in Chapter 11, different provisions apply to suspected terrorists, who can be detained without being charged for 28 days.

Questioning

As in the United States, a suspect is under no legal obligation to answer questions asked by the police. The suspect has an absolute right to silence, and before being questioned used to be told: "You do not have to say anything unless you wish to do so, but what you say may be given in evidence." Problems inevitably arose concerning the voluntariness and reliability of statements made by suspects to the police; to help combat these problems PACE provided for the tape recording of interviews conducted at police stations (Section 60). Controversial provisions of the Criminal Justice and Public Order Act of 1994 allow a court to draw inferences from a suspect's failure to answer police questions (Sections 34, 36, 37). The caution administered by the police has consequently been modified to state: "You do not have to say anything. But it may harm your defence if you do not mention when questioned something which you later rely on in court. Anything you do say may be given in evidence" (Code C, paragraph 10.5).

Investigation of Police Misconduct

Traditionally the English have been less concerned than the Americans about the potential for police abuse of power. For example, although the exclusion of illegally obtained physical evidence has been used in the United States as a mechanism for deterring unlawful police conduct, the English approach has been to allow the introduction of such evidence at trial provided that "it is relevant to the matters in issue" (*Karuma v. R.*, 1955, p. 203 per Lord Goddard C.J.) and to rely far more heavily on civil suits, and in particular police disciplinary procedures, to correct such abuses when they occur.

Traditionally in England, as in much of the United States, complaints against the police have been investigated by other police officers. Historically this has led to concerns about public confidence in the police. Liberty (a civil liberties group that is analogous to the ACLU in the United States) reports that as early as 1929, a Royal Commission had recommended that the Director of Public Prosecutions (DPP) be given investigators who were independent of the police for the purpose of investigating the police (Liberty, 2000, p. 13). By the 1970s there were serious concerns about endemic corruption among detectives in London's Metropolitan Police, and this led to the 1976 Police Act and the establishment of the Police Complaints Board.

A complaint against a police officer was investigated by a member of the officer's force. Unless the chief constable was satisfied that no criminal offense had been committed, the report had to be sent to the DPP for a decision on criminal proceedings. After the DPP made a decision, a copy of the investigator's report was sent to the Police Complaints Board, along with a memorandum from the deputy chief constable indicating whether he had decided to bring disciplinary charges against the officer, and if not, why not. The board could direct a chief constable to institute disciplinary proceedings (Police Complaints Board, 1980, pp. 3–4). Major criticisms about the handling of complaints against the police included concern about the lack of an independent check on the manner in which complaints were investigated and the fact that all complaints, whether minor or serious, were handled in the same fashion (Zander, 1990, p. 215).

Under PACE, a quasi-independent system of handling complaints against the police was implemented with a new Police Complaints Authority (PCA) replacing the old Police Complaints Board. However, this system still involved using police officers to investigate complaints of misconduct by other police officers. Lesser offenses were investigated by officers within the same force, sometimes by a dedicated "internal affairs" department. More serious offenses could be investigated by officers from another force at the behest of the PCA. Certain serious complaints, such as those alleging death or serious injury, had to be referred to the PCA (Section 89). PACE also established an informal process for the resolution of minor complaints (Section 85) and lessened the role of the DPP by requiring that only cases in which the chief constable determines that an officer "ought to be charged" with a criminal offense be forwarded to the DPP (Section 90). The Police Complaints Authority, it should be noted, had the power to direct the chief officer to send the case to the DPP (Section 92).

The Prosecution of Offences Act of 1985

The Prosecution of Offences Act of 1985 set up a new system in which the prosecution of cases was separated from the investigative processes of the police. This represented a departure from the tradition whereby the local constable had been responsible for the arrest, investigation, and prosecution of the offender in magistrate's court. This new system, which is discussed in more detail in Chapter 7, met with initial resistance from the police because it represented a significant reduction in their power. However, research has indicated that by 1990 the negative attitudes exhibited by the police toward the new system had mellowed considerably (Wakefield, Hirschel, & Sasse, 1994), and by the beginning of the 1990s cases appeared to be moving more smoothly and successfully through the prosecution process (House of Commons, 1990, p. 13).

Policing After the Two Acts and into the Twenty-First Century

Despite fears that civil liberties of individuals would be eroded, that the police would be hampered and confined in their abilities to carry out their duties, and that the prosecution of cases would be inhibited due to the separation of the investigation and prosecution functions, the police have come to accept, or at least adhere to, the new codes, which have not hampered effective policing in England. For example, strong initial police opposition to the tape recording of interviews dissipated (Zander, 1990, p. 125), and tape recording "proved to be a strikingly successful innovation providing better safeguards for the suspect and the police officer alike" (Runciman, 1993, p. 26). However, previous allegations of intimidation and the fabrication of admissions by the police were replaced by allegations that statements were unreliable because they had been "obtained in 'informal' interviews outside the PACE safeguards" (Justice, 1994, pp. 2–3). In addition, the percentage of stops and searches that resulted in arrest, though increasing, remained rather low: 13% in 1992 (Home Office, Research and Statistics Department, 1993a), and 14% in 1993 (Home Office, Research and Statistics Department, 1994b). Twenty years later they remained at the same level with 13% of stops and searches in 2002/2003 and 2003/2004 resulting in an arrest (Ayres, Murray, & Fiti, 2003; Murray & Fiti, 2004), and 11% in 2004/2005 (Ayres & Murray, 2005).

By 2004, 70% of complaints against the police from citizens were either withdrawn, dispensed with, or informally resolved. Only 30% were investigated, and only 4% of investigated complaints were substantiated (Cotton, 2004). Overall, from a high of 25,000 complaints registered in 1994 there has been a gradual downward trend to a low of 15,000 in 2002/2003 and a 4% increase to 15,885 in 2004 (Cotton, p. 1). In his overall assessment of then current research on the effects of PACE, Reiner (1992a) suggested that although "bureaucratic procedures tend to be followed faithfully . . . research implies that many of the supposed checks on police powers become mere rubber-stamping routines" (p. 476). However, more current information (Home Office, 2004a) suggests the police have been working diligently to diminish any misperceptions about the public image of police procedures and improve communication with the citizens through the development of new agencies and divisions dealing with police reform (Police Reform Act of 2002; Liberty, 2000; Home Office, 1993b).

In June 1993 the government issued a white paper report with the imposing title, *Police Reform: A Police Service for the Twenty-First Century, the Government's Proposals for the Police Service in England and Wales* (Home Office, 1993b). The

Home Secretary, Michael Howard, issued direct guidelines for the express purpose of building "a strengthened partnership between the police and the public in which we help each other to do our best to make our country a safer place in which to live" (Home Office, 1993b, p. 1).

Although the white paper did not alter the basic aims of the police with traditional statements involving crime prevention, law enforcement, justice, community service, and economic efficiency, it suggested that the number of police forces be reduced. It entrusted to the central government the setting and monitoring of key objectives for the police and focused attention on local needs, providing for a strengthening of local police authorities, and an increased level of consultation with local citizens. Noticeably, the term *police service* is used throughout the report.

The report also articulates the strategies and procedures for achieving the aims and proposals outlined. It concludes by stating: "It is believed that only by working together can the police and the public create and sustain the kind of country we all want" (Home Office, 1993b, p. 47). Although neither drastic nor sweeping in its proposals, it was issued almost simultaneously with the **Sheehy Report** (June 28 and June 30, 1993, respectively), and together the two appeared to be an attempt to bring about more centralization of the police, which had been envisaged since the early days of Sir Robert Peel.

The *Sheehy* Report contained a number of controversial recommendations for police reorganization. The report set out plans for a system of fixed term contracts whereby officers would initially be hired on 10-year contracts and shifted to 5-year renewable contracts thereafter. The *Sheehy* inquiry had concluded that the police service "does not need the whole of the existing rank structure" (Sheehy, 1993, p. 8) and originally recommended the abolition of the ranks of chief inspector, chief superintendent, and deputy chief constable, though the proposals to eliminate the ranks of chief inspector and deputy chief constable were later withdrawn. Another controversial recommendation in the *Sheehy* Report was a change in the pay scales and bonus system, with the new scales appearing to favor the senior ranks over the lower ranks.[7] Response to the *Sheehy* Report was quick and overwhelming. The Police Federation labeled it divisive and reflective of the "fact that the committee understood little or nothing about policing" (Kirby, 1993, p. 1). Others, including the public and police scholars (see, e.g., Waddington, 1993, p. 2), joined in the criticism. Police opposition culminated in an unprecedented gathering of some 21,000 police officers in Wembly Stadium on July 20, 1993, to voice their opposition (in a "peaceful and dignified manner") to the proposals.

As a consequence, the government softened its approach and the Police Act of 1996 did not contain *Sheehy*'s more radical and most unpopular proposals. By 1997 *Sheehy* was buried, along with the Conservative government when Labour and Tony Blair came to power in a landslide election.

PACE had established a new system for examining complaints against the police. However, the 1999 McPherson Report into the police response to the murder of Stephen Lawrence highlighted, once more, a widespread lack of confidence in the system for responding to complaints against the police. In 2000, Liberty published the study *An Independent Police Complaints Commission*. This study was conducted by a number of stakeholders, including representatives of senior and rank and file police officers, criminal defense attorneys, and civil rights lawyers. The Home Office provided "observers." Simultaneously, the government conducted its own consultation, and these various activities led to the 2002 Police Reform Act. A key and consistent recommendation, by McPherson, Liberty, and the consultations, was for the creation of an independent police complaints body with its own dedicated team of investigators. The Police Reform Act of 2002 created the **Independent Police Complaints Commission (IPCC)**, which became operational on April 1, 2004.

The primary distinction between the IPCC and its predecessors is that it is an entirely independent nondepartmental public body. The IPCC has 17 regional commissioners who can never have served as police officers (Section 1). These commissioners are mostly, but not exclusively, drawn from the legal professions, and the initial team has a notably high number of women and members drawn from the minority communities. The IPCC, which has broad investigatory powers, can either supervise police officers or employ its own team of investigators. The IPCC has an express duty to raise public confidence in the police. Whereas the PCA only dealt with incidents involving death or serious injury, the IPCC will accept a broader range of complaints involving police misconduct and will accept complaints not only from victims, but also from those who merely witness misconduct (Section 12).

The IPCC was not universally welcomed by the police, and has already shown an independent mind. Its first high profile investigation was into the shooting of an unarmed Brazilian citizen whom police in London mistook for a suicide terrorist and shot dead on the subway system. The police, including London's most senior police officer, sought to deflect criticism and to justify a secretive "shoot to kill," and "shoot without warning" policy. It is alleged they also sought to block the IPCC from investigating. The IPCC responded by investigating not only the shooting, but also the policies and the commissioner himself for the alleged cover-up.

A final indication of just how radical a change the IPCC is for the English police is found in the senior staff of the IPCC. The first chair of the commission has been Nick Hardwick, the former Director of the Refugee Council, one of Britain's primary immigrant rights nongovernmental organizations (NGOs). The deputy chair has been John Wadham, one of England's most high profile human rights lawyers and the director of Liberty until 2003.

■ The Police in a Correctional Role

One way in which the police in England have differed from their counterparts in the United States has been in their handling of juveniles. As in the United States, police in England always have been and probably always will be the first line of contact in dealing with youth who come within the jurisdiction of the juvenile court. The decisions they make on the street initiate the entire juvenile justice process in both the United States and England, and it is this application of police discretion in dealing with juveniles that makes the job of a police officer quite difficult at times.

From the early 1970s, as a result of the Children and Young Persons Act of 1969, the police in England have been involved in a unique practice of "police cautioning." It involved placing the police in a quasi-correctional role and gained some attention as possibly being useful in the United States (Wakefield, 1983; Wakefield & Hirschel, 1991a, 2005; Wakefield, Kane, & Caulfield, 1984).

Intended as a form of diversion from official court processing, the formal cautioning of a juvenile originally took place after the police youth bureau collected the relevant background information on the juvenile following an arrest.[8] Although juveniles who committed serious offenses were not supposed to receive cautions, this occurred. Provided the juvenile admitted the offense, there was sufficient evidence of guilt, and informed consent had been obtained from the juvenile and the parents or guardians, a time was scheduled for the caution. The caution, which might be classified as a stern verbal lecture, was generally administered at the police station by a senior officer in uniform in the presence of the parents or guardians of the juvenile. The content consisted mainly of a warning about the direction in which the juvenile's behavior was leading and what the possible outcomes (e.g., detention and court processing) would be should the juvenile continue to act delinquently. Cautions could be cited in subsequent proceedings in juvenile court, and each juvenile was allowed a maximum of three cautions (Home Office, 1985, 1990, 1994a).

However, as a result of the Crime and Disorder Act of 1998, offenders under the age of 18 no longer receive cautions. **Reprimands** and **warnings** were implemented as a replacement for cautions. Reprimands are for minor offenses committed by first-time offenders under the age of 18. A second offense results in a final warning. Following a final warning, the youthful offender is to be referred to a **Youth Offending Team (YOT)**, which is responsible for developing a rehabilitation program that is designed to address the juvenile's behavior. Many times the YOT will contact the victim to ascertain if any mediation or reparation is needed. The third offense results in charges being filed, unless the third offense occurred more than two years after the final warning.

However, adults continue to be subject to two types of cautions: simple cautions and conditional cautions. **Simple cautions** consist of simple verbal warnings, with no further police action. Implemented after the Criminal Justice Act of 2003, **conditional cautions** consist of a formal warning that is attached to specific behavioral conditions. Failure of the adult offender to comply with these conditions may result in charges being filed for the original offense. Recently, the cautioning of adult offenders has taken a restorative path, with the aim of encouraging the offender to take responsibility for his or her offense and repair the harm experienced by the victim. Although research suggests that restorative cautioning is not more effective than traditional forms of cautioning, both victims and offenders have benefited from the use of this type of restorative cautioning (Wilcox, Young, & Hoyle, 2004).

Although early research comparing the effectiveness of police cautioning of juveniles was inconclusive, suggesting no significant differences in re-arrests (Farrington & Bennett, 1981, p. 134) or reconvictions (Mott, 1983) between cautioned juveniles and court-processed juveniles, the government felt strongly enough about the potential of cautioning as an alternative to court processing to implement it for adults. Thus, it has been used for several relatively minor adult offenses (e.g., theft and handling stolen goods, drug and alcohol offenses) as a form of diversion from official court processing. In 2000, 239,000 offenders were cautioned for non-motoring offenses, including 60,800 reprimands and warnings for juvenile offenders. About 151,000 of these cautions were for indictable offenses (Johnson, 2001).

Despite concerns over its spreading range of use, the caution appears to be an effective mechanism for helping reduce the clogging of the English courts. However, the United States has not seemed interested in adapting this formal practice for minor criminal offenses, though warning tickets have traditionally been issued for traffic offenses and unofficial warnings are often given to citizens. Perhaps this is an option that the United States, faced with overcrowded dockets in most of its lower courts, might examine more closely.

■ The Public Image of the Police

In order for the police to be effective and efficient, public opinion of the police must be positive. What are the factors that shape an individual's opinion of the police? An individual's own experience with the police is clearly a major factor. In addition, one is influenced by what one hears from others and what one reads about the police. It should be no great surprise that the media have a tremendous effect on how the police are perceived. Reiner (1992b) describes the media image of policing, contrasting factual and fictional presentations of the police and their work.

One of the ways to measure the public image of the police is to examine how often victims of crime report their experiences to the police. As discussed earlier in this book, the British Crime Surveys reveal that crime is rising, but at a slower rate than indicated by data on crimes reported to the police, because the percentage of offenses reported to the police has remained relatively stable since 1997. This may suggest that public confidence in the police has remained stable. Other findings, however, may suggest that public confidence in the police has recently decreased.

In the 1987 British Crime Survey, 59% of those questioned indicated that they had had some contact with the police during the past year. This personal contact seems to have the most impact on individuals' perceptions of the police. In that survey, 85% of those questioned rated the general performance of the police as good or very good; only 4% rated it as very poor (Skogan, 1990, p. 49). However, the good/very good rating had decreased to 49% by 2004/2005 (Nicholas, Povey, Walker, & Kershaw, 2005). Furthermore, both the British Crime Surveys and independent polls suggest that confidence and support have been decreasing: Whereas 43% rated the police as very good in 1982, this figure dropped to about 25% by the end of the decade (Skogan, p. 49). Of course, opinions of those who have had contact with the police vary according to the nature of that contact, be it a criminal matter, a civil case, or a service contact. Increasingly listed in the British Crime Surveys as reasons why the victim did not report the offense to the police was that the "police could do nothing" (71% in 2004/2005), and that "we dealt with the matter ourselves/it was inappropriate for the police" (19% in 2004/2005) (Nicholas et al., 2005).

An ever-widening gap appeared to be developing between what the public expected of the police and the police's ability to deliver services. Some complaints focused on the inability of the police to answer questions and the amount of (or lack of) effort put into a case; others focused on personal matters, such as not being treated politely or with respect (Skogan, 1990, p. 50).

It became apparent that, perhaps because of unrealistic expectations on the part of the public, the image of the police officer was suffering and police-community relations needed to be improved. In the wake of the Brixton riots and the Scarman (1981) report, the Police and Criminal Evidence Act of 1984 established police community consultative committees (Section 106) and in 1986 a system of lay visitors to police stations was introduced (Reiner, 1992a, pp. 484–485). This opened police operations to some outside review, but not generally to those who had regular contact with the police. In 1992, after an International Conference on Police and an outside consultant's report to the Metropolitan Police of London, Sir Peter Imbert, commissioner of the Metropolitan Police, made several changes designed to enhance the public image of the police and police-community relations. These changes included the development of a "common purpose" statement, a change in name from the police being known as a police "force" to a police "service," and an emphasis on **sector policing** (or community policing) programs similar to those used in the United States (The New (and Improved) Scotland Yard, 1992, pp. 10–13).[9]

Another issue related to unrealistic expectations is the fact that sometimes what the public thinks is effective in fighting crime is, in fact, not much help. For instance, Skogan indicates that the public would like a faster response time when the police are summoned, which would require more use of rapid response cars (1990, p. 50). However, as in the United States, reality often does not support the idea that the quicker the police arrive at the scene, the more likely it is that they will apprehend the suspect. Indeed, many of the English forces adopted graded priority response schemes, as is the practice in some forces in the United States. Perhaps more important, as it is in the United States, is how the police treat the public once they have arrived at the scene (Skogan, 1990, p. 50).

As Her Majesty's chief inspector of the constabulary reported in 1992: "The police will have to undergo a fundamental cultural change to win back the faith of a public shaken by the miscarriages of justice (The Guildford Four, Maguire Seven, Birmingham Six etc.), low clear up rates, and perceived incivility and abrasiveness" (Campbell, 1992). Even the Police Federation, which consists of rank-and-file police officers, had to admit that there was "public disquiet" concerning the performance of the police in the last decade of the twentieth century. Additionally, the Steven Lawrence Inquiry added to this negative community perception of the police (McPherson, 1999).

Finally, the results of research examining the attitudes of the public toward the police support the notion that positive attitudes toward the police are being eroded (Reiner, 1992a, p. 471; McPherson, 1999). The English public, as has long been the case in the United States, no longer accept authority unquestioningly and feel that they should be treated much as customers should be treated. Sir John Woodcock, Her Majesty's Inspector of Constabulary, suggested that this is a "cultural change of very considerable magnitude" reflecting the challenge facing the English police as they move into the future (Campbell, 1992).

■ New Directions for the Police in the Twenty-First Century

As previously discussed, the Police Reform Act of 2002 created the Independent Police Complaints Commission and mandated the Home Secretary to prepare and submit to Parliament each year a National Police Plan. With local police chiefs and local police authorities required to produce three-year plans consistent with government policy, this legislation has a significant centralizing influence. This centralizing influence is highlighted by other provisions of the act. Thus, the act permits the Home Secretary to issue codes of practice whenever it is considered necessary to do so "for the purpose of promoting the efficiency and the effectiveness of the police forces" (Section 2). The act also empowers the Home Secretary to mandate that an inspection of a particular police force be carried out (Section 3), and to direct a police authority to take remedial action (Section 4) with appropriate action plans set out (Section 5).

As mentioned, the first three-year National Policing Plan (2003–2006) was issued in November 2002. The overarching aim of the plan was "to deliver improved police performance and greater public reassurance" (p. 3). Four key priorities were presented in this first national policing plan: (1) "tackling anti-social behavior and disorder"; (2) "reducing volume, street, drug related, violent, and gun crime in line with local and national targets"; (3) "combating serious and organized crime operating across force boundaries"; and (4) "increasing the number of offenders brought to justice" (Home Office, 2002, pp. 6–8). As previously discussed, starting with the Police and Magistrates Courts Act of 1994, the Home Secretary has been empowered to set performance targets for achieving policing objectives, and the National Policing Plan of 2003–2006 contains numerical performance targets. Thus, for example, for 2003–2004 every police force was set the target of increasing by 5% the number of offenses "brought to justice" (p. 8), while the police forces participating in the "Street Crime Initiative" were exhorted to "achieve a 14% reduction in robbery from 1999–2000 to 2005" (p. 7). The scheme clearly seeks to centralize strategic management of police forces under the direction of the national government. There is particular focus on crimes that are perceived as being important to ordinary members of the public, namely, so-called "anti-social behavior," which includes petty crime, vandalism, drug use, and noise. Although the centralized approach is very different from that in the United States, the focus is very similar to that pursued by Mayor Rudolph Giuliani as he sought to reduce the levels of crime and incivility in New York City at the turn of the century.

In November 2004, the Home Office produced a white paper titled *Building Communities, Beating Crime: A Better Police Service for the 21st Century*. This white paper discussed the achievements of England's police service and how to prepare for and improve in the twenty-first century. Although the police service is to continue aiming for the "prevention, detection, and reduction of crime," as well as better protection for the public (Home Office, 2004b, p. 6), this paper also called for a number of reforms:

- *Reduction of bureaucracy:* The reduction of bureaucracy is seen as a necessary element of making England's police service more accessible to the public. The goal put forth by this white paper was to free up 12,000 front line officers by 2008 by reducing unnecessary bureaucratic elements and improving efficiency through science and technology.

- *Revitalized community policing:* Improving community policing includes the introduction of neighborhood policing teams responsible for solving specific local problems by working with the community. It also includes becoming more responsive to individual community needs and, in a sense, becoming more customer service oriented. The goal is to make the police a more accessible presence on the streets and in the communities. In fact, this paper reported that a Neighborhood Policing Fund would be available to help put 25,000 new community service officers and wardens on the streets by 2008.

- *Modernization:* One of the most important challenges acknowledged by this white paper is the modernization of England's police forces. Modernization will be achieved through developing the leadership skills of all police officers through improved training and career development.

Following the euphoria surrounding the announcement on July 6, 2005, that London was selected over Paris to host the 2012 Olympics, the terrorist bombings of July 7, 2005, provided a tragic and stark "wake-up call" regarding law enforcement and security concerns in London. The immensity of the prospect of providing security for athletes and citizens from more than 200 countries for two weeks of competition in London in the summer of 2012 became apparent. Along with the "new directions" for the police and the National Policing Plan of 2004 (Home Office, 2004c) came a very practical and real-life exercise for the people of England—more specifically, London. The complex strategy and day-to-day concerns are a challenge to any major police force under the best of circumstances in planning for this type of test of an organization's resolve. Realistically, the performance of the Metropolitan Police during this period will provide the backdrop and foundation for policing in London in the future. It is with these issues that the English police are facing the twenty-first century. It remains for history to record the direction in which they proceed and the results of their efforts for reform.

■ Conclusion

Despite the aforementioned confrontation and controversy, the modern English police continue to occupy a prominent position in the field of law enforcement throughout the world. With its tradition of civility in fighting crime, the English police should continue to occupy this position well into the twenty-first century. However, it will not be without further challenge. Crime prevention and law enforcement are not easy goals to achieve on either side of the Atlantic. The 1992 Rodney King incident in Los Angeles and the 1993 revelations of widespread corruption in the New York City Police Department were testament to the continuing need to improve policing in the United States. Additionally, the response to the bombings in New York on September 11, 2001; the revitalization of the Times Square area in Manhattan, New York; and the national reduction of the crime rate are positive signs for law enforcement in the United States.

The Thatcher government in England "called the bluff" of law enforcement and spent money on the crime problem in the form of increased resources for the police. As the end of the 1980s approached, the crime problem still existed and the police had not been successful in eradicating it. However, as discussed earlier, new measures and practices involving both administration and front-line policing in England in the 1990s and the first part of the twenty-first century are also positive steps.

Will the new police reforms, if implemented, eliminate crime? Will the police have a "leaner, meaner fighting machine" to control crime? Obviously, the answer to this first question is "no" because crime and criminality are subject to many more variables than effective law enforcement. However, the English have clearly made some strides and are making an effort, just as the United States is doing, to continue the assault on crime as we move forward in the twenty-first century.

CHAPTER SPOTLIGHT

- After World War II, England had 143 police forces (police departments). By 2006 this had been reduced to 43. The boundaries of 41 forces roughly follow the boundaries of local governments. The other two forces are located in London.
- The forces outside London are accountable through a tripartite system of the local police authority, the Home Office, and the chief constable. In London, the Metropolitan Police is headed by the Commissioner of the Police of the Metropolis who is appointed by the crown, on the advice of the Home Secretary. The commissioner reports to the Metropolitan Police Authority, part of the Greater London Authority. The City Police has its own Commissioner and is supported by the City of London Council.
- Outside London the chief constables are appointed by the individual police authorities, but this power is subject to the approval of the Home Secretary.
- Central government, through the Home Secretary, has considerable power to impose central government policies and to determine objectives for policing upon the police authorities. Police authorities must "have regard" for the views of the Home Secretary and comply with any central government directives.
- The government centrally establishes a National Policing Plan, which sets "strategic policing priorities" for the next three years.
- In addition to the 43 forces, there are a number of regional crime squads intended to fight specific crimes including child pornography, drugs, and soccer violence. There is also the Serious Organised Crime Agency (SOCA), which was created in 2006 to target organized crime.
- Centralization means that recruitment qualifications for police officers are uniform. No formal educational requirements exist; applicants must understand English and pass a numeracy exam.
- Officers identified to have the desired leadership skills are eligible for the High Potential Development Scheme (HPD). This provides an accelerated promotion scheme. Another such scheme is aimed at university graduates and provides for entry into the police Leadership Academy.
- New recruits receive a 14–17-week training course that includes training in diversity and the elimination of discriminatory behavior and the use of sound decision making. This is in addition to courses on law, community relations, driving, and first aid. New officers go on to have a two-year probationary period.
- The Metropolitan Police has its own training center, the Peel Centre. A proposal exists at the time of writing to close this facility and decentralize training.
- English police officers are not routinely armed and do not receive firearms training as part of their basic training. Specialized units provide armed response vehicles as back-up to patrol officers, protection to VIPs, and security at locations such as airports, embassies, and government buildings.

- Armed response vehicles (ARVs) are fast armed response vehicles containing two or three heavily armed officers who are on call to attend incidents at the request of other officers.
- Police training in England and Wales has moved more towards "human relations" training and the proper use of discretion. In particular the 1981 Brixton riots and the murder of Stephen Lawrence have caused training to focus more on racial and cultural awareness.
- Promotion is from constable to sergeant to inspector to chief inspector to superintendent to assistant chief constable to deputy chief constable to chief constable. Upward mobility is the product of service experience, interview, and examination.
- The upper ranks are filled by officers who have successfully completed courses at the Leadership Academy for Policing, a national police academy providing centralized, standardized training for senior officers. The academy has an international reputation and has provided training for officers from around the world.
- Police training had been centralized though the Central Police Training and Development Authority or Centrex until April 2007 when it was replaced with the National Police Improvement Agency.
- Women and minorities are under-represented in the police forces. Since the 1980s and particularly since the 1999 McPherson Report, efforts have been made by central government to address this.
- The number of minority officers doubled between the late 1980s and 1991, and has increased since, but they are still significantly under-represented.
- Affirmative action has not been employed in England and, in fact, is illegal. This is in contrast to the last 40 years in the United States and the resulting representation of women and minorities at all ranks.
- Centralization of authority in the office of Home Secretary has been a major theme in English policing. However, this has led to a deterioration in police/community relations. This is particularly the case in urban areas.
- A major shift in police accountability and powers came with the Police and Criminal Evidence Act of 1984 (PACE). Through its Codes of Practice, police powers of stop, search, seizure, road checks, warrantless searches, and detention were, for the first time, codified. Such powers are still much broader and less subject to restriction than in the United States.
- In 1985, the police lost their traditional power to prosecute cases. Investigation and prosecution were separated and the latter transferred to the Crown Prosecution Service.
- In 2002 the Independent Police Complaints Commission (IPCC) was created to provide an independent body to investigate and prosecute allegations of police misconduct.
- In contrast to the United States, English police have more discretion in dealing with juvenile offenders. From 1969 until the late 1990s there was a practice of police cautioning. This placed the police in a quasi-correctional role and was a form of diversion from official court processing. Cautions required sufficient evidence to convict and an admission of guilt. It took the form of a stern verbal lecture from a senior officer in front of the juvenile's parents.

- Since 1998, cautions have been replaced by a system of reprimands and warnings. The former for first time offenders, the latter being a second and final warning. Conditional warnings have existed since 2003 and involve specific behavioral conditions intended to direct offenders towards a restorative path.
- Statistics show that public confidence in the police is in decline. There is also a widening gap between public expectations and the ability of the police to deliver. The Home Office has attempted to address this through greater centralization such as the National Police Plan and codes of practice under the 2002 Police Reform Act, and empowering the Home Secretary to set performance targets for police forces.
- There is a disconnection in public expectations and what the public perceives as an effective way to fight crime, and what is actually effective.
- The Home Office continues to look for ways to more effectively fight crime and build public confidence in the police through modernization, reduced bureaucracy, and revitalized community policing. Again, these approaches will be centrally dictated and monitored.

KEY TERMS

Central Police Training and Development Authority (Centrex): Created as a result of a 1999 Home Affairs Report, which suggested deficiencies in police training and was responsible for improving police training and police training facilities.

Codes of Practice: Comprehensively changed and codified the law relating to police powers to stop and search, conduct roadchecks, effect warrantless arrests, obtain search warrants, and detain suspects.

Conditional cautions: Consist of a formal warning that is attached to specific behavioral conditions.

Cultural and Communities Resource Unit: Created by the MPS to better equip police officers with specialized knowledge of different neighborhoods and cultures.

Guidance circulars: Distributed periodically to all divisions of the police outlining existing/proposed policies and procedures.

Her Majesty's Chief Inspector of Constabulary Office: Responsible for the formal inspection and assessment of all 43 police forces, and acts in a key advisory role to the Home Office, the chief constables, and the local police authorities.

High Potential Development Scheme (HPD): Program designed to provide guidance and support to those police officers who possess desired leadership traits.

Independent Police Complaints Commission (IPCC): An entirely independent nondepartmental public body.

Local police authority: A police agency responsible for the local area.

Met Careers Team: Promotes careers in the police service to currently underrepresented groups. Also known as Positive Action Team.

National Policing Plan: Local police chiefs and local police authorities are required to produce three-year plans consistent with government policy.

National Policing Improvement Agency (NPIA): Recently created to support self-improvement of police agencies and to improve various areas of policing.

Reprimands: A verbal disciplinary action for first-time offenders under the age of 18 who commit minor offenses; implemented as a replacement for cautions.

Sector policing: A form of policing similar to community policing in the United States, which was implemented to improve police-community relations and enhance the public image of the police.

Sheehy **Report:** A report that set out plans for a system of fixed term contracts whereby officers would initially be hired on 10-year contracts and shifted to 5-year renewable contracts thereafter.

Simple cautions: Consist of simple verbal warnings, with no further police action.

Warnings: A verbal disciplinary action used after a second offense and the use of the reprimand; implemented as a replacement for cautions.

Youth Offending Team (YOT): Responsible for developing a rehabilitation program that is designed to address a juvenile's behavior.

PUTTING IT ALL TOGETHER

1. What steps must an individual take in order to become a police constable? How does this compare to the United States?

2. How do the policies and procedures in England compare to those in the United States?

3. In what ways do the United States and England differ in the handling of juveniles?

ENDNOTES

1. This was only the fifth Royal Commission appointed to examine the police in England. The others, occurring in 1839, 1855, 1905, and 1929, covered areas ranging from police complaints to policing powers and procedure (Critchley, 1967, p. 267).

2. In the early to mid 1990s, female police chiefs were represented by Elizabeth M. Watson (Austin, Texas), Mary Ann Viverette (Gaithersburg, Maryland), and Linda Weaver (Johnstown, Pennsylvania). Many of the major cities across the United States (e.g., New York, Detroit, Los Angeles, Atlanta, and Portland) have had minority police chiefs. As might be expected, this number continues to increase in the twenty-first century for both women and minorities. It is estimated that in 2005, there were 200 female law enforcement officers in top-ranked positions (National Center for Women and Policing, 2002; Sabalow, 2005).

3. In addition, the Police Act of 1964 gave statutory recognition to many agencies and initiatives that had developed, such as the police college, district training centers, forensic science laboratories, and the Police Advisory Board set up to advise the Home Secretary (Sections 41, 46). The act also established the statutory basis for the formation of the Research

and Planning Branch of the Home Office (Section 42). This unit has come to be a leader in criminal justice research in England and worldwide, lending its expertise to all aspects of criminal justice research. For a comprehensive review of early police research in England, see Reiner (1992a). An up-to-date review of police research in England can be found at the Home Office web site (http://www.homeoffice.gov.uk) under the Research, Development and Statistics Directorate link.

4. For an intensive discussion of the provisions of these acts and an in-depth examination of police powers after passage of PACE, see Robilliard and McEwan (1986) and Zander (1990).

5. Prohibited articles include offensive weapons or articles connected with burglary or theft (PACE 1984, Section I(7)). Drugs are covered under separate statutory provisions.

6. U.S. Constitution, Amendment IV. For a comparison with the evidentiary standards in the United States, compare the definition of "probable cause" given in *Draper v. U.S.* (358 U.S. 307, 313 (1958)) with the standard of "reasonable grounds to believe" required for a stop and frisk (*Terry v. Ohio* 392 U.S. 1 (1968)).

7. By 2005, after completion of training, a new police constable would start at 29,103 pounds a year; sergeants at 31,092; inspectors at 41,586; chief inspectors at 45,852; superintendents at 53,046; and chief superintendents at 63,345 (Metropolitan Police Service, 2004c).

8. In the 1980s a streamlined version of the caution, known as the instant or early caution, was instituted. Geared to juveniles arrested for the first time for a minor offense, it does not involve a full investigation and is administered very soon after the arrest. See "Police in a Correctional Role: Cautioning By the English Police and Its Viability as an Option for Offenders" (Wakefield & Hirschel, 2005).

9. Other changes included the implementation of a "Plus Program" of participatory management and bureaucracy reduction; a revision of the old shift system, which had been in place (out of a sense of tradition) since 1829, to a system more responsive to crime patterns; and a stepped-up, aggressive minority-recruitment program (The New (and Improved) Scotland Yard, 1992, pp. 10–13).

REFERENCES

Adler, E., Mueller, G., & Laufer, W. (1994). *Criminal justice*. New York: McGraw-Hill.

Ayres, M., & Murray, L. (2005). *Arrests for notifiable offenses and the operation of certain police powers under PACE: England and Wales 2004/05*. London: Home Office Research and Statistics Department.

Ayres, M., Murray, L., & Fiti, R. (2003). *Arrests for notifiable offenses and the operation of certain police powers under PACE: England and Wales 2002/03*. London: Home Office Research and Statistics Department.

Chapter Resources

Banton, M. (1964). *The policeman in the community*. London: Tavistock.

Barclay, G. C. (1993a). *Criminal justice system in England and Wales* (2nd ed.). London: Home Office.

Barclay, G. C. (1993b). *Digest 2: Information on the criminal justice system in England and Wales*. London: Home Office Research and Statistics Department.

BBC News. (2005, July 23). *Shooting watershed for UK security*. Retrieved May 4, 2007, from http://news.bbc.co.uk/1/hi/uk/4708373.stm.

BBC News. (2006, September 11). *Axed police mergers cost millions*. Retrieved May 4, 2007, from http://news.bbc.co.uk/1/hi/uk/5334230.stm.

BBC News. (2007). *1992: Top policewoman suspended from duty*. Retrieved May 30, 2007, from http://news.bbc.co.uk/onthisday/hi/dates/stories/january/9/newsid_4063000/4063817.stm.

Bennett, T. (1990). *Evaluating neighborhood watch*. Brookfield, VT: Gower.

Benyon, J. (1986). Policing in the limelight: Citizens, constables and controversy. In J. Benyon & C. Bourn (Eds.), *The police: Powers, procedures, and proprieties* (pp. 3–42). Oxford: Pergamon.

Benyon, J., & Bourn, C. (1986). *The police: Powers, procedures, and proprieties*. Oxford: Pergamon.

Bibi, N., Clegg, M., & Pinto, R. (2005). *Home Office statistical bulletin: Police service strength*. London: Home Office Research and Statistics Department.

Bobbies Rally against Service Reforms. (1993, July/August). *Law Enforcement News*, 14.

Bourn, C. (1986). The police, the acts, and the public. In J. Benyon & C. Bourn (Eds.), *The police: Powers, procedures, and proprieties* (pp. 281–298). Oxford: Pergamon.

Brogden, M. (1982). *The police: Autonomy and consent*. London: Academic.

Brogden, M. (1991). *On the Mersey beat*. Oxford: Oxford University Press.

Campbell, D. (1992, June 18). Police chief admits public faith shaken. *The Guardian*, 21.

Campbell, D. (1994, May 17). Some police to get guns for routine patrol. *The Guardian*, 1.

Centrex. Retrieved December 22, 2005, from http://www.centrex.police.uk:8080/cps/rde/xchg/SID-3E8082DF-BF893508/centrex/root.xsl/home.html.

Christophersen, O., & Cotton, J. (2004). *Police service strength*. London: Home Office Research and Statistics Department.

Cotton, J. (2004). *Home Office statistical bulletin: Police complaints and discipline*. London: Home Office Research and Statistics Department.

Critchley, T. A. (1967). *A history of police in England and Wales*. London: Constable.

Farrington, D., & Bennett, T. (1981). The police cautioning of juveniles in London. *British Journal of Criminology, 21*(2), 123–135.

Fisher, Sir H. (1977). *The Confait case: Report by the Hon. Sir Henry Fisher*. London: H.M.S.O.

Gay Police Association. About the GPA. Retrieved December 22, 2005, from http://www.gay.police.uk/about.html.

Gorer, G. (1955). *Exploring English character*. London: Cresset.

Gould, R. W., & Waldron, M. L. (1986). *London's armed police: 1829 to the present*. London: Arms and Armour.

Grant, L. (1992, January 26). Not one of the boys in blue! Did sexism end a top policewoman's career, or did she throw it away? *The Independent (London)*, 20.

Hansen, O. (1986). A balanced approach? In J. Benyon & C. Bourn (Eds.), *The police: Powers, procedures, and proprieties* (pp. 103–111). Oxford: Pergamon.

Her Majesty's Chief Inspector of Constabulary (HMIC). (2006). *Annual report 2004–2005*. London, England: TSO.

Her Majesty's Inspector of Constabulary. (2005). *Closing the gap: A review of the "fitness for purpose" of the current structure of policing in England and Wales*. London: Home Office.

Home Affairs. (1999). *First special report*. Retrieved December 22, 2005, from http://www.publications.parliament.uk/pa/cm199900/cmselect/cmhaff/77/7702.htm.

Home Office. (1985). *Circular 14/1985: The cautioning of offenders*. London: H.M.S.O.

Home Office. (1988). *British crime survey report*. London: H.M.S.O.

Home Office. (1990). *Circular 59/1990: The cautioning of offenders*. London. H.M.S.O.

Home Office. (1992). *Surveying crime: Findings from the 1992 British crime survey*. London: H.M.S.O.

Home Office. (1993a). *Criminal statistics: England and Wales 1992*. London: H.M.S.O.

Home Office. (1993b). *White paper: Police reform: A police service for the twenty-first century, the government's proposals for the police service in England and Wales*. London: H.M.S.O.

Home Office. (1994). *Circular 18/1994: The cautioning of offenders*. London: H.M.S.O.

Home Office. (2001). *Race Relations (Amendment) Act 2000: New laws for a successful multi-racial Britain*. London, England: H.M.S.O.

Home Office. (2002). *The national policing plan 2003–2006*. London, England: H.M.S.O.

Home Office. (2003). *The national policing plan 2004–2007*. London, England: H.M.S.O.

Home Office. (2004a). *Annual police use of firearms statistics 2003/2004 for forces in England and Wales*. London, England: H.M.S.O.

Home Office. (2004b). *Building communities, beating crime: A better police service for the 21st century*. London, England: H.M.S.O.

Home Office. (2004c). *The national policing plan 2005–2008*. London, England: H.M.S.O.

Home Office. (2004d). *The National Policing Plan 2005-2008*. London, England: H.M.S.O.

Home Office. (2005). *Improving opportunity, strengthening society: The government's strategy to increase race equality and community cohesion*. London, England: H.M.S.O.

Home Office. (2006). *Police reform: National Policing Improvement Agency: Frequently asked questions*. Retrieved June 13, 2006, from http://police.homeoffice.gov.uk/police-reform/policing-improvement-agency.

Home Office, Equal Opportunity Monitoring, FI Division. (1992). *Minority ethnic and female members of all 43 police forces in England and Wales*. London: H.M.S.O.

Chapter Resources

Home Office, Research and Statistics Department. (1993a). *Operation of certain police powers under PACE: England and Wales, 1992.* Croydon, England: H.M.S.O.

Home Office, Research and Statistics Department. (1993b). *Police complaints and discipline: England and Wales, 1992.* Croydon, England: H.M.S.O.

Home Office, Research and Statistics Department. (1994a). *The criminal histories of those cautioned in 1985, 1988 and 1991.* London, England: H.M.S.O.

Home Office, Research and Statistics Department. (1994b). *Operation of certain police powers under PACE: England and Wales, 1993.* Croydon, England: H.M.S.O.

House of Commons. (1990). *Committee of Public Accounts, 2nd report: Review of the Crown Prosecution Service.* London: H.M.S.O.

House of Commons. (1999). Home Affairs—Fourth Report. London: H.M.S.O.

Independent Police Complaints Commission. Our values. Retrieved May 30, 2007, from www.ipcc.gov.uk/index/about_ipcc/ourvalues.htm.

Johnson, K. (2001). *Cautions, court proceedings and sentencing.* London: Home Office.

Johnston, L. (1992). British policing in the nineties: Free market and strong state? *International Criminal Justice Review, 2,* 1–18.

Joint Central Committee of the Police Federation of England and Wales. (1994). The best possible start. *Police: The Voice of the Service, 26*(7), 12–14.

Jones, T., & Newburn, T. (2001). *Widening access: Improving police relations with hard to reach groups.* London: Home Office.

Justice. (1994). *Annual report 1994.* London.

Karuma v. R. (1955). A.C. 197, 203(J.C.).

Kennedy, D. (1986). *Neighborhood policing: The London Metropolitan Police Force. Kennedy School of Government Case Program.* Cambridge: Harvard University.

Kirby, T. (1993, July 1). Police jobs for life culture swept away by inquiry. *The Independent,* 1.

Liberty. (2000). *An independent police complaints commission.* London, England.

Loveday, B. (1992). Book review essay: English criminal justice in crisis. *International Criminal Justice Review, 2,* 129–133.

MacDonald, W. (1982). *Police prosecutor relations in the United States.* Washington, D.C.: U.S. Department of Justice.

Marzouk, L. (2006). Police college to close. *This is local London.* Retrieved May 29, 2006, from http://www.thisislocallondon.co.uk/news/topstories/display.var.774258.0.police_college_to_close.php.

Mayhew, P., Maung, N. A., & Mirrlees-Black, C. (1993). *The 1992 British crime survey.* London: H.M.S.O.

McKenzie, I. K. (1993, March 17). Separate but equal: Developments in non-discriminatory policy and practice in British and American policing. Paper presented at the Academy of Criminal Justice Sciences Meeting, Kansas City, MO.

McPherson, W. (1999). *The Stephen Lawrence inquiry.* London: The Stationary Office. Retrieved March 3, 2006, from http://www.archive.official-documents.co.uk/document/ cm42/4262/4262.htm.

Meet the Future (1989). *Journal of International Criminal Justice 5*(5) 3. Oxford Press, NY.

Merricks, W. (1986). An independent prosecution service: Principles and practice. In J. Benyon & C. Bourn (Eds.), *The police: Powers, procedures, and proprieties* (pp. 243–250). Oxford: Pergamon.

Metropolitan Police Service. (2004a). *Cultural and communities resource unit.* Retrieved May 30, 2007, from http://www.met.police.uk/scd/specialist_units/serious_and_organised_crime.htm.

Metropolitan Police Service. (2004b). *High potential development scheme.* Retrieved December 22, 2005, from http://www.metpolicecareers.co.uk/default.asp?action=article&ID=202.

Metropolitan Police Service. (2004c). *Met careers team.* Retrieved May 29, 2007, from http://www.metpolicecareers.co.uk/default.asp?action=article&ID=24.

Metropolitan Police Service.(2004d). *Working together for a safer London.* Retrieved June 10, 2006, from http://www.met.police.uk.

Metropolitan Police Service. (n.d.). *Structure of policing in London.* Retrieved May 30, 2007, from http://www.met.police.uk/about/organisation.htm.

Metropolitan Police Service. (2007). *New constable/selection process.* Retrieved May 30, 2007, from http://www.metpolicecareers.co.uk/default.asp?action=article&ID=207.

Morgan, J. B. (1990). *The police function and the investigation of crime.* Brookfield, VT: Gower.

Mott, J. (1983). Police decisions for dealing with juvenile offenders. *British Journal of Criminology, 23*(3), 249–262.

Murray, L., & Fiti, R. (2004). *Arrests for notifiable offenses and the operation of certain police powers under PACE: England and Wales 2003/04.* London: Home Office Research and Statistics Department.

National Center for Women & Policing. (2002). Equality denied: The status of women in policing: 2001. Retrieved May 29 2007, from http://www.womenandpolicing.org/PDF/2002_Status_Report.pdf.

National Policing Improvement Agency (NPIA). (2007). Retrieved May 29, 2007, from http://www.centrex.police.uk.

The New (and Improved) Scotland Yard. (1992, September 15). *Law Enforcement News, 18*(364), 10–13.

Nicholas, S., Povey, D., Walker, A., & Kershaw, C. (2005). *Home Office statistical bulletin: Crime in England and Wales, 2004/2005.* London: Home Office Research and Statistics Department.

Oxford, K. (1986). The powers to police effectively. In J. Benyon & C. Bourn (Eds.), *The police: Powers, procedures, and proprieties* (pp. 61–74). Oxford: Pergamon.

Phillips, C. (1981). *Royal Commission on criminal procedure: Report.* London: H.M.S.O.

Police Complaints Board. (1980). *Triennial review report 1980.* London: H.M.S.O.

Police—Could You? (2007a). *Detailed eligibility requirements.* Retrieved May 30, 2007, from http://www.policecouldyou.co.uk/officers/eligible_details.htm.

Police—Could You? (2007b). *Other roles.* Retrieved May 30, 2007, from http://www.policecouldyou.co.uk/other/overview.html

Police—Could You? (2007c). *Detailed High Potential Development Scheme.* Retrieved May 30, 2007, from http://www.policecouldyou.co.uk/officers/hpds/overview.html.

Price, C., & Caplan, L. (1977). *The Confait confessions*. London: Marion Boyars.

Reiner, R. (1985). *The politics of the police*. Brighton: Wheatsheaf.

Reiner, R. (1991). *Chief constables*. Oxford: Oxford University Press.

Reiner, R. (1992a). Police research in the United Kingdom: A critical review. In M. Tonry & N. Morris (Eds.), *Modern policing, crime and justice: A review of research* (vol. 15, pp. 435–508). Chicago: University of Chicago Press.

Reiner, R. (1992b). *The politics of the police* (2nd ed.). London: University of Toronto Press.

Roberts, R. (1873). *The classic slum*. London: Penguin.

Robilliard, J., & McEwan, J. (1986). *Police powers and the individual*. Oxford: Basil Blackwell.

Rowe, D. (1986a). On Her Majesty's service: Criminal justice in England and Wales: Part I. *Criminal Justice International, 2*(5), 10–16.

Rowe, D. (1986b). On Her Majesty's service: Criminal justice in England and Wales: Part II. *Criminal Justice International, 2*(6), 9–16.

Runciman, Viscount. (1993). *The royal commission on criminal justice: Report*. London: H.M.S.O.

Sabalow, Ryan. (2005, November 29). New police chief breaks ground for female cops. *Auburn Journal*, p. 2.

Scarman, Lord. (1981). *The Brixton disorders 10–12 April 1981: Report of an inquiry by the Rt. Hon. the Lord Scarman, OBE*. London: H.M.S.O.

Serious Organised Crime Agency (SOCA). (2006). Retrieved May 30, 2007, from http://www.soca.gov.uk.

Sheehy, Sir P. (1993). *Inquiry into police responsibilities and rewards: Executive summary*. London: H.M.S.O.

Skogan, W. (1990). *The police and public in England and Wales: A British crime survey report*. London: H.M.S.O.

Smith, D. L., & Gray, J. (1985). *Police and people in London*. Aldershot, Hants, England: Gower.

Stead, P. (1985). *The police of Britain*. New York: Macmillan.

Stone, R. (1986). Police powers after the act. In J. Benyon & C. Bourn (Eds.), *The police: Powers, procedures, and proprieties* (pp. 53–60). Oxford: Pergamon.

Struck, Doug (2006, June 6). Mounties set for force's first same sex wedding. *The Washington Post*, A.11.

Tonry, M., & Morris, N. (1992). *Modern policing: Crime and justice: A review of research*. Chicago: University of Chicago Press.

Uglow, S. (1988). *Policing liberal society*. New York: Oxford.

Waddington, P. (1991). *The strong arm of the law*. Oxford: Clarendon.

Waddington, P. (1993, July 1). The case of the hidden agenda. *The Independent*, 29.

Wakefield, W. (1983, March). Gimme that ol' time religion: Metropolitan Police handling of juveniles in London. Paper presented at the annual Academy of Criminal Justice Sciences meeting, San Antonio, TX.

Wakefield, W., & Hirschel, J. D. (1991a, November). The police caution of juveniles by the English police: Its viability as an option for other offenders. Paper presented at the American Society of Criminology meeting, Baltimore, MD.

Wakefield, W., & Hirschel, J. D. (1991b, November). Public prosecution in England: Resistance to change. Paper presented at the American Society of Criminology meeting, Baltimore, MD.

Wakefield, W., & Hirschel, J. D. (2005). Police in a correctional role: Cautioning by the English police and its viability as an option for offenders in the United States. In C. Fields & R. Moore, Jr. (Eds.), *Comparative criminal justice: Traditional and nontraditional systems of law and control* (2nd ed., pp. 209–224). Prospect, IL: Waveland Publishing.

Wakefield, W., Hirschel, J. D., & Sasse, S. (1994). Public prosecution in England: Resistance to change in a major police force. *American Journal of Police, 13*(3), 169–189.

Wakefield, W., Kane, J., & Caulfield, D. (1984, March). Attitudes of London Metropolitan Police officers toward the practice of police cautioning of juveniles. Paper presented at the annual Academy of Criminal Justice Sciences meeting, Chicago, IL.

Waldren, M. (1994). Arming the police: An historical appraisal. *Criminal Justice Matters, 17*, 11.

Wilcox, A., Young, R., & Hoyle, C. (2004). *An evaluation of the impact of restorative cautioning: Findings from a reconviction study*. London: Home Office Research and Statistics Office.

Willink, Sir H. (1962). *Royal commission on the police: Final report*. London: H.M.S.O.

Zander, M. (1990). *The Police and Criminal Evidence Act 1984* (2nd ed.). London: Sweet and Maxwell.

Chapter Resources

The Judicial System of England

III

The Criminal Courts, Judges, and Lawyers

7

Chapter Objectives

After completing this chapter, you will be able to:

- Outline the structure and jurisdiction of the English criminal courts.
- Describe the general qualifications, mode of selection, and duties of the judges who preside in the various criminal courts in England.
- Compare and contrast the qualifications and training required to become a solicitor with the qualifications and training required to become a barrister.
- Compare and contrast how solicitors and barristers are organized and the types of legal work they perform.
- Describe how cases were prosecuted prior to, and how they are being prosecuted after, implementation of the Prosecution of Offenses Act of 1985.
- List the circumstances in which an accused in England is entitled to obtain the assistance of counsel.
- Outline both the circumstances and the manner in which counsel is provided in England to indigent defendants.

■ Introduction

In this chapter we will examine the structure of the English court system and the qualifications, duties, and powers of the various officials who appear in the courts: the judges, lawyers, and other court officials. The court process itself will be investigated in the following chapter.

■ Structure and Jurisdiction of the Courts

In England there are two levels of courts of original jurisdiction, courts that actually hear and decide cases—the Magistrates' Courts and the Crown Courts. There are four levels of courts of appellate jurisdiction; they are, in ascending order, the Crown Courts, the Divisional Court of the Queen's Bench, the Criminal Division of the Court of Appeal, and the House of Lords. The judicial functions currently exercised by certain members of the House of Lords, the Lords of Appeal in Ordinary, however, are to be transferred to the judges of a new Supreme Court of the United Kingdom (Constitutional Reform Act of 2005, Section 23).

Overall responsibility for the administration of the court system has rested with the **Lord Chancellor,** who has acted as (1) head of the judiciary, (2) speaker of the House of Lords, and (3) a member of the Prime Minister's Cabinet. However, until 2003 there was no unified system of court administration. In particular, the Magistrates' Courts were administered at the local level. This situation changed in 2003 as a result of passage of the Court Act, which established a system of central administration for all of the courts except the House of Lords.

Magistrates' Courts

The **Magistrates' Courts** are the courts of inferior jurisdiction, which are responsible for trying the less serious offenses, the summary offenses. As indicated in Chapter 3, in England crimes are categorized as being: either (1) summary, in which case they are tried in the Magistrates' Court; (2) indictable, in which case they are tried in the Crown Court; or (3) either summary or indictable, in which case they can be tried in either court.

With regard to cases that can be tried in either court (cases "triable-either-way"), it is the magistrates who make the initial determination of whether the case is more suitable for summary trial or for trial on indictment (Magistrates' Courts Act of 1980, Section 19). A defendant must, however, in most circumstances consent to the case being tried in summary fashion (Section 20, as amended by Schedule 3 of the Criminal Justice Act of 2003). If the defendant does not consent, the case will be tried on indictment in Crown Court. A defendant may believe that there are advantages to the case being tried in Crown Court. First, in Crown Court, unlike in the Magistrates' Court, the case is heard by a jury. Second, the delay that inevitably accompanies a trial in Crown Court may result in prosecution witnesses being unavailable. However, this belief may be tempered by the fact that heavier penalties may be imposed in Crown Court. Thus, for example, if a defendant is convicted in a Magistrates' Court of assault occasioning actual bodily harm (Offences against the Person Act of 1861, Section 47), the maximum penalty the magistrates can impose is 6 months' imprisonment. If, however, the defendant has been convicted of the same offense on indictment in Crown Court, a sentence of up to 5 years imprisonment may be imposed.

The situation is further complicated by the fact that a defendant who has been convicted in a Magistrates' Court of an offense that is triable-either-way can be sent to the Crown Court for sentencing if the magistrates who have tried the case believe that the defendant deserves a more severe sentence than that which they are legally empowered to impose (Powers of Criminal Courts [Sentencing] Act of 2000, Section 3). Until 2003, the maximum amount of active time that could be given for a single offense in a Magistrates' Court was 6 months. With the aim of reducing the number of cases sent to Crown Court for sentencing, the maximum sentence was increased to 12 months by the Criminal Justice Act of 2003 (Section 152). In the Crown Court the defendant may receive whatever sentence could have been imposed if the case had been processed as an indictable offense (Powers of Criminal Courts [Sentencing] Act of 2000, Section 5).

Much like the equivalent courts of inferior jurisdiction in the United States, the Magistrates' Courts have other general responsibilities in criminal matters. They are the courts from which arrest and search warrants and criminal summonses (orders directing defendants to appear in court) are issued. They are the courts in which defendants make their initial appearances and have the question of pretrial release decided. And, they are the courts in which the preliminary examination of persons charged with indictable offenses is conducted.

Crown Courts

The **Crown Courts** are the superior courts of original jurisdiction that try the more serious offenses, the indictable offenses. Strictly speaking there is one single Crown Court, which is divided between 78 centers in the 6 judicial districts into which England and Wales are divided. Crimes such as murder, rape, robbery, and causing death by dangerous driving must be tried in Crown Court. The Crown Courts also exercise appellate jurisdiction over cases that have been decided in the Magistrates' Courts. A defendant who has been convicted at trial in a Magistrates' Court can appeal to the Crown Court against the conviction or sentence. A defendant who has pled guilty in Magistrates' Court can appeal against the sentence (Magistrates' Courts Act of 1952, Section 83).

Appellate Courts

Appeals from the Magistrates' Courts on points of law are sent to the Divisional Court of the Queen's Bench (**Figure 7-1**). This court also hears appeals from the Crown Courts on points of law. These appeals are by way of "case stated" and may be made by either the defense or the prosecution. The court receives a concise statement of the law involved in the case upon which it rules. Appeals from the Divisional Court of the Queen's Bench go directly to the House of Lords, which as previously noted, will be replaced by the new Supreme Court of the United Kingdom.

Appeals against a conviction or sentence given in Crown Court go to the Criminal Division of the Court of Appeal. This court is the final arbiter on matters of fact. The Criminal Justice Act of 1988 (Section 36) gives the prosecution, through the Attorney General, the ability to submit cases in which "unduly lenient" sentences have been imposed to the Court of Appeal for resentencing.

Appeals on matters of law also lie from the Crown Court to the Criminal Division of the Court of Appeal. The Attorney General may refer a point of law to the Court of Appeal for clarification. The Court of Appeal's decision will not, however, affect the acquittal. From the Criminal Division of the Court of Appeal, an appeal on a point of law of general public importance can be made to the House of Lords.

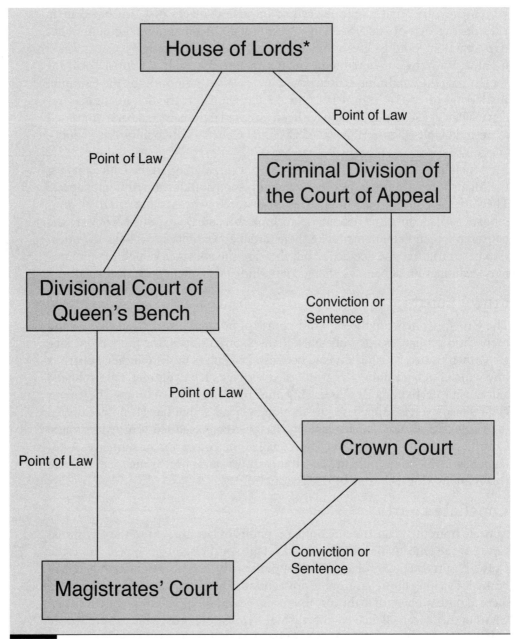

Figure 7-1 Structure of the English Criminal Courts
*To be replaced by the Supreme Court of the United Kingdom

As a result of the numerous miscarriages of justice that had occurred, a **Criminal Cases Review Commission (CCRC)** was established by the Criminal Appeal Act of 1995. This commission functions as an independent public body that reviews suspected miscarriages of justice and decides whether they should be referred to an appellate court because the commission believes that the defendant either has been wrongly convicted or unjustly sentenced (Criminal Cases Review Commission [CCRC], 2006b). Between its inception in March 1997 and December 31, 2005, the CCRC received a total of 8,343 applications for review. A total of

308 of the 7,590 cases that had been completed by December 31, 2005, were referred to the Court of Appeal for review of either conviction or sentence. A total of 178 of the 252 cases decided by the Court of Appeal by that date resulted in either the conviction or the sentence being quashed (CRCC, 2006a).

■ The Judges

A wide variety of judges with different types of background qualifications and modes of selection preside over these courts. These judges are, in ascending order, lay magistrates, District Judges, recorders, circuit judges, High Court judges, Lord Justices of Appeal, and Lords of Appeal in Ordinary.

Magistrates

The Magistrates' Courts form the backbone of the English judicial system. It has been estimated that these courts deal with 95% to 98% of criminal matters referred to court (Cross & Ashworth, 1981, p. 3; Department for Constitutional Affairs, 2004a, Section M1; Fitzgerald & Muncie, 1983, p. 82; Skyrme, 1979, p. 5). The vast majority of these officials are **lay magistrates** who are also justices of the peace. These lay magistrates are prominent citizens whose names have been submitted to the Lord Chancellor by local advisory committees, one third of whom are currently serving magistrates (Department for Constitutional Affairs, 2003, para. 12). They are appointed by the Lord Chancellor on behalf of the Crown to serve in one of the counties of England or Wales or in one of the London commission areas. They must reside within 15 miles of the commission area to which they have been appointed and have no jurisdiction outside that area. All newly appointed magistrates receive training prior to assuming duties on the bench. This training focuses on such topics as court procedure, evidence, and sentencing and includes visits to penal institutions. When first on the bench they sit with experienced magistrates. All magistrates appointed after January 1, 1980, are required to attend approved refresher training totaling at least 12 hours every 3 years. In order to chair a court hearing a magistrate must, since 1996, have completed a chairmanship course.

As indicated in the National Strategy for the Recruitment of Lay Magistrates, the prime consideration in appointing lay magistrates is not prior legal training. Rather, it is that the prospective magistrate possess the "six key qualities" that are taken to define "personal fitness, namely: good character; understanding and communication; social awareness; maturity and sound temperament; sound judgment; and, commitment and reliability" (Department for Constitutional Affairs, 2003, para. 19). Although the desirability of obtaining representation from all sections of the community has long been recognized (see, e.g., Skyrme, 1979, p. 53), older, white, conservative-oriented males of the upper socioeconomic classes have tended to be appointed magistrates (Fitzgerald & Muncie, 1983, pp. 110–112). Concern about minorities being underrepresented among the lay magistracy has led, since the 1980s, to increases in minority appointments (Department for Constitutional Affairs, 2004a, Section M4; Lord Chancellor's Department, 1988, p. 77). In 2005, 6.5% of lay magistrates were members of minority ethnic groups, not far short of their representation of 7.8% in the general population (Department for Constitutional Affairs, 2005). Because magistrates are not barred from participating in politics, and indeed are often highly active in politics, an attempt is generally made to obtain a reasonably representative balance among those affiliated with the different political parties.

A very small number of magistrates, some 124 out of over 25,000 (Department for Constitutional Affairs, 2004b, p. 127), are professional, qualified lawyers who are paid for sitting on the bench. These judicial officials, formerly known as stipendiary magistrates, are now called District Judges. Most of these District Judges are based in London. The rest are located in other major metropolitan areas. District Judges are appointed by the Crown on the recommendation of the Lord Chancellor, and must have held the right of audience for at least 7 years in relation to any class of proceedings in any part of the Supreme Court, or all proceedings in County or Magistrates' Courts.

Although District Judges can be dismissed by the Lord Chancellor only for misbehavior or inability, lay magistrates hold office at pleasure, and legally may be removed at any time by a Lord Chancellor without citing any reason. In practice, however, a magistrate will not be dismissed without due cause. Problems posed by infirm and incompetent magistrates are facilitated by the ability of the Lord Chancellor to place such magistrates on a supplemental list. Persons on this list remain justices of the peace, but are not summoned to court. The retirement age for magistrates is 70.

The District Judges nearly always sit alone in court. The lay magistrates, on the other hand, sit as a bench of two to seven (but normally two or three) under the guidance of a chairperson they have elected. They must be available for 26 court sittings (half court days) a year and indicate to the chief clerk of the court to which they have been appointed when they will be available to sit. Just prior to court sessions, benches of available magistrates are finalized by the chief clerk. It is unlikely that a magistrate will know in advance with which other magistrate(s) he or she will sit on a particular day.

The lay magistrates undertake their duties "at some inconvenience to themselves and with no expectation of material reward" (Skyrme, 1979, p. 6). They can, however, receive allowances for "traveling, subsistence and financial loss" (Justice of the Peace Act of 1979, Section 12). The major advantages of the lay magistrate system lie in its low comparative costs and its flexibility. If the workload increases, more magistrates can be called in and extra court sessions held, provided that courtroom space is available.

The chief clerk, who effectively had to be a barrister or solicitor of 5 years' standing (Justice of the Peace Act of 1979, Section 26) traditionally had a dual role to fulfill. The chief clerk was responsible both for the administration of the court (including overseeing legal aid) and for giving the magistrates "advice about law, practice or procedure" (Richman & Draycott, 1993). These functions were, however, often delegated to court clerks, who often were not attorneys: In 1987 about one quarter were barristers or solicitors (Zander, 1992, p. 341). The Police and Magistrates Courts Act of 1994 created the position of justices' chief executive, and these officials took over some of the administrative responsibilities of the chief clerk. A 1998 government report recommended that there be greater separation of these administrative and judicial functions, with the chief clerk, who from January 1, 1999, would have to be a professionally qualified lawyer, able to act as "senior legal advisor" to the magistrates (Department for Constitutional Affairs, 1998).

Unlike the judges in the higher courts, who are solely the arbiters of law, the magistrates are judges of both law and fact. In determining matters of law and procedure, the lay magistrates are assisted by the chief clerks, who may in practice wield considerable authority in the courtroom. However, although chief clerks advise the lay magistrates on such issues, they do not participate in deciding cases.

When a case is sent from Magistrates' Court on appeal to Crown Court, between two and four magistrates must sit with the Crown Court judge (Supreme Court Act of 1981, Section 74[1]). The decision is made by majority vote.

Crown Court Judges

There are three different types of judges who sit regularly in Crown Court. They are High Court judges, circuit judges, and recorders. All these judges are currently appointed by the Crown on the recommendation of the Lord Chancellor. Considerable prior legal experience is required of all of these appointees. A **High Court judge** must have had 10 years' advocacy experience in the High Court or been a circuit judge for at least 2 years. A **circuit judge** must have been entitled to appear as an advocate for 10 years in either the Crown Court or the County Court, or been a recorder. A **recorder**, finally, must have been entitled to appear as an advocate for 10 years in Crown or County Court (Courts and Legal Services Act of 1990, Section 71).

High Court and circuit judges hold office during good behavior until they reach their respective retirement ages. The retirement age for High Court judges is 75, for circuit court judges 72. High Court judges may be removed by the Crown on an address by both Houses of Parliament. Since this procedure was instituted by the Act of Settlement in 1701 only one judge has been removed in this fashion, an Irish judge who was removed in 1830 after embezzling money and fleeing to France. Circuit court judges may be removed by the Lord Chancellor for incapacity or misbehavior. Recorders are appointed on contracts for fixed terms and sit in Crown Court on a part-time basis. They may be reappointed until they reach the age of 65.

Offenses tried in Crown Court are divided into four categories. Offenses in the first category, the most serious of the offenses such as murder and offenses against the Official Secrets Act, must be tried by a High Court judge. Offenses in the second category (such as manslaughter and rape) are also tried by High Court judges unless the presiding judge of the circuit releases the case to another judge. Offenses in the third category (the less serious indictable offenses) may be tried by High Court judges, circuit judges, or recorders. Offenses in the fourth category (offenses triable either summarily or on indictment) are normally tried by circuit judges or recorders. All cases held in Crown Court are, as mentioned before, tried before a jury.

Appellate Judges

The Divisional Court of the Queen's Bench, which hears appeals on points of law from both the Magistrates' and Crown Courts, sits in London. It is composed of two or more (usually three and occasionally five) High Court judges who have been assigned to that court. The Lord Chief Justice presides over the Divisional Court. The Lord Chief Justice must be either a judge of the Court of Appeal or qualified for appointment as a Lord Justice of Appeal, and is appointed by the Crown on the recommendation of the Prime Minister.

The **Lord Justices of Appeal** who sit in the Court of Appeal must be lawyers of 10 years' standing or High Court judges. They are appointed by the Crown on the recommendation of the Prime Minister and hold office during good behavior until they reach the compulsory retirement age of 75. There are currently 37 Lord Justices of Appeal. The Criminal Division of the Court of Appeal is presided over by the Lord Chief Justice. Appeals are heard by a panel of three judges.

The highest court in the land currently is the House of Lords. A judge of the House of Lords is known as a **Lord of Appeal in Ordinary** (or more informally as a Law Lord) and is appointed by the Crown on the recommendation of the Prime Minister. To be eligible for appointment a candidate must have been a lawyer of at least 15 years' standing or have held high judicial office for at least 2 years. Traditionally, appointment has carried with it a life peerage that entitled the holder to sit and vote in the legislative chamber of the House of Lords. Again, tenure is during good behavior until attainment of the age of 75. There are 12 Lords of Appeal in Ordinary. Appeals have generally been heard by five Lords of Appeal in Ordinary.

The Constitutional Reform Act of 2005, as noted earlier in this chapter, establishes a new Supreme Court of the United Kingdom (Section 23). The 12 judges who sit on this court are to be appointed by the Crown on the recommendation of the Prime Minister (Section 26) and the new court is to take over the role formerly exercised by the Lords of Appeal in Ordinary. To provide continuity with the present system, those who are Lords of Appeal in Ordinary at the time of the establishment of the new Supreme Court are to become judges of the Supreme Court, with the person who was the senior Lord of Appeal in Ordinary becoming the President of the Court (Section 24). The judges of the new Supreme Court are to hold office during "good behavior," but may be removed on address of both Houses of Parliament (Section 33).

Until passage of the Constitutional Reform Act of 2005, the Lord Chancellor acted as the head of the judiciary, and as such played a major role in the selection of judges. This act modifies the position of Lord Chancellor and provides that judicial duties previously exercised by the Lord Chancellor be entrusted to the Lord Chief Justice, who serves as the President of the Courts of England and Wales and the Head of the Judiciary of England and Wales (Section 7). The result is that the Lord Chancellor's previous judicial functions are minimized and that office's executive functions, and to a lesser extent legislative functions, are given more prominence.

The Lord Chancellor is now entitled "Secretary of State for Constitutional Affairs and Lord Chancellor." The Lord Chancellor's Department, which had previously provided the Lord Chancellor with a staff, has been subsumed into the Department of Constitutional Affairs (DCA). This new department was established in 2003 and is responsible for running the courts, improving the justice system, and ensuring the provision of human rights.

The Constitutional Reform Act of 2005 also established an independent Judicial Appointments Commission, composed of a lay chairperson and 14 commissioners

(Section 61 and Schedule 12). This commission is taking over the responsibility for selecting candidates for judicial appointment. The establishment of such a commission promotes the doctrine of separation of power and helps further the independence of the judiciary, an issue that is stressed in the Constitutional Reform Act itself (see, e.g., Section 3).

■ The Lawyers

There are two types of lawyers who practice in England: solicitors and barristers. Whereas the **solicitor** has been compared to the general practitioner in the medical profession, the **barrister** has been likened to the medical specialist. The solicitor has traditionally handled cases in the lower courts, namely the Magistrates' Courts, whereas the barrister has enjoyed the right of audience in the higher courts. However, recent changes in the law have resulted in solicitors being able to acquire all of the rights of audience previously exercised only by barristers and being eligible for judicial appointments previously available only to barristers. A client can hire a solicitor directly, but until 1989 could hire a barrister only through a solicitor. Starting in 1989 this rule started to be relaxed, first allowing professional clients, such as accountants, and now members of the public to have some direct contact with barristers (Zander, 2003, pp. 760–764; The Bar Council, 2004, Sections 204–205, 401, 501–502).

Barristers

The historical origins and the development of the professions of solicitors and barristers are rather different. Barristers came into existence in the thirteenth century when Roman Catholic priests were forbidden from practicing law. Barristers' association with what are now known as the **Inns of Court** began in the fourteenth century when the buildings previously occupied by the crusading order of the Knights Templar were vacated and taken over by lawyers who wished to be near to the Royal Courts of law. The lawyers formed different societies, and in the seventeenth century these societies, or Inns of Court, obtained the exclusive right to practice law in the Royal Courts. Although for many years these lawyers, or barristers, as they became known, dealt directly with clients, it became firmly established in the nineteenth century that clients could approach barristers only through solicitors. Prospective barristers joined one of the Inns of Court where they learned the law under the pupilage of an experienced barrister, took part in moots (mock trials), and ate the required number of dinners, where knowledge was imparted by seasoned members of the profession. In 1852 a Council of Legal Education was set up by the Inns of Court, and in 1872 examinations were introduced.

Today there are four Inns of Court in existence: Inner Temple, Middle Temple, Lincoln's Inn, and Gray's Inn. In order to become a barrister an applicant must be admitted by one of the Inns. Normally a law degree is required. If, however, the applicant does not have a law degree, he or she will have to take a one-year law course and pass the Common Professional Examination shared with solicitors. Training has two major additional components: Taking the 30-week course and passing the bar examination established by the Council of Legal Education, and reading for one year as a pupil in the chambers of a practicing barrister. Unlike their counterparts in the United States, all English lawyers must serve an apprenticeship before being allowed to practice law on their own. The Bar Council now requires all barristers to fulfill continuing education requirements. The new practitioners' program established in 1997 requires a minimum of 45 hours of continuing education during the first 3 years of practice. After that, 12 hours of continuing education are required each year (The General Council of the Bar, 2005)

Barristers are not permitted to form partnerships. Such a rule, it is believed, promotes a client's ability to select a particular barrister for a case. However, all practicing barristers are required to join a set of "chambers," which consists of offices shared with other barristers. The senior member of the group is known as the head of chambers. A shortage of space prompted the Bar to relax the rules, and this resulted in more chambers being set up in London outside the Inns. As of December 2004 there were 14,364 barristers practicing in 355 sets of chambers in England, with 210 of these in London (The Bar Council, 2006). Until the 1990s, all but one of the London chambers were located in the Inns of Court.

Working in chambers enables barristers to share expenses incurred for rent, equipment, and services. With the increased concern for economy, efficiency, and profitability, the size of chambers doubled between the 1970s and 1990s. The average set of chambers in London in the early 1990s had 15 to 20 members, elsewhere about 12 (Zander, 1992, p. 634). Sharing chambers also provides a setting for guidance and mutual support. In addition, the constant proximity to other barristers can act as a barrier to the infringement of professional rules, such as the rule against touting for clients. The chambers employ a clerk who is responsible for obtaining work for the members of the group. This is an outgrowth of the stipulation that barristers may not advertise for clients, and may not normally obtain a client except through a solicitor. The clerk, who works on a commission basis, also acts as the office administrator for the group, and as the business manager of each individual member of the group. Most chambers do, in fact, employ more than one clerk, with one of them designated as the senior clerk.

The barrister's role consists of presenting cases in court, drafting legal documents, and giving legal advice. Generally a barrister will have an area of specialization, and will work primarily in that area. Except when appearing in a Magistrates' Court or when representing a minor in Crown Court, all barristers must wear robes and wigs in court.

There are two classes of barrister. There are the ordinary barristers, known as junior counsel, and there are the senior barristers, Queen's Counsel (QCs), who are called leading counsel or "silks" (in recognition of the fact that they are entitled to wear silk gowns, not the wool/polyester gowns worn by junior counsel). Until recently, a junior barrister had to have been in practice for 10 years in order to become a silk. With the extension to solicitors of rights of audience previously enjoyed only by barristers, it was argued that certain solicitors should be entitled to be appointed silks, and since July 1995 solicitor advocates (solicitors with the right of audience in the higher courts) have held that right (Zander, 2003, p. 764). Until 2004 QCs were appointed by the Crown on the advice of the Lord Chancellor. Since then, in order to promote selection on the basis of merit and introduce peer review, recommendations for appointment have been made by an independent selection panel, composed of both lawyers and nonlawyers. Queen's Counsel do not draft documents and do not normally appear in court without the assistance of junior counsel. Traditionally, about 10% of barristers have been QCs (Zander, 1992, p. 629). Very few solicitors have as yet been appointed QCs.

The position of silks has been under attack in recent years, with the Office of Fair Trading suggesting that the existence of this rank artificially raises, without sufficient benefit to the public, the price of services rendered (Office of Fair Trading, 2001). Some of the recent changes, such as the new appointment process, have been in response to the concerns raised by the Office of Fair Trading.

Since 1997, complaints about barristers have been submitted to a Complaints Commissioner who conducts the initial investigation and is authorized to refer substantiated complaints to the Professional Conduct and Complaints Committee, which is composed of barristers and members of the Bar Council's panel of lay representatives. Until 2000, barristers enjoyed immunity for negligent work performed both in court and in preparation for court. However, in *Arthur JS Hall and Co v. Simons* (2000) the House of Lords ruled 4–3 that they should no longer enjoy this immunity.

Solicitors

The precursors of the modern-day solicitors were the attorneys, solicitors, and proctors who practiced law in the old common law, chancery, and ecclesiastical and admiralty courts by helping litigants with the formalities required for bringing actions in those courts. By 1739 these three types of lawyers had formed the Society of Gentlemen Practisers in the Courts of Law and Equity (Price, 1979, p. 9). In 1831 this society was incorporated as the Law Society with its members known as solicitors.

The traditional training for a solicitor has been an apprenticeship known as "articles" with a practicing solicitor. From 1836 an examination has also been required, and since 1877 the Law Society has been responsible for administering this examination. In 1980 another prerequisite was added: possession of a college law degree. Those who did not possess such a degree had to pass an initial professional examination.

The current guiding regulations are the Training Regulations of 1990. These regulations stress the twin aspects of (1) academic and (2) vocational training. The academic component is satisfied by either successfully completing a law degree or passing the Common Professional Examination and then taking the Legal Practice Course. The vocational component is satisfied by entering into and completing a training contract for 2 years' full-time work or the equivalent of 2 years' full-time work spread over a period of up to 4 years. After fulfilling these requirements, the newly qualified solicitor is entitled to have his or her name added to the Roll of Solicitors by the Master of the Rolls and to apply to the Law Society for a practicing certificate. This certificate has to be renewed each year. Since August 1985 the Law Society has required continuing education courses, first for solicitors in their first 3 years of practice, and then for all solicitors. Currently, 16 hours of continuing professional development are required each year (Law Society, 2005a). In addition to overseeing the training of solicitors, the Law Society is the body responsible for hearing complaints against solicitors and taking disciplinary action when deemed appropriate.

Solicitors may practice law individually or together with others in a firm. Most law firms are fairly small. Of the 9,198 firms in England in 2003, 84.7% had 4 or less partners and only 5.5% more than 10 partners. Though only 1.6% of the firms had 26 or more partners, they employed 37.3% of the 92,752 practicing solicitors (The Law Society, 2004). It is highly common for solicitors in a firm to have areas of specialization. Very often the senior solicitors in a firm will be partners, working for a share of the firm's profits, while the junior members are employees working for fixed salaries.

As mentioned earlier, the solicitors are the lawyers who handle the criminal cases in Magistrates' Court. Until the 1990s their ability to present cases in the higher courts (Crown Court and above) was highly limited. With permission from the Lord Chancellor they could present cases in Crown Court. Provided they or other members of their firm handled the case in Magistrates' Court, they could appear in Crown Court on appeals or committals for sentence from Magistrates' Court. Usually, however, they used to appear in Crown Court as assistants to the barristers who had the right of audience there. Unlike the barristers who wear gowns and wigs, the traditional attire for a solicitor is a business suit.

For over a century there has been debate over whether the two branches of the legal profession should be merged. Such fusion, it has been argued, would save time, effort, and money, and lessen duplication and increase continuity in handling cases. On the other hand, it has been suggested fusion would diminish the general availability of legal specialists and result in lawyers being less likely to refer clients to specialists. The Royal Commission on Legal Services considered the issue and in 1979 issued a report that unanimously supported continuation of the two branches. Fusion of the profession, it was believed, would reduce the quality of advocacy. In an era when there was a greater call for specialization, merging the two branches would lead to specialists being less readily available.

The Royal Commission's report did not, however, dispose of the matter. Indeed, the Conservative government once again in January 1989 brought the issue of merger to the forefront by suggesting, among other proposals, that duly licensed solicitors be allowed to present cases in the higher courts, that clients be able to approach barristers directly, and that solicitors be eligible for appointment to the higher judicial posts (Lord Chancellor's Department, 1989). These proposals posed a radical change for the English legal profession and, it has been suggested, created "fission rather than fusion" (*Manchester Guardian Weekly*, 1989, p. 12). They promoted intense debate and were modified in the ensuing government white paper and legislation (The Courts and Legal Services Act of 1990). The act provided for solicitors both to obtain the right of audience in the higher courts (Sections 17 and 27) and to be appointed judges in the higher courts (Section 71).

A cumbersome scheme was initially established for vetting solicitors who sought the right of audience in the higher courts. The Access to Justice Act of 1999, however, facilitated the process by entrusting the Law Society with establishing appropriate qualifications and regulations. These are to be found in the publication *Higher Rights of Audience* (The Law Society, 2005b). By April 2003 a total of 955 solicitors had qualified for rights of audience in the higher criminal courts (Zander, 2003, p. 751).

Lawyers for the Prosecution

In England, under common law, any person could commence a criminal prosecution. Thus, traditionally, the prosecution of cases was considered a private matter and it was up to the citizen to institute criminal proceedings, produce witnesses, and handle the case from the initial prosecution to the trial under the "direction" of a magistrate (DeGama, 1988, p. 342). Needless to say, in such an environment many cases were discontinued or not even prosecuted. Criminals went free and corruption was apparent.

This approach was followed for many years until Robert Peel established the police and they developed an image of a "neutral" civil force somewhat independent from the state. Despite the fact that a public prosecution office was created in 1879 (the Office of the Director of Public Prosecutions), the office itself was relatively small and handled few cases. There was fear that if public prosecutors were to become a powerful reality, the courts and the criminal justice system would become biased, incompetent, uncaring, and corrupt. There was also the sentiment that such prosecutors would constitute a significant encroachment by the state on the citizenry.

Creating a balance between the government and the people that was acceptable, the police filled the void between the state and the citizenry. Police officers, who were viewed as citizens in uniform, undertook the majority of prosecutions (DeGama, 1988). Private nonpolice prosecutions were, in fact, rare, with estimates that such cases constituted less than 1% of all prosecutions (Saunders, 1985, p. 5) or 2.4% of nonpolice prosecutions (Lidstone, Hogg, & Sutcliffe, 1980, p. 15). The offenses most commonly prosecuted by private individuals were shoplifting and common assault (Zander, 1984, p. 199). In addition, there was a time when "banks and other business houses often preferred to prosecute for offences that involved their internal workings" (Jackson, 1971, p. 85).

Under the prosecutorial schema in operation in England prior to passage of the Prosecution of Offenses Act of 1985, the police fulfilled the primary prosecution function in addition to their traditional policing responsibilities. Because a police prosecution was considered a private prosecution, the police could not be forced to prosecute a case. As Lord Denning M. R. summed it up in *R. v. Metropolitan Police Commissioner ex p. Blackburn*:

> *No Minister of the Crown can tell him (a chief constable) that he must, or must not, keep observation on this place or that; or that he must, or must not, prosecute this man or that one. Nor can any police authority tell him so. The responsibility for law enforcement lies on him. He is answerable to the law and to the law alone. (1986, p. 894)*

Only if a police department adopted a policy of wholesale nonenforcement of a law, as the Metropolitan Police did in 1966 with regard to illegal gambling in gaming clubs, could that department be ordered by the courts to enforce the law. And even here the police could not be ordered to initiate prosecution in a specific case.

Although minor offenses, such as traffic and theft offenses, were often prosecuted by police officials, the more serious offenses were handled by lawyers employed by the police. In employing such lawyers the police enjoyed a traditional client-lawyer relationship. As the Royal Commission on Criminal Procedure pointed out, "The solicitor may offer advice but the final decision on who shall be prosecuted and for what offence rests with the police" (1981a, p. 52). To assist in the prosecution of cases, most of the 43 English police departments established prosecuting solicitors' departments.

A limited amount of supervision over the prosecution of cases was exercised by the central government through the office of the Attorney General (AG) and the officials working under the Attorney General, most notably the Director of Public Prosecutions. The **Attorney General,** who is appointed by the party in power, is the chief law officer of the Crown, the titular head of the English Bar, and a member of either the House of Commons or the House of Lords. The Attorney General represents the government in court in important cases, answers questions in Parliament, and provides legal advice to government departments. The Attorney General is the official who exercises the Crown's power to stop a criminal prosecution by entering a "nolle prosequi" (a latin term meaning "to be unwilling to pursue"), and must consent before certain types of criminal prosecution (e.g., some of the offenses against official secrets) are commenced. The deputy of the Attorney General is the Solicitor General. This official is also politically appointed, and sits in the House of Commons or Lords.

The **Director of Public Prosecutions (DPP),** who must be a barrister or solicitor of at least 10 years' standing, was until 1986 appointed by the Home Secretary. Now the DPP is both appointed by, and works under, the general supervision of the Attorney General. The DPP has been responsible for prosecuting particularly important or difficult cases and for advising chief constables. The DPP's consent is required before prosecution can be commenced for some 60 statutory offenses (Zander, 1992, p. 213).

This form of prosecution was so successful that critics were virtually nonexistent for almost a century. It was only in the latter part of the twentieth century, as concerns about the quality of British policing grew, that cries for reform of the prosecution process began to resound. It was suggested that the police could not be objective in prosecuting cases, that it was no longer apparent that the system was cost-effective, and that an independent prosecuting body would be in both the government's and the people's interests. In 1970 the Justice Committee recommended that an independent system similar to those operating in Scotland and the United States be implemented. In 1978 the issue was referred to a Royal Commission on Criminal Procedure, which after an exhaustive investigation, issued its report in 1981. In evaluating the current system and assessing the potential of proposed modifications, the commission placed considerable emphasis upon the criteria of "fairness, openness and accountability, and efficiency" (1981a, p. 127). It concluded that the current system fell short when measured by these criteria, and proposed that "the police should retain control of the procedures up to and including the point of accusation" (p. 194), but that the responsibility for the prosecution of cases should be entrusted to an independent Crown Prosecution service. Although the commission recognized that the new system would cost more, it believed that the added expense was "worth paying to achieve a fairer, more open and more efficient system, which will have the confidence of the public in whose name it will operate" (p. 170).

The government accepted the commission's recommendations in principle and set up a working party for advice "on what would be the best model for the organization of such a service" (Home Office, 1983, p. 7). In a white paper presented to Parliament in October 1983, the government endorsed the working party's recommendation of a national prosecution service headed by the Director of Public Prosecutions and under the supervision of the Attorney General (Home Office, 1983). In a second white paper presented to Parliament in December 1984, the government outlined its proposals for the distribution of functions between the Director of Public Prosecutions and his or her headquarters and the Crown Prosecutors and their local staffs. The guiding principle to be applied in distributing function was "maximum delegation consistent with proper accountability" (Home Office, 1984, p. 1).

In late 1984, the enabling legislation was introduced in Parliament. Entitled the Prosecution of Offenses Act, it was passed in 1985 and implemented in 1986. With this statute, the power of prosecution, which had been the domain of the police for the previous 150 years, was passed over to the new prosecution service.

The Prosecution of Offenses Act of 1985

The Prosecution of Offenses Act of 1985 is divided into four parts, with part one focusing on the structure and operation of the new Crown Prosecution Service, part two dealing with the issue of costs in criminal cases, and parts three and four presenting various miscellaneous and supplemental matters.

The Act established the **Crown Prosecution Service (CPS)** under the Director of Public Prosecutions (DPP), who is appointed by and works under the Attorney General (Sections 1–3). The CPS has the duty to take over the conduct of all criminal proceedings instituted by the police. The right of private individuals to prosecute offenses, subject to the director's power to take over the conduct of proceedings, is, however, preserved (Section 6).

The DPP and the headquarters' staff are located in London. Initially, the country was divided into 31 prosecution areas, each headed by an area chief appointed by and answerable to the DPP. Under a reorganization plan announced in 1993, the number of prosecution areas was reduced to 13 (Runciman, 1993, p. 71). A 1998 report, however, concluded that this reorganization had been a mistake because it had resulted in overly centralized and excessively bureaucratic management (*Review of the Crown Prosecution Service*, 1998). It was recommended that the CPS be reorganized into 42 areas that paralleled the jurisdictions of the respective police forces with one area for both the Metropolitan and City of London police forces. These 42 areas would be "bound by central policies and procedures," but with "a large degree of autonomy in carrying out their professional duties and managing their local offices" (*Review of the Crown Prosecution Service*, p. 3).

Under the Prosecution of Offences Act of 1985 the Director of Public Prosecutions is responsible for issuing the central policies and procedures that provide the lawyers working for the CPS with guidance on the general principles to be applied when making decisions about prosecutions (Section 10). In making these decisions, two major criteria are applied: (1) whether there is sufficient evidence to warrant prosecution (the "realistic prospects of conviction" test), and (2) whether prosecution is deemed to be in the public interest.

The English Crown Prosecution Service in Operation

Although the police no longer enjoy the authority to decide whether to prosecute a case, they still control initial entry into the court system. They initiate legal proceedings by preferring charges against defendants, and it is only after this step that the CPS becomes involved. The police may dispose of a case by issuing a caution, or they may completely drop a case. Thus, their response is extremely important because in essence the prosecutors represent only a second set of filters with the power only to continue or discontinue a case. This differs from the situation in the United States where the prosecutor makes the decision to prefer the charges after arrest, and quite often plays a major part in the investigative function. The English Crown Prosecution Service does not have the option of going to a grand jury to initiate criminal investigations, because the English grand jury was abolished in 1933.

One option exercised by the CPS is to drop a case that the police have presented to them. This practice has prompted widespread criticism by the police for several reasons. First, the police may believe they have not been properly consulted or have received proper feedback when cases are dropped. The police position is that if they had been consulted prior to dropping cases, they would have had the opportunity to pursue further evidence. Discussions with officers have indicated that this practice affects police morale, particularly in cases where the officer has been assaulted or has a personal interest in the case. Secondly, there have been complaints from both the police and the National Audit Office (1989, p. 2) that the Crown Prosecution Service has not carried out a systematic or consistent method of discontinuing cases. Since the inception of the CPS, discontinuance rates have varied widely throughout the country. In the first 5 years they ranged from a low of 4% to a high of 19% (*Crown Prosecution Service, Memoranda of Evidence*, 1990, p. 10). Although the reduction of the number of prosecution areas to 13 was designed to "achieve consistency of practice across the country" (Runciman, 1993, p. 71), local variations in discontinuance rates continued to exist (National Audit Office, 1997).

In order for the CPS to meet its statutory obligations and to be successful, it is imperative that its prosecutors are supplied with both the correct evidence and paperwork in the right form at the right time. This requires both the CPS and the police to make sincere efforts to exchange records and other forms of paperwork. A lack of efficiency in this area causes delays and unnecessary adjournments of cases. According to a CPS survey of adjournments, 21% of the adjournments were occasioned by the CPS itself, and a further 26% were "police inspired" (*Crown Prosecution Service, Memoranda of Evidence*, 1990, p. 13). This may be taken as an indication of communication problems between the CPS and the police.

Concern about these issues has led to the development of closer working relationships between the CPS and the police. Prosecution teams, composed of police and CPS representatives, have been established, and through CPS Direct, an off-hours telephone service, prosecutors now provide the police with charging advice. The observed decline in the national discontinuance rates in Crown Court, which fell steadily from 16.2% in 2002–2003 to 12.5% in 2004–2005, has been attributed to these reforms (Crown Prosecution Service, 2005).

A related problem arose from the relatively low pay received by employees of the Crown Prosecution Service. The cost savings that some anticipated would accrue from the new service (see, e.g., Jones, 1983, p. 23) did not materialize, and the need to keep costs low, coupled with the need to handle the backlog of cases, initially resulted in inexperienced back-up teams being hired, compounding all the existing problems and further compromising the efficiency and quality of the judicial system. However, it appears that the CPS was a beneficiary of the economic recession of the early 1990s as the dearth of available positions attracted a higher caliber of attorneys to the CPS, and reliance on part-time legal assistance decreased.

It has been claimed that the police have benefited from the new prosecution scheme in that they have been protected from allegations of prosecuting overzealously in the face of insufficient evidence, and have received advice from criminal attorneys on evidential and public interest questions. It has also been stated that the prosecuting body has been more objective in its pursuits (*Crown Prosecution Service, Memoranda of Evidence*, 1990, p. 9). Whether this is the case is still somewhat open to question.

■ Lawyers for the Defense

In England, as in the United States, the accused is entitled to the assistance of an attorney, either privately retained or provided at public expense, at various stages of the criminal justice process.

Right to the Assistance of Counsel

The right to the assistance of counsel begins when a suspect has been arrested and is being held in custody at the police station. The governing provision is Section 58 of the Police and Criminal Evidence Act of 1984, which is noteworthy for a number of reasons. First, despite announcing at the outset that the arrestee "shall be entitled, if he so requests, to consult a solicitor privately *at any time*" (Section 1, emphasis added), the section then describes the circumstances under which an arrestee may be denied access to a solicitor for progressively longer periods of time to a maximum of 48 hours. Second, following a crime control model of criminal justice, and contrary to the law in the United States, the more serious the offense the longer access to the solicitor may be denied—up to 36 hours for "serious arrestable offences" (Section 58[6]) and 48 hours for acts of terrorism (Section 58[13]). Third, those suspected of terrorism may be denied the right to consult *privately* (Section 58[14]). Finally, it may be noted that though a person who goes to the police station voluntarily and asks about entitlement to legal advice enjoys the same right as someone who has been arrested (Home Office, 1991, Code C, para. 3.16; Home Office, 2005, Code C, para. 3.21–22), suspects (whether arrested or not) do not have to be informed of their rights to the assistance of counsel until they arrive at the police station (Zander, 1992, p. 106).

Right to the Assistance of Counsel at Public Expense

Prior to the twentieth century, attorneys were not supplied at public expense to indigent defendants in any systematic manner, although the judge could "invite" any attorney who was present in court to assist the defendant. In addition, there was a custom that defendants on trial on indictment could produce a certain sum of money and select any of the barristers who was robed and present in court, who then had to conduct the defense (known as the "dock brief" system, which was only finally abolished in 1980).

Until very recently the English favored this system of supplying individual private attorneys to indigent defendants on a case by case basis rather than establishing a government-sponsored public defender office. The Poor Prisoners' Defence Act of 1903 provided legal aid for defendants on trial on indictment and the Poor Prisoners Defence Act of 1930 extended legal aid to those on summary trial. However, the system did not operate well because the magistrates who decided on grants of legal aid were reluctant to provide aid and the lawyers were poorly paid (Spencer, 1989, p. 474). In 1945 the Rushcliffe Committee declined to establish a centrally or locally administered legal aid department, preferring a scheme that emanated from the courts and utilized private solicitors (*Report of the Committee on Legal Aid*, 1945, pp. 23–24). Attempts to improve the situation with passage of the Legal Aid and Advice Act of 1949 did not succeed because the necessary funding was not provided. A committee was appointed in 1964 (the Widgery Committee) to review the whole system of legal aid, and the next two decades saw the passage of a series of acts outlining the organization and administration of legal aid. In 1980 the supervision of legal aid was transferred from the Home Office to the Lord Chancellor's Department. In 1988 the newly constituted Legal Aid Board took over administrative functions previously exercised by the Law Society.

The decision of whether to grant legal aid is made by the court in which the defendant appears. An application for aid may be made orally, but in practice is nearly always made on the prescribed forms to the clerk of the court. It is the clerk who makes the initial decision, but if the clerk is for some reason unable to make the decision, the judge(s) must decide. Until 1983 a defendant denied legal aid did not have a right of appeal. However, the Legal Aid Acts of 1982 and 1988 gave a defendant a "right of recourse" to a legal aid committee composed of practicing barristers and solicitors.

The decision to grant aid is determined by two tests: a means test and a merits test. The merits test questions whether it is in the interests of justice for the defendant to be granted legal aid, and follows criteria originally laid down in the 1966 Widgery Committee report and placed on a statutory basis in the Legal Aid Act of 1988 (Section 22). Any defendant who is being tried in Crown Court is automatically considered to pass the merits test. In some cases the law requires that a defendant be granted legal aid before a particular decision can be made (e.g., to send the defendant to prison for the first time). Prior to 1967 if a person was granted legal aid all of that person's legal expenses were paid. Since 1967 there has been a flexible part payments scheme with a means test that determines a defendant's fair share. A convicted defendant can also be ordered to pay part of the prosecution costs (see Section 18 of the Prosecution of Offences Act of 1985).

Research indicated that many eligible defendants did not apply for legal aid, and that there was a lack of uniformity in the granting of legal aid (Zander, 1992, p. 519). Concerns about a lack of availability of solicitors to defendants at their first court appearance led in 1972 to the establishment in Bristol of a duty solicitor scheme whereby solicitors were routinely available for consultation by defendants. Success with this program led to the introduction of similar schemes in many parts of the country. The Legal Aid Act of 1982 set up the **National Duty Solicitor Scheme**.

The duty solicitor scheme is available to all defendants. There is no means test to determine a defendant's eligibility to receive services and there is no provision for reimbursement by defendants. The solicitors, who are paid for their time at court, whether or not they work with clients, provide a safety net for criminal defendants. There are, however, limits on what they are allowed to do. They may, for example, give a defendant general advice or apply for legal aid or bail for a defendant, but cannot argue a defendant's case on a "not guilty" plea.

The duty solicitor scheme was extended by the Police and Criminal Evidence Act of 1984, which allowed for the provision of duty solicitors at police stations. In 1986 a national scheme was established to provide such assistance. Although the percentage of suspects seeking legal advice at police stations increased, it remained low (overall about 25%, with great variations among the police stations surveyed [Brown, 1989; Sanders & Bridges, 1990]). In 1991 there were duty solicitors available in 97% of England's police stations and Magistrates' Courts. Between April 1, 1991, and March 31, 1992, 549,083 clients were assisted at police stations at a cost of £71.37 per case, and 232,588 clients were served by court duty solicitors at a cost of £23.04 per case (Legal Aid Board, 1992, pp. 71–73).

This system of legal aid was revamped by the Access to Justice Act of 1999, which established a new Criminal Defence Service operating under a Legal Services Commission. The old legal aid system had been extremely expensive to run, and a major concern had been that a very small percentage of cases had consumed a large percentage of the budget. In 1996–1997, for example, 1% of the cases in Crown Court had consumed 42% of the budget (Zander, 2003, p. 577). Under the new system, potentially highly expensive cases are handled through individually negotiated contracts. Another concern with the legal aid system had been the uneven quality of service clients had received. Under the new system, solicitors' offices have to be reviewed by, and enter into a contract with, the Legal Services Commission before they can provide clients the legal services covered by the scheme.

The Criminal Defence Service began operation on April 2, 2001. The duty solicitor scheme continues to operate as before, free of charge both in police stations and at Magistrates' Courts. The scheme is, however, under the supervision of the Legal Services Commission instead of the Law Society. The decision as to whether to grant legal aid continues to be made by the court in which the defendant appears. In practice, virtually all defendants (about 95%; see Zander, 2003, p. 580) in Crown Court continue to be represented by counsel at public expense. Although the Criminal Defence Service essentially utilizes the model of providing individual private attorneys to defendants, it has established a pilot program of public defenders' offices with staff directly employed by the commission. Although some suggest that public defenders' offices can provide better quality services at a lower cost than individual attorneys, others are concerned about a lack of independence on the part of state-salaried employees. It remains to be seen to what extent this model of salaried state-employed defense counsel replaces the traditional model of using private attorneys to provide legal assistance to indigent defendants.

■ Conclusion

From the above description of the organization of the legal profession and court system in England, it can be seen that there are many differences between England and the United States.

The division of the legal profession into barristers and solicitors, with clients until very recently only allowed to approach barristers through solicitors, may appear odd. It is a system that, in theory at least, permits highly specialized expertise to be available to all types of clients, unlike in the United States where such expertise may be concentrated in highly prestigious and expensive law firms accessible to only the rich. However, it is a system that provides some duplication of work and is somewhat expensive to run. As we have seen, there has been momentum toward fusion of the legal profession, and one could hardly advocate with justification the creation of such a bifurcated system of legal representation in the United States.

The requirement that both solicitors and barristers serve a period of apprenticeship as part of the qualification process is, however, quite another matter. Such a requirement prohibits newly qualified attorneys from hanging out their shingles and working with clients without ever having worked under the guidance and supervision of experienced attorneys.

The English have opted to provide legal representation to indigent defendants by employing private attorneys to conduct such work rather than setting up, as is common in urban areas in the United States, state-administered public defenders' offices. The English have been particularly concerned about state involvement in the defense of someone who is being prosecuted for offenses against the state. The provision of adequate remuneration for these private attorneys has, however, been a constant problem. The duty solicitor scheme represents an effort of somewhat limited success to provide initial legal representation to all defendants. The legal authority that police enjoy to keep those suspected of committing serious offenses from consulting with an attorney runs contrary to the law in the United States, and provides an interesting example of how in England crime control interests sometimes take precedence over due process interests.

The historical development of the prosecution process in England left the prosecution function for many years almost entirely in the hands of the police. This concentration of power in the police caused increasing concern in the latter part of the twentieth century as faith in the reliability and integrity of the police decreased. This led in 1985 to the establishment of an independent prosecution service, the Crown Prosecution Service (CPS). Although plagued with horrendous initial problems, the CPS appears to be functioning on a more even keel, and hopefully will eventually fulfill the stated objectives of "fairness, openness and accountability, and efficiency" (Royal Commission on Criminal Procedure, 1981a, p. 127).

The English rely on the process of appointment for selection of their state prosecution and judicial officials. Concerns have been raised about the prevalence of older, white, conservative-oriented males of the upper socioeconomic classes among those appointed lay magistrates, and the highly select group from whom the higher court judges are selected. In many ways an appointment process can more readily meet these concerns than the election process so common in the United States. Indeed, it would appear preferable to improve upon the current appointment procedures, as the English are currently seeking to do, rather than move to selection by election with all its attendant ills (e.g., selection by an uninformed electorate, qualified candidates unwilling to run for office).

The overall structure of the English courts is very similar to that of the criminal courts in the United States. In some ways, however, the structure is far more flexible. The category of cases "triable-either-way" allows for certain cases to be officially resolved in either the lower or higher criminal courts. In addition, it is possible for a defendant to be tried in a Magistrates' Court and sent to the Crown Court to receive a more severe sentence than could be imposed in the Magistrates' Court. Finally, the prosecution has the ability to appeal against "unduly lenient" sentences. The extent to which these features affect the processing of cases will be examined in the following chapter.

The extensive use of lay magistrates in the lower courts is a distinctive feature of the English court system. Low comparative costs and flexibility are two major advantages of the system. Whether the quality of justice is adversely affected by such a heavy infusion of lay judges is another issue for further examination in the next chapter.

CHAPTER SPOTLIGHT

- In England there are two courts with original jurisdiction: the Magistrates' Courts and the Crown Courts.
- There are four courts with appellate jurisdiction. In ascending order they are: Crown Courts, the Divisional Court of the Queen's Bench, the Criminal Division of the Court of Appeal, and the House of Lords.
- The House of Lords' role is to be transferred to the new Supreme Court of the United Kingdom.
- Magistrates' Courts are the courts of inferior jurisdiction. They are the courts from which search warrants and summonses are issued and pre-trial bail is considered, and are responsible for trying "summary" offenses and determining whether an "either-way" offense is tried there or sent to the Crown Court.
- The maximum sentence a Magistrates' Court may impose is 12 months in jail.
- Crown Courts are the superior courts of original jurisdiction. They are responsible for trying serious indictable offenses. All Crown Court cases are tried by juries.
- In addition to the ascending levels of appeal, there is the Criminal Cases Review Commission, created in 1995 to review and investigate suspected miscarriages of justice.
- Magistrates are "lay persons," or nonlawyers. They are prominent citizens who live within their commission area, are appointed by the Crown through the Lord Chancellor, and sit in benches of at least two.
- Lay magistrates sit with a chief clerk, an experienced solicitor or barrister who provides advice on law and procedure.
- A small number of paid lawyers sit in Magistrates' Courts, primarily in London and other metropolitan areas. They hold the title of District Judge.
- Three types of judges sit in the Crown Court: High Court judges, circuit judges, and recorders.
- The most serious offenses are tried only by a High Court judge. Some second category offenses may be released to circuit judges by the High Court judge. Third category offenses may be tried by a recorder. Fourth category offenses are tried only by a circuit judge or recorder.
- The Divisional Court of the Queen's Bench sits in London and hears appeals on points of law from the Magistrates' Court and Crown Court.
- The Court of Appeal hears cases from the Divisional Court and is composed of the Lord Justices of Appeal.
- The highest court in England is the House of Lords, composed of the Lords of Appeal in Ordinary. There are 12 law lords and they typically sit on a bench of 5.
- The new Supreme Court of the United Kingdom will have 12 justices, appointed by the Crown on the recommendation of the Prime Minister.
- The Constitutional Reform Act of 2005 created an independent Judicial Appointments Commission to select judicial candidates and to promote the independence of the judiciary.

- There are two types of lawyers in England: solicitors and barristers.
- In addition to receiving academic training and having to pass qualifying examinations, all English lawyers must serve an apprenticeship prior to practicing law on their own.
- Traditionally, solicitors have handled cases in the lower courts and barristers cases in the higher courts. However, since 1990 solicitors have been able to obtain the right to appear in all courts, rather that just the Magistrates' Courts as was previously the case.
- Senior barristers and solicitor advocates may be appointed Queen's Counsel (QC). Since 2004, promotion to QC has been on merit through peer review and an independent selection panel.
- For 150 years prior to 1986, prosecutions were primarily the responsibility of the police. In 1986 the Crown Prosecution Service (CPS) was created to take over this role.
- The CPS is headed by the Director of Public Prosecutions (DPP), who is appointed by the Attorney General and is responsible for setting central policies and procedures for CPS lawyers to apply in prosecuting cases.
- The police do, however, retain the power to initiate charges against defendants, drop a case, or issue a caution, a formal, recorded warning.
- Grand juries do not exist in England. They were abolished in 1933.
- The right to counsel begins when a suspect is arrested and detained at a police station. Defendants may have either the counsel of their choice or a duty solicitor.
- Access to a solicitor may be denied in certain circumstances.
- Indigent defendants may receive financial aid to pay for an attorney, usually of their choice, through a legal aid scheme funded by the state.
- Legal aid is granted subject to a means and merits test.
- There is also a duty solicitor scheme covering all police stations and Magistrates' Courts. This is not subject to a means or merits test.
- The Legal Aid and Duty Solicitor schemes were modified in 1999 by the Access to Justice Act, which created the Criminal Defence Service operating under the Legal Services Commission. Solicitors must now meet designated criteria, enter into a contract to provide services, and be reviewed.

KEY TERMS

Attorney General (AG): Appointed by the party in power, is the chief law officer of the Crown, the titular head of the English Bar, and a member of either the House of Commons or the House of Lords. The A.G. represents the government in court in important cases, answers questions in Parliament, and provides legal advice to government departments.

Barrister: An English lawyer who has traditionally handled cases in the higher courts where he or she appears in gown and wig. Until recently, a barrister could only be hired through a solicitor.

Circuit judge: The middle ranked of High Court judges; he or she must have been entitled to appear as an advocate in either the Crown Court or the County Court, or been a recorder. Generally tries category three or four offenses.

Criminal Cases Review Commission (CCRC): Established by the Criminal Appeal Act 1995, this Commission functions as an independent public body that reviews suspected miscarriages of justice and decides if they should be referred to an appellate court because the Commission believes that the defendant either has been wrongly convicted or unjustly sentenced.

Crown Courts: The superior courts of original jurisdiction that try more serious offenses. There is, technically, one Crown Court for England and Wales, divided between 78 centers, in 6 judicial districts. The Crown Court also has appellate jurisdiction over cases decided in the Magistrates' Courts.

Crown Prosecution Service (CPS): An independent authority established by the Prosecution of Offences Act of 1985 to take over the conduct of all criminal proceedings instituted by the police.

Director of Public Prosecutions (DPP): Must be a barrister or solicitor of at least ten years' standing. Until 1986 was appointed by the Home Secretary. Now the DPP is both appointed by, and works under, the general supervision of the Attorney-General. The DPP heads the Crown Prosecution Service and is responsible for prosecuting particularly important or difficult cases and for advising chief constables.

High Court judge: The most senior of High Court judges; he or she must have had at least 10 years' advocacy experience in the High Court and have been a circuit judge for at least 2 years. Only one of the three High Court judges empowered to try category one offenses. Also tries category two offenses unless the case is released to another judge.

Inn of Court: Professional society to which a prospective barrister must be admitted. There are four Inns of Court: Inner Temple, Middle Temple, Lincoln's Inn, and Gray's Inn.

Lay magistrates: Prominent citizens appointed to administer justice in "benches" of two to seven (but normally two or three) in one of the counties of England and Wales or in one of the London Commission Areas.

Lord of Appeal in Ordinary/Law Lord: A judge of the House of Lords who is appointed by the Crown on the recommendation of the Prime Minister. Traditionally, appointment has carried with it a life peerage which has entitled the holder to sit and vote in the legislative chamber of the House of Lords. Tenure is during good behavior until age seventy-five. There are twelve Lords of Appeal in Ordinary. Law Lords will become the Justices of the Supreme Court of the United Kingdom, when this new court becomes operational.

Lord Chancellor: Senior government minister who formerly played multiple roles as: (1) head of the judiciary; (2) speaker of the House of Lords; and, (3) a member of the Prime Minister's Cabinet. As of April 2006, the Lord Chancellor is no longer head of the judiciary.

Lord Justice of Appeal: Judge who sits in the Court of Appeal. Must be a lawyer of ten years' standing or a High Court judge. Appointed by the Crown on the recommendation of the Prime Minister and holds office during good behavior until the compulsory retirement age of seventy-five. There are currently thirty-seven Lord Justices of Appeal.

Magistrates' Courts: The inferior courts of original jurisdiction, responsible for trying less serious offenses, summary offenses, and either-way offenses. They are also the courts that issue arrest and search warrants, and criminal summonses.

National Duty Solicitor Scheme: Free legal service available to defendants at police stations and Magistrates' Courts.

Recorder: The most junior of High Court judges; often employed part-time, must have been entitled to appear as an advocate for at least 10 years in the Crown Court or County Court. Normally tries category three or four offenses.

Solicitor: An English lawyer who has traditionally handled cases in the lower courts. He or she could always be hired directly by a client.

Triable-either-way: Offenses that can be tried in either the Magistrates' court or Crown Court, depending upon the severity of the specific case and the sentence likely to be imposed.

PUTTING IT ALL TOGETHER

1. What background qualifications should be required to become a judge? Should a law degree be required?

2. Should judges be appointed or elected?

3. Is it good policy to have unpaid lay judges decide cases?

4. Should practical work experience be required to qualify as a lawyer?

5. Should the two branches of the English legal profession be merged?

6. What have been the advantages and disadvantages of the English creating the Crown Prosecution Service?

7. Should defendants enjoy broader access to legal assistance in more or less serious cases?

8. For what types of cases, and at what stages of the criminal justice process, should indigent defendants be provided with legal assistance at public expense?

9. Is it desirable to establish public defenders' offices rather than relying on private attorneys to provide legal services to indigent defendants?

REFERENCES

Arthur JS Hall and Co v. Simons. (2000). 3 All ER 673.

Ashworth, D. (1987). The public interest element in prosecutions. *Criminal Law Review*, 595–607.

Bailey, S. H., & Gunn, M. J. (1991). *Smith and Bailey on the modern English legal system* (2nd ed.). London: Sweet and Maxwell.

The Bar Council. (2004). *Code of conduct.* (8th ed.). Retrieved January 29, 2006 from http://www.barcouncil.org.uk/document.asp?documentid=2811.

The Bar Council. (2006). *Bar statistics—December 2004.* Retrieved January 29, 2006, from http://www.barcouncil.org.uk/documents/BarStatsDec04.doc.

Barnard, D. (1988). *The criminal court in action* (3rd ed.). London: Butterworths.

Brown, D. (1989). *Detention at the police station under the PACE Act 1984.* London: H.M.S.O.

Criminal Cases Review Commission (CCRC). (2006a). *Case statistics.* Retrieved January 16, 2006, from http://www.ccrc.gov.uk/cases/case_44.htm.

Chapter Resources

Criminal Cases Review Commission (CCRC). (2006b). *Our role*. Retrieved January 16, 2006, from http://www.ccrc.gov.uk/canwe/canwe_27.htm.

Cross, Sir R., & Ashworth, A. (1981). *The English sentencing system*. (3rd ed.). London: Butterworths.

Crown Prosecution Service, Memoranda of Evidence. (1990). Ordered by the House of Commons, Home Affairs Committee, 1990 Session, 1989–1990, Vol. 1. London: H.M.S.O.

Crown Prosecution Service. (2005). *Annual report 2004–2005*. London: H.M.S.O.

Darbyshire, P. (1989). *English legal system in a nutshell*. London: Sweet and Maxwell.

DeGama, K. (1988). Police powers and public prosecutions: Winning by appearing to lose? *International Journal of the Sociology of Law, 16*, 339–357.

Department for Constitutional Affairs. (1998). *The future role of the Justices' clerk: A Lord Chancellor's Department consultation paper*. Retrieved January 18, 2006, from http://www.dca.gov.uk/consult/general/jc-rolfr.htm#part4.

Department for Constitutional Affairs. (2003). *National strategy for the recruitment of lay magistrates*. Retrieved January 16, 2006, from http://www.dca.gov.uk/magist/recruit/magrecruit.htm#back.

Department for Constitutional Affairs. (2004a). *Judicial appointments annual report 2003–2004*. Retrieved January 16, 2006, from http://www.dca.gov.uk/judicial/ja-arap2004/parttwo.htm.

Department for Constitutional Affairs. (2004b). *Judicial statistics: Annual report 2004*. London: TSO.

Department for Constitutional Affairs. (2005). *Regulatory impact assessment: Supporting magistrates' courts to provide justice*. Retrieved January 26, 2006 from http://www.dca.gov.uk/risk/suppmcsjust.pdf.

Department for Constitutional Affairs. (2006). *Judicial appointments in England and Wales: Policies and procedures*. Retrieved January 26, 2006, from http://www.dca.gov.uk/judicial/appointments/japp_ch2.htm.

Fitzgerald, M., & Muncie, J. (1983). *System of justice: An introduction to the criminal justice system in England and Wales*. Oxford, England: Basil Blackwell.

The General Council of the Bar. (2005). *Continuing professional development: Information pack 2005*. London: General Council of the Bar.

Home Office. (1983). *An independent prosecution service for England and Wales*. London: H.M.S.O.

Home Office. (1984). *Proposed Crown Prosecution Service*. London: H.M.S.O.

Home Office. (1991). *Police and Criminal Evidence Act 1984 codes of practice A–G: 1991 edition*. London: H.M.S.O.

Home Office. (2005). *Police and Criminal Evidence Act 1984 codes of practice A–G: 2005 edition*. London: TSO.

Jackson, R. M. (1971). *Enforcing the law*. Harmondsworth, England: Penguin Books.

Jones, P. (1983). *The costs of an independent prosecution service*. Home Office Research and Planning Unit, *16*, 21–23.

The Law Society. (2004). *Key facts 2003: The solicitors' profession*. London: The Law Society.

The Law Society. (2005a). *Continuing professional development: Law society requirements*. London: The Law Society.

The Law Society. (2005b). *Higher rights of audience.* London: The Law Society.

Legal Aid Board. (1992). *Report to the Lord Chancellor on the operation and finance of the Legal Aid Act 1988 for the year 1991–2.* London: H.M.S.O.

Lidstone, K. W., Hogg, R., & Sutcliffe, F. (1980). *Prosecutions by private individuals and non-police agencies.* (Research Study No. 10 for the Royal Commission on Criminal Procedure.) London: H.M.S.O.

Lord Chancellor's Department. (1988). The ethnic composition of the magistracy. *The Magistrate, 44,* 77–78.

Lord Chancellor's Department. (1989). *The work and organization of the legal profession.* London: H.M.S.O.

Manchester Guardian Weekly. (1989, February 5), p. 12.

National Audit Office. (1989). *Review of the Crown Prosecution Service.* London: H.M.S.O.

National Audit Office. (1997). *Press notice: Crown Prosecution Service.* London. Retrieved January 20, 2006, from http://www.nao.org.uk/pn/9798400 .htm

Office for Fair Trading. (2001). *Competition in professions.* London: H.M.S.O.

Price, J. P. (1979). *The English legal system.* Plymouth, England: MacDonald and Evans.

R. v. Metropolitan Police Commissioner ex p. Blackburn. (1986). 2 W.L.R.

Report of the Committee on Legal Aid and Advice in England and Wales. (1945). London: H.M.S.O.

Review of the Crown Prosecution Service (The Glidewell Report). (1998). London: TSO.

The Royal Commission on Criminal Procedure. (1981a). *The investigation and prosecution of criminal offences in England and Wales: The law and procedure.* London: H.M.S.O.

The Royal Commission on Criminal Procedure. (1981b). *Report.* London: H.M.S.O.

The Royal Commission on Legal Services. (1979). *Report.* London: H.M.S.O.

Runciman, Viscount. (1993). *The Royal Commission on Criminal Justice: Report.* London: H.M.S.O.

Saunders, A. (1985, January). Prosecution decisions and the Attorney General's guidelines. *Criminal Law Review,* 4–19.

Sanders, A., & Bridges, L. (1990). Access to legal advice and police malpractice. *Criminal Law Review,* 494–509.

Skyrme, T., Sir. (1979). *The changing image of the magistracy.* London: MacMillan Press.

Spencer, J. R. (1989). *Jackson's machinery of justice.* Cambridge, England: Cambridge University Press.

Richman, J., & Draycott, A.T. (1993). *Stone's Justices' Manual.* London: Butterworths.

Timmons, J. (1986). The crown prosecution service in practice. *Criminal Law Review,* 28–32.

Zander, M. (1984). *Cases and materials on the English legal system.* (4th ed.). London: Weidenfeld and Nicolson.

Zander, M. (1992). *Cases and materials on the English legal system* (6th ed.). London: Weidenfeld and Nicolson.

Zander, M. (2003). *Cases and materials on the English legal system* (9th ed.). London: Butterworths.

The Court Process

8

◼ Introduction

Now that we have examined the structure of the English court system and the officials who work in the courts, we will investigate how the courts actually operate. However, before we embark upon a detailed examination of the different stages of the court processes for both summary and indictable offenses, a brief outline will be provided.

◼ Outline of the Court Process

A defendant can be brought into the English criminal court system in one of three ways: as a result of being served a summons, after being arrested on an arrest warrant, or after being arrested without a warrant. A summons will specify the date on which a defendant is to make his or her initial appearance in court. An offender who has been arrested may be held for a limited amount of time by the police before being brought to court. For nonarrestable offenses this can be up to 36 hours (Criminal Justice Act of 2003, Section 7, which increased the previous limit of 24 hours). For arrestable offenses the period of detention may be as long as 96 hours, provided that a magistrate's approval has been obtained after 36 hours of detention (Police and Criminal Evidence Act of 1984, Sections 41–44). After being charged, a defendant is then to be brought before a magistrate "as soon as practicable" (generally within 24 hours, but possibly only within 48 or 96 hours depending on holidays and court sittings; Police and Criminal Evidence Act of 1984, Section 46).

Summary Offenses

The case may be decided at the initial appearance before the magistrates if it involves a summary offense and the defendant decides to plead guilty. If the maximum penalty for the offense is 3 months' imprisonment or less, the defendant may arrange not to appear in court and plead guilty by mail. If the offense is a summary offense and the defendant decides to plead not guilty, a trial date will be set. The magistrates will decide on the pretrial status of the defendant, namely whether the defendant will remain free on bail or be incarcerated prior to trial. If the offense is one that is **triable-either-way**, and (1) the magistrates determine that it is appropriate to try the case summarily and (2) the defendant consents to a summary trial, the case is processed as a summary case. If either of these conditions is not satisfied, then the case is processed as an indictable offense.

The trial of a summary offense takes place before the magistrates, who are the arbiters of both fact and law. If convicted, the defendant is sentenced by the magistrates, subject to the provision that the magistrates may send a case involving a triable-either-way offense to the Crown Court for sentencing if they believe that the defendant deserves a more severe sentence than they are legally empowered to impose. A presentence report (previously called a social enquiry report) is required before certain sentences (e.g., a community service order) can be imposed.

Indictable Offenses

If the case involves an indictable offense, or a triable-either-way offense which is to be tried on indictment, the case will have to be sent to Crown court for resolution. At the initial appearance before the magistrates a date used to be set for committal proceedings. At these proceedings the prosecution had to prove that there was a **prima facie** case for the defendant to answer. This meant that the prosecution had to provide some evidence to show that it could prove all of the elements required for conviction and that the defendant would have to rebut this evidence. Nowadays, however, as discussed later in this chapter, committal proceedings have been abolished for indictable offenses and reduced to paper proceedings for triable-either-way offenses. At Crown court the case will be tried before a jury unless the defendant decides to plead guilty. If convicted, the defendant will be sentenced by the judge.

■ Special Features of the English Court Process

Perhaps the most striking aspect of the English court process to an observer from the United States lies in its formality and the special attire (robes and wigs) worn by the judges and barristers in Crown Court. It is believed by some that the effect is to provide distance between the defendant and the court officials and underscore the solemnity and seriousness of the proceedings. To others the robes and wigs represent archaic theatrical attire that may exert a prejudicial effect on the interests of justice by unduly intimidating the naive by, for example, making truthful falteringly delivered testimony appear false, and appearing ridiculous to the criminally sophisticated. To the critics there is irony in the fact that in the House of Lords the Lords of Appeal in Ordinary dispense justice dressed in lounge/business suits. As more solicitors gain the right of audience in the Crown Court, it may be anticipated that the momentum for change will grow.

Tied in with this formality is the total absence of cameras and sketch artists from the English courtroom. Indeed, the English news media are entitled to present only the barest of details of cases that are *sub judice* (currently under trial): the names of the defendants and the charges. Thus, the English avoid the high level of media attention afforded in the United States to high profile cases, like those involving O. J. Simpson and Michael Jackson, and do not face the types of problems of prejudicial pretrial publicity that are encountered in the United States.

Another noticeable feature of the English court process arises from the courtroom layout. Unlike the defendant in the United States, who is seated next to his or her attorney at a table parallel to the prosecution table, the defendant in an English court is placed alone in the dock. In addition to hampering effective client-attorney interaction, the setup highlights the defendant, who is distanced from the other participants in the courtroom, and may exert a psychological impact on both the defendant and courtroom observers. The Runciman Commission recommended that in the future the dock be made as comparable as possible to other parts of the court, not be "unnecessarily large," and "be positioned as close as possible" to the defendant's attorneys (Runciman, 1993, p. 143). These recommendations have not been implemented. However, it should be noted that when the defendant is a juvenile, she or he sits with the defense team and neither the barrister nor the judge wears robes or wigs.

■ Pretrial Release

As a signatory to the European Convention on Human Rights, England must abide by its provisions. As Ashworth and Redmayne note, there are three fundamental principles in those provisions and the related case law that apply to the decision of whether to release a defendant prior to trial (**pretrial release**): (1) that there always be "a presumption of liberty and a presumption of innocence"; (2) that the decision-makers "avoid stereotypical reasoning and [to] assess each case individually"; and (3) that there should be "care to impose the least restrictive regime on a defendant pending trial" (2005, pp. 208–209).

As discussed earlier in the chapter, anyone arrested by the police can be held in police custody for only a limited period of time. If the arrest has been made pursuant to a warrant, the warrant will state whether the arrestee is to be held in custody or released on police bail. If the arrest has been made without a warrant, the custody officer must decide whether the arrestee should be released. The presumption is in favor of release except in certain circumstances such as when the arrestee's name or address cannot be ascertained, or the custody officer has reasonable grounds to believe the arrestee will fail to appear in court, or if released will interfere in the administration of justice (Police and Criminal Evidence Act of 1984, Section 38). Until 1994, the police could not place any conditions on release. The Criminal Justice and Public Order Act of 1994, however, empowered them to do so (Section 27). Available data had shown that about 5% of defendants are held in police custody until the first court appearance (Home Office, 1992a, p. 194). Research conducted some 6 months after passage of the Criminal Justice and Public Order Act of 1994 indicated a 2.5% decrease in the percentage of defendants maintained in police custody, but a far greater decrease (9.6%) in the percentage granted unconditional release (Raine & Willson, 1997, p. 597). The Criminal Justice Act of 2003 granted the police the additional power to place suspects on bail without first taking them to the police station ("street bail"; Section 4). Although suspects placed on "street bail" must be required to attend a police station, no other condition can be imposed on their release.

An arrestee who has been denied police bail must be brought before a Magistrates' Court as soon as is practicable, and in any case no later than the first sitting after he or she was charged (Police and Criminal Evidence Act of 1984, Section 46).

At the first appearance in Magistrates' Court, the issue of pretrial release is raised once more unless the case is resolved by the magistrates at that sitting. Available data show that over 80% of summary offenses, and about 40% of indictable offenses, are resolved at this stage (Ashworth & Redmayne, 2005, p. 219). As with police bail, the presumption is in favor of pretrial release. Under the 1976 Bail Act, an accused is to be granted bail unless "the court is satisfied that there are substantial grounds for believing that the defendant, if released on bail . . . would—(a) fail to surrender to custody, or (b) commit an offense while on bail, or (c) interfere with witnesses or otherwise obstruct the course of justice . . ." (Sched. 1, Part 1, Section 2).

The bail decision is made by the magistrates, who must consider the nature of the offense, the defendant's background, the defendant's performance under previous grants of bail, the strength of the prosecution's case, and the likely sentence. Bail may be granted subject to conditions (e.g., that the defendant reside at a particular address or report daily to a police station), but a rationale must be given for the imposition of any condition. Unlike the situation in the United States, bail does not involve the defendant either posting money or forfeiting money if he or she fails to appear. However, a surety can still be required (Bail Act of 1976, Section 3), in which case a third party may forfeit money if the defendant fails to appear. Failure to appear constitutes an arrestable offense (Sections 6–7). A limited number of bail hostels have been established, generally under the management of the Probation Service, to accommodate defendants who are homeless or require some supervision while on bail. Concerns have been raised, however, that these hostels may be used to accommodate those who would otherwise have been bailed, rather than as an alternative to pretrial incarceration (Pratt & Bray, 1985).

A great deal of concern has been voiced about crimes committed by defendants who are on bail. Research conducted in England has shown that about 10% to 21% of those released on bail commit further offenses while on bail (Morgan & Henderson, 1998, p. 44; Morgan & Jones, 1992, p. 43). This is in line with the rearrest rate of 16% reported for felony defendants released in 2000 in the 75 largest counties in the United States (Maguire & Pastore, 2001, p. 456). Whether these reoffending rates are acceptable depends upon one's philosophy on the balance to be struck between releasing defendants who reoffend while on bail and incarcerating defendants who could be safely released. Concern about reoffending led to legislation that requires magistrates to give reasons when they grant bail against police objections to defendants facing murder, manslaughter, or rape charges (Criminal Justice Act of 1988, Section 153). Additional legislation in 1994 provided that those charged with these offenses not be granted bail if they had already been convicted of such an offense (Criminal Justice and Public Order Act of 1994, Section 25). The Criminal Justice Act of 2003, however, softened this provision a little and allows bail to be granted if "the court is satisfied that there is no significant risk of his committing an offence while on bail" (Section 14).

The vast majority of those granted bail turn up for court. Available data show that only 8% of those bailed for summary offenses, and 17% of those bailed for indictable offenses, fail to appear at subsequent proceedings (Home Office, 2004, p. 84). These failure-to-appear rates compare favorably to those in the United States (22%, for example, for felony defendants released in 2000 in the 75 largest counties in the United States; Maguire & Pastore, 2001, p. 456).

If the magistrates deny bail they must state the reasons for doing so. A defendant who has been denied bail may appeal to another bench of magistrates if there has been a change in circumstances, to a High Court judge in chambers, or to the Crown Court.

Those denied bail in England suffer the same disadvantages as those denied bail in the United States: disruption of employment and family ties, lessened access to defense attorneys with potential detrimental consequences to the preparation of a defense, and confinement in correctional institutions under worse conditions than those experienced by convicted offenders.

Keeping a defendant in custody prior to trial is costly for the state and exerts a heavy demand on prison resources because a great deal of time has to be spent in processing these short-term prisoners and transporting them to court. In

1993 a private firm, Group Four, was employed for the first time to transport prisoners. Unfortunately a spate of escapes during the first weeks of operation marred this attempt at privatization. The private firm Reliance currently provides prisoner transport in Southwest England and Wales.

Analysis of available data shows that the percentage of defendants denied bail remained reasonably constant between 1993 and 2003, with 1% to 2% of those charged with summary offenses, and 7% to 13% of those charged with indictable offenses denied bail (Home Office, 2004, p. 79). Particularly likely to be denied bail are, not surprisingly, those charged with serious violent offenses. Thus, about 50% of robbery offenders are committed for trial in custody (Home Office, 2003, p. 134). Of concern are the substantial percentages of defendants denied bail who are subsequently acquitted or given noncustodial sentences upon conviction: 21% and 22% respectively in 2003 (Home Office, 2004, p. 83). It should be noted, however, that the 21% acquittal rate includes a large percentage of cases where the judge ordered acquittal because the prosecution was unable or unwilling to proceed (Ashworth & Redmayne, 2005, p. 267).

■ Cases Triable-Either-Way

The category of offenses triable-either-way adds an officially imposed flexibility to the English court process that is unknown in the United States. A large number of offenses are classified as triable-either-way. This category allows for a middle range type of offense to be tried as a summary offense, unless the magistrates decide that the case is too serious to be heard by them or the defendant elects to be tried by a jury at Crown Court (Magistrates' Court Act of 1980, Sections 17–22, as amended by Criminal Justice Act of 2003, Sched. 3). In addition, if after trying the case the magistrates decide, on learning of the defendant's character and prior record, that the defendant deserves a more severe sentence than they are authorized to impose, they may send the case to Crown Court for sentencing (Powers of Criminal Courts [Sentencing] Act of 2000, Section 3). Court data from 1998 through 2003 indicate that 88% of triable-either-way cases are resolved in Magistrates' Court (Home Office, 2004, p. 69).

Research conducted at eight courts in 1986 indicated that, with marked geographical variation, an average of 40% of triable-either-way cases were committed to Crown Court because magistrates declined jurisdiction and 60% because the defendant elected for a jury trial (Riley & Vennard, 1988). A 1989 study of convicted offenders conducted in five areas found these percentages reversed (Hedderman & Moxon, 1992, p. 7). The prime motivating factor behind a defendant electing for jury trial appears to be the belief that the likelihood of acquittal is greater if the case is tried by a jury at Crown Court. Ironically, although the prime reason cited by defendants whose cases were sent by magistrates for trial at Crown Court, but who would have preferred trial at Magistrates' Court, was the prospect of a lighter sentence at Magistrates' Court, almost as high a percentage of those who elected Crown Court trial thought they would receive a lighter sentence at Crown Court—a sentiment not supported by the data (Hedderman & Moxon, pp. 20–21, 27–37). Defendants clearly appear to lack trust in the magistrates, who are seen to be on the side of the police.

A large percentage of the triable-either-way cases are committed to Crown Court because the magistrates decline jurisdiction. However, many of the sentences handed out in triable-either-way cases in Crown Court are sentences that could have been imposed in Magistrates' Courts (Riley & Vennard, 1988, p. 35; Runciman, 1993, p. 85). Guidelines issued in 1990 by judges of the Queen's Bench Division suggested that either-way offenses be tried summarily unless the court found the existence of one or more aggravating factors and considered its sentencing powers to be insufficient. Examples of aggravating factors included burglary accompanied by vandalism, unlawful wounding committed with a weapon likely to cause serious injury, and a wide disparity in age between the victim and offender in an unlawful sexual intercourse offense (Practice Note, 1990).

The trend has been to reclassify triable-either-way offenses downwards to summary offenses in order to lighten the load on the Crown Court and keep down court costs. It has always been considerably more expensive to try a case in Crown Court than in Magistrates' Court (£3,100 as opposed to £295 in 1988–89, for example; Hedderman & Moxon, 1992, p. 38). In the Criminal Justice Act of 1988, common assault, unauthorized taking of a motor vehicle, criminal damage above £400 and below £2,000, and driving while disqualified were all reclassified as summary offenses. It was estimated that this reclassification resulted in a 5% reduction of persons tried in Crown Court (Home Office Research and Statistics Department, 1992, p. 3).

To further the likelihood that magistrates would try cases themselves, unless they considered their sentencing powers to be inadequate, a new procedure (**plea before venue**) was introduced in 1997. This procedure allowed defendants to enter their plea before the magistrates made the mode of trial decision (Criminal Procedure and Investigations Act, Section 49). This resulted in an immediate reduction in the number of defendants committed for trial at Crown Court, but an increase in the number committed for sentencing (Home Office, 2003, p. 62). This procedure was strengthened by a 2002 Practice Direction that created a presumption in favor of summary trial unless the case contained an aggravating factor, such as the infliction of substantial injury or high property loss (currently £5,000 or more), and the magistrates considered their sentencing powers to be insufficient (Practice Direction, 2002, pp. 943–945). In addition, an incentive to accept summary trial has been provided for defendants. The Criminal Justice Act of 2003 allows a defendant to request the magistrates to indicate whether they are likely to impose a custodial or noncustodial sentence if the defendant pleads guilty (Sched. 3). In cases where the magistrates indicate they are likely to impose a noncustodial sentence, the defendant is likely to agree to summary trial rather than risk conviction and a custodial sentence in Crown Court. If the defendant does go ahead and plead guilty, the magistrates are bound to impose a noncustodial sentence.

A matter of continuing concern has been the high percentage of defendants who opt for trial at Crown Court and then decide to plead guilty when the case reaches the Crown Court (about 70%; Hedderman & Moxon, 1992, p. 231; Justice, 1993b, p. 5).[1] To save the state time and money it was recommended that those who plead guilty at later stages of the proceedings should be given less of a sentencing discount than those who plead guilty earlier on (Runciman, 1993, pp. 111–112; Seabrook, 1992, p. 38). Some, however, have suggested that the loss of time and money occasioned by these late guilty pleas was not so much the result of a lack of incentive for the defendant to plead guilty, as the result of inefficiency in the system, evidenced by barristers and clients meeting for the first time on the morning of the scheduled Crown Court trials and a lack of effective liaison between prosecution and defense prior to the day of the trial (Justice, 1993b, pp. 5–9). As discussed later in this chapter, the Criminal Justice and Public Order Act of 1994 stated that in passing sentence the court should take into account the stage at which the defendant indicated the intention to plead guilty (Section 48).

In 1993 the Runciman Commission recommended that the defendant should no longer have the right to insist on a jury trial in cases triable-either-way. The commission suggested that the initial decision on the mode of trial should rest jointly with the prosecution and defense. If they disagreed, the decision would be made by the magistrates (1993, pp. 87–89). This recommendation, however, encountered considerable opposition and, despite being supported by Sir Robin Auld in his 2001 report on the criminal courts, has not been enacted into law.

■ Committal Proceedings

In the United States, felony cases may be processed through both felony (probable cause) and grand jury hearings before they reach the higher courts for resolution. It has been suggested that these two stages of the court process, which are intended to test the sufficiency of the prosecution case, duplicate each other and that the closed nonadversarial grand jury hearings considered to rubberstamp the decisions of prosecutors should be abolished in favor of the open court felony hearings.

In England, the grand jury originally provided some protection against unfair prosecution, with juries refusing to indict when they considered prosecution unwarranted. However, as the prosecution process changed and committal proceedings developed in such a way that they duplicated the filtering functions of the grand jury, the grand jury became superfluous. The grand jury was abolished in 1933, and the system appears to have continued functioning without any adverse effects.

Like the U.S. felony hearings, the modern English **committal proceedings** serve as a filter on serious cases (indictable offenses) that must be sent to the higher court (the Crown Court) for trial. Committal proceedings are held in open court and provide the defense an opportunity to hear the prosecution's case and cross-examine prosecution witnesses. However, the evidentiary hurdle that the prosecution must overcome—the establishment of a prima facie case— is lower than the standard of probable cause required in U.S. felony hearings, and is nowhere near the standard of proof beyond a reasonable doubt required for conviction at trial. Consequently, the prosecution does not have to present all of its evidence. In recognition of this fact, and to save time and expedite proceedings, a number of changes were made. In 1967, a shortened form of committal proceedings, known as **paper committal** or Section 6(2) proceedings, was instituted. Under this version, a legally represented defendant who was given copies of all prosecution witness statements could agree to be committed for trial at Crown Court without the magistrates considering the evidence (Magistrates' Courts Act of 1980, Section 6[2]). Even when there were full committal proceedings, provided the defense did not object, a written statement could be submitted in lieu of oral testimony (Magistrates' Court Act of 1980, Section 102).

The introduction of paper committal resulted in the vast majority of committal proceedings being conducted in this format. A 1981 national study calculated that 92% of committal proceedings were paper committals (Jones, Tarling, & Vennard, 1985). Although this figure apparently decreased to 87% in 1986 (Lord Chancellor's Department, 1989), it rose to 93% in 1991 (Runciman, 1993, p. 89). Concerns were raised, however, about the effectiveness of committal proceedings as a filter. The 1981 study found that 88% of full and 99% of paper committals resulted in the defendant being committed to Crown Court. Of those cases committed for trial, 34% of the cases where there had been a full and 15% of the cases where there had been a paper committal resulted in the defendant being acquitted of all charges (Jones et al., p. 358).

Arguably, committal proceedings were not serving their function well and there were proposals to modify or abolish them (see, e.g., Lord Chancellor's Department, 1989; Runciman, 1993). Since 1988, committal proceedings could be bypassed in serious and complex fraud cases and sent to Crown Court by a notice of transfer, subject to the defendant's right to apply to the Crown Court for the case to be dismissed on the grounds that the evidence that has been disclosed is not sufficient for a jury to convict (Criminal Justice Act of 1987, Sections 4–6; Criminal Justice Act of 1988, Section 144). The Runciman Commission recommended that committal proceedings be abolished, but that the defendant retain the right before trial to make a submission of "no case to answer." This motion would be decided by the Crown Court if the offense were an indictable one, and by a stipendiary magistrate if it were an offense triable-either-way (Runciman, 1993, p. 90).

Subsequent legislation was enacted that abolished committal proceedings for indictable offenses (Crime and Disorder Act of 1998, Section 51), but retained them for offenses triable-either-way. However, as a result of passage of the Criminal Procedure and Investigations Act of 1996 (Section 47 and Sched. 1), since 1997 only paper committals have been conducted in these cases. Thus, in addition to abolishing the grand jury, the English have effectively removed committal proceedings as a filter of the strength of the prosecution's case.

■ The Exclusion of Illegally Obtained Evidence

One way in which the English courts have long differed from their counterparts in the United States has been in the handling of illegally obtained evidence, particularly illegally obtained physical evidence (see, e.g., Hirschel, 1984). In the United States the 1960s saw a growing trend against allowing evidence to be used in court if it was obtained in violation of any of a variety of offenders' rights.[2] In England, the rule has been that such evidence should be admitted into court provided that "it is relevant to the matters in issue" (*Karuma v. R.*, 1955, p. 203; per Lord Goddard, Chief Justice). The English have been concerned not so much about the manner in which evidence has been obtained as its probative value and whether its introduction in court is likely to unfairly prejudice the defendant. The basic rule has been that the "trial judge in a criminal trial has always a discretion to refuse to admit evidence if in his opinion its prejudicial effect outweighs its probative value" (*R. v. Sang*, 1979, p. 272).

Not surprisingly, illegally obtained verbal statements have always been far more likely to be excluded in England than illegally obtained physical evidence. Involuntary confessions—that is, confessions "obtained as a result of a threat or promise held out by a person in authority"—have been excluded for centuries (Zander, 1992, p. 405). Under the Police and Criminal Evidence Act of 1984, a confession is admissible if "relevant," provided that it was not obtained by "oppression" or in circumstances that render it "unreliable" (Section 76).

Until passage of the Police and Criminal Evidence Act of 1984, except for confessions, illegally obtained evidence was very rarely excluded. Illustrative of the operation of English law until this act is the case of *Jeffrey v. Black* (1977). In that case, after arresting Mr. Black for stealing a sandwich from a public house, the police conducted an illegal search of his lodgings and discovered cannabis. At Mr. Black's trial for possession of illegal drugs the justices excluded the illegally obtained evidence. On appeal, however, it was held that although the evidence had been "irregularly obtained," "the test whether evidence was admissible was whether it was relevant" and that "the justices had erred in law in refusing to admit the evidence" (pp. 490–491).

Under the Police and Criminal Evidence (PACE) Act of 1984 judges were given statutory authority to exclude evidence if its admission "would have such an adverse effect on the fairness of the proceedings that the court ought not to admit it" (Section 78[1]). Contrary to expectations, this provision has exerted an impact on the English criminal justice system, with the courts more likely than before to exclude illegally obtained evidence.

The police role in numerous miscarriages of justice in England, such as the Guildford Four, the Birmingham Six, and the Maguire Seven, had raised questions about whether the courts should more closely scrutinize the methods by which the police secure evidence. In order to succeed in having evidence excluded it appears that the defense needs to "establish that a significant and substantial breach of the [PACE] rules or other impropriety has occurred" and that "it affects the fairness of the proceedings which is sufficiently serious as to require that the court exclude the evidence" (Zander, 2003, p. 459). In *R. v. Nathaniel* (1995), for example, DNA identification evidence, which had been instrumental in securing the defendant's conviction for rape, was excluded on appeal because it had been taken 4 years earlier in connection with another case and should have been destroyed. Despite this change in orientation, the English courts still lean toward the inclusion of illegally obtained evidence in court, provided that it is reliable and its introduction would not unduly affect the fairness of the proceedings. In *R. v. Khan* (1996) the House of Lords allowed the introduction of improperly obtained tape recordings, noting that: "Under English law, there was in general nothing unlawful about a breach of privacy and the common law rule that relevant evidence obtained by the police by improper or unfair means was admissible in a criminal trial" (p. 558). Interestingly, it has been suggested that the courts are particularly disinclined to exclude evidence in cases involving "professional" criminals (Doherty, 1999). Many in the United States might applaud this approach because "the criminal is (not) to go free because the constable has blundered" (*People v. Defore*, 1926, p. 587).

■ Guilty Pleas and Plea Bargaining

As in the United States, the majority of criminal cases in England are resolved by defendants pleading guilty. Court data have indicated that some 70% of defendants in Crown Court plead guilty (Zander, 1992, p. 280), and it has been calculated that an even higher percentage (79% to 93%) plead guilty in Magistrates' Court (Bottoms & McClean, 1976, p. 105; McConville, Sanders, & Leng, 1991, p. 149; Seifman, 1980, p. 182). Regional variations in guilty plea rates have been consistently observed, with London and the South East returning the lowest rates (Ashworth & Redmayne, 2005, p. 267; Zander, 1992, pp. 280–281).

The extent to which guilty pleas are tendered as a result of a plea bargain is open to debate. Many criminal court cases involve defendants who are clearly guilty. There is little point in these defendants contesting the charges, and they may themselves be eager to have their cases settled. Moreover, if not already aware, they will undoubtedly be made aware by their lawyers of the sentence discount generally given to those pleading guilty. This discount has been justified on the grounds that a guilty plea is an indication of remorse justifying a lesser sentence (a highly questionable proposition). It has been estimated that the discount is in the range of a quarter to a third of the expected sentence (Baldwin & McConville, 1977, p. 50; Thomas, 1979, p. 52; Thomas, 1992, A8-2c), and it has been recognized on appeal that a defendant who has pled guilty should not receive the maximum sentence (see, e.g., *R. v. Barnes* [1984] where the maximum sentence of seven years for attempted rape was reduced on appeal to six years). As in the United States, the incentive to plead guilty may take the form of a charge reduction, with the defendant pleading guilty to a lesser charge (to manslaughter instead of murder, for example), or the prosecution dropping one or more of multiple charges. Whereas in the United States a **plea bargain** is openly acknowledged as an agreement between the prosecution and defense whereby the defendant agrees to enter a guilty plea in exchange for the prosecution dropping or reducing a charge or recommending a particular sentence, in England the process has been more covert with no clear agreement made between the prosecution and defense.

Prior to the introduction of the Crown Prosecution Service (CPS) in 1986, the potential role of the police in inducing guilty pleas was formidable. Responsible for charging and prosecuting as well as investigating criminal offenses, the police could keep a suspect in custody and use the threat of denial of police bail accompanied by denial of access to legal representation, as well as the promise of a more lenient outcome to the case if the defendant pled guilty, in order to induce a defendant to plead guilty. Even after the establishment of the CPS it was observed that the CPS usually did not question deals obtained by the police (McConville et al., 1991, p. 166).

Although the English had thus far refused to formalize the plea-bargaining process, they had long insisted that the judiciary oversee discussions and decisions regarding a defendant's plea. Guidelines issued in the case of *R. v. Turner* (1970) stated that there must be freedom of access between attorneys and the judge, but that because justice should be administered as much as possible in open court, attorneys should use such right of access only when "really necessary" (p. 1098). All discussions had to take place with both defense and prosecution attorneys present, and the content of the discussions had to be disclosed to the defendant. Finally, although the judge should never indicate the specific sentence he or she was inclined to impose, or that a more severe sentence was likely to be imposed as a result of the defendant pleading not guilty, the judge could state that whether the defendant pled guilty or not guilty, the sentence imposed upon conviction would take a particular form (e.g., imprisonment or probation).

It was observed that the *Turner* guidelines discouraged many judges from seeing attorneys. Moreover, interviews with defendants indicated that, despite the *Turner* guidelines, many defendants were unaware of the existence and/or the nature of negotiations being carried out on their behalf (Seifman, 1980). Defense attorneys were also criticized for pressuring clients to plead guilty (Baldwin & McConville, 1977). Finally, it was clear that informal plea-bargaining still continued between defense attorneys and police and prosecution officials (Seifman, p. 186).

To cope with the pressures exerted by rising caseloads, and to avoid the inconvenience occasioned by last minute guilty pleas, many Magistrates' Courts set up pretrial case reviews. In these reviews the prosecution and defense examined areas of dispute and very often settled the case: 48% of anticipated contested hearings in Nottingham Magistrates' Court, for example, resulted in either a guilty plea or the prosecution offering no evidence (Baldwin, 1985). This represented a step in the direction of formalizing plea-bargaining in England, a modification of the judicial process later recommended by the Seabrook report (1992, pp. 44–45), and apparently supported by a majority of Crown Court judges and barristers (Justice, 1993b, p. 1). In 1993, the Runciman Commission recommended that the *Turner* guidelines be modified and that judges be allowed to indicate the maximum sentence they would impose on the basis of the facts available to them (1993, p. 113).

The 1994 Criminal Justice and Public Order Act codified the practice of giving defendants sentence discounts by providing that, in passing sentence on a defendant who has pled guilty, the court should take into account both the stage at which the defendant indicated the intention to plead guilty and the circumstances in which this indication was given (Section 48; replaced by Section 152 of the Powers of Criminal Courts [Sentencing] Act of 2000). In addition to replacing these earlier provisions (Section 144), the Criminal Justice Act of 2003 established a Sentencing Guidelines Council (Section 167), which was mandated to issue guidelines for courts to consult when sentencing defendants (Section 172). In November 2003, the Sentencing Guidelines Council issued the consultation paper *Reduction for Guilty Plea* (2004). This paper recommended that a reduction from one third to one tenth of the sentence be given in exchange for a guilty plea. The extent of the reduction would be affected by the stage of the proceedings at which the plea was tendered, with those pleading guilty at the first reasonable opportunity receiving a sentence discount of one third and those pleading guilty after the trial has begun receiving a discount of only one tenth of the sentence (pp. 4–5).

■ Juries

We have seen that about 95% to 98% of English criminal cases are dealt with at Magistrates' Court and that some 70% of cases that go to Crown Court are resolved by a guilty plea. Thus, less than 1% of cases are heard by a jury. However, the right to trial by jury is one of the most hallowed of rights, and provides important lay involvement in the criminal justice process.

Until 1974, eligibility for jury service was governed by property qualifications, which were originally highly restrictive in nature but by then were out of date. Under the Jury Act of 1974 people are eligible for jury service if: (1) they are between the ages of 18 and 65 (later raised to 70 by the Criminal Justice Act of 1988); (2) they are on the electoral register; and (3) they have been a resident of the United Kingdom for at least 5 years since the age of 13. Though these qualifications were intended to be highly comprehensive, it was estimated that they missed about 20% of those from the Commonwealth countries and 20% of those between the ages of 21 and 24 (Zander, 1992, p. 419).

As in the United States, certain categories of persons have been disqualified from jury service, while others have been deemed ineligible or excused of right. Disqualified from serving are those who have been imprisoned or received a suspended sentence or community service order within the past 10 years, those who have been placed on probation within the last 5 years, and mentally disordered persons. Permanently disqualified are those who have been sentenced to a term of imprisonment of 5 years or more (Juries Act of 1974, Sched. 1; Criminal Justice Act of 2003, Section 321). From 1988 until 2001 the police made random inspections of jury lists to check for the presence of disqualified persons. In 2001 a Jury Central Summoning Bureau (JCSB) was established to coordinate the jury process, and this agency now conducts record checks of prospective jurors. It is a criminal offense for a person to serve on a jury knowing that he or she is disqualified (Criminal Justice Act of 1967, Section 14[3]).

Until passage of the Criminal Justice Act of 2003, judges, persons involved in the administration of justice, and members of the clergy were ineligible for jury service. In addition, members of the Houses of Commons and Lords, full-time members of the armed forces, doctors, dentists, and other members of the medical profession were excused of right. In 1993, the Runciman Commission recommended that clergymen no longer be ineligible (p. 132), and in 2001 Lord Justice Auld, in his review of the criminal courts, recommended that no one be automatically ineligible or excusable from jury service. The Criminal Justice Act of 2003 responded to these recommendations by removing the provisions granting the right to be automatically ineligible or excusable from jury service. Only full-time members of the armed forces retain the right to be excused, but this is subject to the proviso that the commanding officer certify that "it would be prejudicial to the efficiency of the service if that member were to be required to be absent from duty" (Section 321).

If summoned, a prospective juror can apply to the JCSB for deferral or to be excused for personal reasons (e.g., illness, child care needs, forthcoming examinations). If service is deferred, the prospective juror will be offered another date within a year. From April 2002 to March 2003 the JCSB summoned approximately 480,000 prospective jurors. A total of 25% were excused from jury service (Department for Constitutional Affairs, 2003). A prospective juror denied a deferral or excusal from jury service may appeal the decision to the head of the JCSB and ultimately to a judge at the appropriate court.

Since 1981, jury panels have been randomly selected by computer from the electoral register. The jury for a particular case is selected by ballot in open court from the jury panel that has been summoned to court (Jury Act of 1974, Section 11). About 20 prospective jurors are brought into court for each case. Unlike the situation in the United States, there is no standard questioning of prospective jurors (i.e., the process of voir dire does not exist). The panel may be challenged on the grounds that it was unfairly chosen (a more likely possibility prior to random computer selection), but such challenges have been "extremely rare" (Spencer, 1989, p. 396) and there has not, in fact, been a successful challenge since the nineteenth century (Enright & Morton, 1990, p. 54).

The attorneys in the case are given only the names and addresses of the prospective jurors, though until 1973 they also were informed of the prospective jurors' occupations. Both the prosecution and defense may challenge a prospective juror for cause. However, a juror can only be questioned to establish a challenge for cause if prima facie grounds for such a challenge have been made, and it has been alleged that this rule has "rendered the challenge for cause almost obsolete" (Enright & Morton, 1990, p. 56).

In addition, the prosecution may, without giving any reason, ask a prospective juror to "stand by for the Crown." If this is done, the prospective juror stands to the side and does not serve unless a jury cannot be empaneled without his or her inclusion, a situation that seldom arises. According to guidelines issued by the Attorney General, this right is exercised sparingly and is used to obtain a competent jury, not one that is well disposed to the prosecution (Practice Note, 1988). Occasionally the prosecution engages in the controversial practice of **jury vetting**, in which more detailed checks are made of potential jurors' backgrounds. Guidelines issued by the Attorney General limit such vetting to cases involving national security and terrorists, and require the consent of the Attorney General (Practice Note, 1988).

Until 1988, the defense had a complementary right to a certain number of peremptory challenges whereby a prospective juror was dismissed without the defense having to show cause. These challenges could be used to increase the likelihood of obtaining minority representation on the jury. However, over the years the number of challenges to which the defense was entitled was gradually reduced until the right was completely abolished by Section 118 of the Criminal Justice Act of 1988.

The overall objective is to obtain an impartial jury, and the ultimate responsibility for achieving this goal rests with the judge, who is bound to excuse anyone who is unfit or incompetent to serve. Occasionally judges themselves have undertaken some questioning of prospective jurors or asked them to disqualify themselves if they are members of certain (e.g., racial) groups or hold strong (e.g., anti-black) views. However, the judge does not have the legal authority to intervene in the selection of a jury in order to obtain a racial balance (*R. v. Ford*,

1989). The view is that, in general, impartiality is to be assumed. The Runciman Commission, however, recommended that "in the exceptional case where compelling reasons can be advanced" the prosecution or defense be able to apply to the judge for "the selection of a jury containing up to three people from ethnic minority communities" (1993, p. 133).

Until 1968 there had to be unanimity among the 12 members of the jury for a defendant to be convicted. Concerns about "jury nobbling" (bribery) and the time and expense occasioned by hung juries led to the introduction of majority verdicts under the Criminal Justice Act of 1967. Now a defendant can be convicted if 10 of the 12 jurors vote for conviction. Because alternate jurors are not used in England, there is provision for conviction if 10 of the remaining 11 jurors, or 9 of the remaining 10, vote for conviction (Section 13). If there is a hung jury, then, as in the United States, the defendant can be retried.

It has been estimated that from 1984 through 1989 about 70% of cases heard in Crown Court were resolved through a guilty plea. Great regional variations were noted in guilty plea rates, with London consistently having the lowest, and the North East, Midland, and Oxford circuits the highest rates. Of the contested cases, about 50% (or 15% of all Crown Court cases) resulted in conviction. The 15% of Crown Court cases that resulted in acquittal were somewhat evenly divided between jury and nonjury judge-initiated acquittals. Ironically, the percentage of nonjury acquittals, generally an indication of weak prosecution cases, increased since the introduction of the CPS, constituting 53% of such acquittals in 1988 and 54% in 1989. Most noticeable was the increase in the percentage of cases in which judges discharged defendants, usually after the prosecution offered no evidence. Finally, between 12% and 14% of jury verdicts were calculated to be majority verdicts (Zander, 1991).

The jury system has been under attack on a number of grounds. It has been alleged that juries are prone to unjustly acquit guilty defendants. It has been suggested that jurors often serve unwillingly, do not pay close attention to the details of the case, and may be confused by legal technicalities. Concern has also been raised about the lack of representativeness of juries, in particular the underrepresentation of minorities. On the other hand, it has been suggested that it is inappropriate for randomly selected juries to decide certain types of cases. Thus, it has been proposed that complex fraud cases be heard by a special tribunal composed of a judge and two lay members with expertise in complex business transactions (Roskill, 1986, pp. 147, 149), a proposal reiterated by Lord Auld in his 2001 report.

Reforms have been instituted to meet some of the concerns raised about the jury system. The introduction of majority verdicts in 1967 was, for example, in large measure a response to the concern about jury nobbling. Both the Jury Act of 1974 and the Criminal Justice Act of 2003 broadened the population base of those eligible for jury service.

Research has shown that some of these concerns may not be as warranted as they appear. There is some evidence that juries may be more inclined than judges to acquit defendants. Baldwin and McConville, for example, in their 1975–1976 study of cases in Birmingham Crown Court found that 32% of the judges had serious doubts, and 6% some doubts, about the justice of the jury acquittals in the cases over which they presided (1979, p. 46). Sympathy toward the defendant or antipathy toward the victim was a salient factor in about a quarter of the acquittals (1979, p. 49). In her study of contested wounding, assault occasioning actual bodily harm, shoplifting, and other theft/handling cases in Magistrates' and Crown Court, Vennard (1985) likewise found that, even after allowing for the nature and apparent strength of the prosecution case, the likelihood of acquittal was significantly higher in Crown Court. Juries, who were less trusting than magistrates of confessions, eyewitness accounts, and prosecution testimony, might, she surmised, "demand a stronger body of evidence from the prosecution in order to be convinced of the defendant's guilt" (p. 146). Her study would appear to support the findings of studies conducted with mock juries (see, e.g., McCabe & Purves, 1974) that indicate that juries take their responsibilities seriously and decide on the basis of the evidence and their sense of equity. In their study of 361 jurors who had completed jury service at six Crown Court centers, Matthews, Hancock, and Briggs (2004) found that, after serving on a jury, 57% had a more positive view of the jury system, 62% considered juries "very important, essential, integral or necessary," and 32% considered juries "quite important" to our system of justice (p. 4). The overall conclusion would appear to be that there is "little evidence of perverse verdicts" (Zander, 1992, p. 449). And the involvement of lay participants in the decision-making process of the criminal justice system would seem to be warranted on the grounds that despite, or perhaps because of, its imperfections, as in the United States, it substitutes the judgment and biases of 12 people for one, and provides protection against the overzealous prosecutor and/or the compliant or eccentric judge.

Sentencing

If the defendant has pled guilty, or been found guilty after trial, he or she will have to be sentenced accordingly. In Magistrates' Court the sentencing decision is made by the bench of lay magistrates or by the District Judge, in Crown Court by the High Court judge, circuit court judge, or recorder. The defendant may be sentenced on the day on which he or she has been convicted, or sentencing may be deferred.

The English tradition has been to leave a great deal of discretion to the sentencer, with statutes only setting the maximum penalties that can be imposed. In choosing the sentence the sentencer may be assisted by a presentence report prepared by a probation officer, and must have obtained one before imposing a custodial sentence for summary or triable-either-way offenses (Criminal Justice Act of 1991, Sections 3[1]–[2]). Unlike the situation in the United States, the prosecutor makes no formal sentence recommendation to the court. However, the defendant is given the opportunity to ask the court to take mitigating factors into consideration.

Under English law, the defendant is also entitled to ask the court to "take other offenses into consideration." This practice was instituted at the beginning of the twentieth century to prevent the police from being able to hold offenses over a defendant's head, and arrest a defendant for another offense just as he or she was being released from a correctional institution. The effect of asking for other offenses to be taken into consideration is that the defendant can be given a more severe sentence than would have been imposed as a result of conviction of the original offense. However, the sentence imposed cannot exceed the maximum penalty permissible for the original offense. It is possible for this practice to be abused by the police in a desire to clear offenses or by the defendant to escape more severe punishment or to shield another person who is in fact guilty of the offense the defendant has asked be taken into consideration. A defendant may ask for a large number of offenses to be taken into consideration. There is one reported case of 513 other offenses, and 80 or 90 "are by no means rare" (Cross, 1981, p. 88).

Reasons for the choice of a particular sentence have not normally been given unless the sentence imposed has been unduly harsh. When reasons have been given they have usually been rather general in nature. Under the Criminal Justice Act of 1991 the court is required to state reasons when "a custodial sentence for a term longer than is commensurate with the seriousness of the offense" is imposed (Section 2[3]). The court has the authority to vary the sentence within 28 days after it has been imposed, with the Crown Court, but not the Magistrates' Court, entitled to increase the sentence.

A defendant may appeal a sentence. From Magistrates' Court this is to the Crown Court, or by way of case stated on a point of law to the High Court. From Crown Court it is to the Criminal Division of the Court of Appeal. The prosecution may, through the Attorney General, appeal cases in which "unduly lenient" sentences are considered to have been imposed. On appeal the court may impose any sentence that could have been given by the trial court (Criminal Justice Act of 1988, Section 36).

The sentencing decision in England, as in the United States, is complicated by the competing sentencing rationales of retribution/just deserts, rehabilitation, deterrence (both specific and general), and incapacitation. Although the first of these rationales focuses on making the sentence commensurate with the crime, the others are in different ways more concerned with decreasing the general level of crime in society, and allow for the sentence to be tailored more to characteristics of the defendant.

Numerous studies have found regional variations in sentencing practice that cannot be attributed solely to the characteristics of the offenses or the offenders appearing before the different courts (see, e.g., Hood [1962, 1972]; Tarling, Moxon, & Jones [1985]; and Barclay [1991] with regard to Magistrates' Courts, and Barclay [1991] and Hood [1992] with regard to Crown Court). Magistrates appear to develop individual bench sentencing practices (see, e.g., Hood, 1972; Tarling et al., 1985). From the mid 1960s until passage of the Criminal Justice Act of 2003, the Magistrates' Association issued nonbinding guidelines for the Magistrates' Courts. However, because these guidelines "had no official or legal status" (Thomas, 2002, p. 486), and were in no way binding on the magistrates, they appear to have had little impact.

The **<u>Criminal Division of the Court of Appeal</u>**, through the appellate process, provides judicial overview of sentencing decisions. The court will generally only modify a sentence if it is not within the appropriate range. Annually about

7% of defendants sentenced in Crown Court appeal, and in about 25% of the appeals the sentence is quashed (set aside) or varied. This compares with less than 1% who appeal from Magistrates' Courts, with about 50% of those sentences quashed or varied on appeal (Home Office, 1992b, p. 142; Home Office Research and Statistics Department, 1994a, Tables 3, 4). The decisions of the Court of Appeal are reported in the Criminal Appeal Reports (Sentencing). Periodically, the court issues sentencing guidelines for specific offenses. In *R. v. Billam* (1986), for example, the court set out a series of guiding principles and enunciated what it considered appropriate sentences for rapes committed in a variety of circumstances. David Thomas's *Current Sentencing Practice* used to provide regularly updated synopses of sentencing law and practice. In 1998, a Sentencing Advisory Panel was established by the Crime and Disorder Act of 1998 to assist the Court of Appeal in developing its sentencing guidelines (Section 81).

Concerns about the sentencing disparity that had resulted from the individualized sentencing model associated with the rehabilitative rationale led to a renewed focus on proportionality between the offense and the sentence. **Sentencing disparity** occurs when different sentences are handed out to different offenders who have similar criminal backgrounds, have been convicted of similar offenses, and would be expected to receive similar sentences. A 1990 government white paper urged that proportionality be the guiding criterion in setting sentences (Home Office, 1990, p. 5). The 1991 Criminal Justice Act embodied this principle in law by declaring that, when imposing a custodial sentence, the court should impose a term that "is commensurate with the seriousness of the offence" (Section 2[2][a]). Concerns about prison overcrowding provided the rationale for a presumption against imposition of a custodial sentence unless "the offence, or the combination of the offence and one other offence associated with it, was so serious that only such a sentence can be justified for the offence" (Section 1[2][a]). In the case of violent or sexual offenses, however, the guiding criterion was whether such a sentence was needed to "protect the public from serious harm" (Section 1[2][b]). In 1993 the limitation to considering only "one other offence" before imposing a custodial sentence was removed (Criminal Justice Act of 1993, Section 66[1]).

In the United States, a major determinant of the sentence an offender receives has been the offender's prior record of criminal convictions. In England, an offender's prior record has not generally had as great an impact on the sentence. The Criminal Justice Act of 1991 (Section 29) did not allow the court to take previous convictions into account unless they presented an aggravating factor with regard to the current offense (e.g., showed racial motivation in unprovoked assaults). This statutory provision came under immediate attack, and the Criminal Justice Act of 1993 replaced it with a new provision that allows the court to "take into account any previous convictions of the offender or any failure of his to respond to previous sentences" (Section 66[6]).

A thorough review of the English sentencing system was undertaken in the late 1990s, and in 2001 the government issued its recommendations in *Making Punishments Work: Report of a Review of the Sentencing Framework for England and Wales* (Home Office, 2001a). The overall goal of the proposed sentencing framework was to establish a sentencing system that would "improve outcomes, especially by reducing crime, at justifiable expense" (p. ii). Of prime concern was the "unclear and unpredictable approach to persistent offenders" (p. ii). Although the proportionality principle was to be retained, and prison was to be reserved for those for whom no other sentence would be severe enough, the panel recommended clearer and more predictable accounting of previous convictions in passing sentence. "Simply requiring them to be 'taken into account' was 'not enough'" (p. iii). Sentencing guidelines were to be developed that took into account both the severity of the current offense and prior convictions. Because short custodial sentences did not give the Prison Service sufficient time to work effectively with offenders, noncustodial sentencing options were to be considered when a custodial sentence of 12 months or more was not required to "meet the assessed needs of crime reduction, punishment and reparation" (p. iii).[3]

The Criminal Justice Act of 2003 provides the general structure of the current English sentencing system. At the outset, the statute recognizes the competing rationales that affect sentencing, instructing courts to consider "punishment," the "reduction of crime," "the reform and rehabilitation of offenders," "the protection of the public," and "reparation by offenders to persons affected by their offences" when passing sentence (Section 142). In assessing the seriousness of the current offense, the court is instructed to consider both the offender's culpability and the harm that was inflicted, that was intended to be inflicted, or that might "forseeably" (sic) have been inflicted (Section 143). Although, as discussed earlier, guilty pleas would result in a reduction in sentence (Section 144), aggravating factors motivating the commission of the offense, such as a racial bias (Section 145) or hostility based on the sexual orientation of the victim (Section 146), would result in an increase in sentence. Previous convictions are to be considered aggravating factors justifying a more severe sentence (Section 143[2]).

In order to promote greater consistency in sentencing, the Criminal Justice Act of 2003 established a **Sentencing Guidelines Council** (Section 167). This eight-person council is composed of the Lord Chief Justice, who serves as the council's chair, and three judicial and four nonjudicial members. Although the judicial members must include a circuit judge, a District Judge from the Magistrates' Court, and a lay justice, the four nonjudicial members must "have experience" in policing, criminal prosecution, criminal defense, or "the promotion of the welfare of victims of crime," with at least one member having experience in each of the listed areas (Section 167). Taking over the roles previously exercised by the Court of Appeal and the Magistrates' Association, the Sentencing Guidelines Council is mandated to issue guidelines for all courts to consult when sentencing defendants (Section 172). The **Sentencing Advisory Panel,** established by the Crime and Disorder Act of 1998 to assist the Court of Appeal in developing its sentencing guidelines (Section 81), now acts in an advisory capacity to the council. The council must consult with the Sentencing Advisory Panel before issuing or modifying sentencing guidelines.

Sentencing Patterns

There are some marked differences in the extent to which different sentences are imposed in England and in the United States. **Table 8-1** presents information on the sentences imposed in English courts for the years 2000 through 2002.

| Table 8-1 | Percentage of Offenders Receiving Various Sentences for Indictable and Non-Motoring Summary Offenses in England by Court, 2000–2002 |

Magistrates Courts

Sentence	2000 Offense Indict:	2000 Sum	2001 Offense Indict:	2001 Sum	2002 Offense Indict:	2002 Sum
Absolute Discharge	1	1	1	1	1	1
Conditional Discharge	19	9	18	10	17	9
Probation[a]	15	3	15	3	14	3
Fine	31	80	30	78	29	78
Community Service Order[b]	9	2	8	2	8	2
Other Community Penalties[c]	8	2	10	4	11	3
Young Offender Institution/Detention	4	*	4	*	4	*
Imprisonment Fully Suspended	*	*	*	*	*	*
Unsuspended	10	1	10	1	11	1
Otherwise Dealt With	3	2	4	2	4	2
N (000s)	743.3		693.9		748.3	

Crown Courts

Absolute Discharge	*	*	*	*	*	*
Conditional Discharge	3	20	3	21	3	21
Probation	10	13	12	15	11	14
Fine	3	16	3	14	3	13
Community Service Order	11	19	11	21	11	19
Other Community Penalties	4	4	4	5	7	6
Young Offender Institution/Detention	14	6	14	5	14	6
Imprisonment Fully Suspended	3	1	2	1	2	1
Unsuspended	49	18	49	17	49	17
Otherwise Dealt With	2	3	2	3	2	4
N (000s)	73.6		71.2		75.4	

* = Less than 1%.

[a]Since April 2001 called community rehabilitation orders. Includes supervision orders (probation orders for juveniles).

[b]Since April 2001 called community punishment orders.

[c]Includes curfew orders, reparation orders, action plan orders, drug treatment and testing orders, attendance center orders, referral orders, and combination orders.

| Table 8-1 | Percentage of Offenders Receiving Various Sentences for Indictable and Non-Motoring Summary Offenses in England by Court, 2000–2002—continued |

Sentence	All Courts					
	2000		2001		2002	
	Offense		Offense		Offense	
	Indict:	Sum	Indict:	Sum	Indict:	Sum
Absolute Discharge	1	1	1	1	1	1
Conditional Discharge	16	9	15	10	14	9
Probation	13	3	14	3	14	3
Fine	25	80	24	77	23	78
Community Service Order	9	2	9	2	8	2
Other Community Penalties	6	2	9	4	11	2
Young Offender Institution/Detention	6	1	6	*	6	*
Imprisonment Fully Suspended	1	*	1	*	1	*
Unsuspended	19	1	19	2	20	1
Otherwise Dealt With	3	2	3	2	3	2
N (000s)	816.9		765.1		823.7	

* = Less than 1%

Sources: Home Office, 2001b, Table 7.1; 2002, Table 7.1; 2003, Table 4.1.

Perhaps the most striking feature of the table is the high percentage of convicted offenders who receive fines as their sentence. Around 80% of offenders convicted of summary offenses and about 25% of those convicted of indictable offenses are fined. As would be expected, fines are imposed far more frequently in Magistrates' Courts than in Crown Court for both summary and indictable offenses. In the United States, in contrast, fines are imposed on convicted offenders far less frequently. In federal courts in fiscal year 2001 only 4.1% of sentenced defendants received fines as the sole sentencing disposition (Maguire & Pastore, 2003, Table 5-19). Even when combined with other sentences, it is estimated that only about 13% of federal defendants are fined (Vigorita, 2002). In state courts, it has been estimated that in the United States fines are imposed alone in 36% of the cases in courts of limited jurisdiction and in less than 10% of the cases in courts of general jurisdiction, though they are imposed with another type of penalty in 86% of the cases in courts of limited jurisdiction and in 42% of the cases in courts of general jurisdiction (Hillsman, Mahoney, Cole, & Auchter, 1987, p. 2). Although fines are used far more frequently in England than in the United States, their use has been diminishing. Analysis of court sentences for 1987 to 1991, for example, revealed that over four fifths of offenders convicted of summary offenses and over one third of those convicted of indictable offenses were fined (Hirschel & Wakefield, 1995, pp. 146–148).

The Criminal Justice Act of 1991 (Section 18) introduced a new system of unit fines in Magistrates' Courts. Under this system the amount of the fine was calculated by multiplying the number of units ascribed to the offense by a figure representing the offender's disposable weekly income. Though this system had been piloted successfully in a number of jurisdictions prior to introduction by statute nationwide, the mathematical formula used to calculate the amount of the fine was modified (with the top level on the income scale raised from £20 or £25 to £100), and a number of well-publicized cases highlighted problems with the scheme (Dyer, 1993). In one case, for example, a man was fined £1,200 for a minor littering offense. (He had, in fact, failed to disclose his income, and the magistrates had assessed the maximum amount possible, rather than cite him for contempt of court.) Facing mounting public criticism and continuing opposition from a number of magistrates, about 30 of whom had resigned in protest at the new scheme, the government suddenly announced its intention to scrap the scheme altogether, but still require magistrates to consider offenders' means when imposing fines (Travis & White, 1993). This intention was enacted into law by the Criminal Justice Act of 1993 (Section 65).

Although fines are imposed far more often in England than in the United States, probation, which since April 2001 has been called **community rehabilitation**, is used far more sparingly. Whereas about 14% of those convicted of indictable offenses are placed on probation, only 3% of those convicted of summary offenses are given probation. Interestingly, whereas those convicted of indictable offenses in Magistrates' Courts are far more likely than those convicted of summary offenses to receive probation, in Crown Court those convicted of summary offenses are more likely to receive probation than are those convicted of indictable offenses (see Table 8-1). In the United States, in contrast, in 2002 31% of those convicted of felonies in state courts, and in fiscal year 2001 13.1% of those convicted of felonies and 45.4% of those convicted of misdemeanors in federal courts, received probation (Maguire & Pastore, 2003, Tables 5-19, 5-47).

Part of the discrepancy between the percentage of offenders receiving probation in England and the United States may be attributable to the English using discharge as a sentencing option. With an absolute discharge, no further action is taken after the finding of guilt. With a conditional discharge, likewise no sentence is imposed. However, if the offender is convicted of a further offense during a specified period (which may be up to 3 years), he or she may also be sentenced for the offense for which a conditional discharge was given (Home Office, 1991, p. 19). In the years 2000 through 2002, about 20% of those convicted of indictable offenses and 10% of those convicted of summary offenses were given absolute or conditional discharges in Magistrates' Courts. In Crown Court about 20% of those convicted of summary offenses were given absolute or conditional discharges and only 3% of those convicted of indictable offenses received this sentence (Table 8-1).

As was noted above with regard to fines, there has been a decrease in the use of absolute and conditional discharges in recent years. Data depicting the sentencing of defendants convicted of indictable offenses show that from 1992 to 2002 there was a gradual decline in the use of discharges. Whereas 21% of these defendants received discharges in 1992, in 2002 this figure was down to 15% (Home Office, 2003, p. 95). The decrease in the use of fines and discharges has been compensated by increases in the use of community sentences, such as probation and community service orders, which accounted for 23% of sentencing dispositions in 1992 and 33% of sentencing dispositions in 2002 (Home Office, p. 95), and in the use of incarceration. This shift in sentencing practice, the Home Office has noted, is "reflecting a general shift upward in sentencing tariffs" (p. 80). This increase in sentencing tariffs is, in fact, evident in the sentencing guidelines. Whereas the 1993 guidelines issued by the Magistrates' Association indicated that a community sentence was the presumptive sentence for assault occasioning actual bodily harm, the 1997 guidelines indicated imprisonment as the presumptive type of sentence (Davies, Croall, & Tyrer, 2005, p. 314).

Among Europeans, the English have earned a reputation for being quick to use imprisonment as a sentencing option. Imprisonment data have consistently shown England to have one of the highest per capita imprisonment rates in Western Europe (see, e.g., Barclay & Tavares, 2000; National Association for the Care and Resettlement of Offenders, 1991). Concerns about the rising prison population, coupled with awareness that imprisonment might be being overused, led to a change in sentencing philosophy toward the end of the twentieth century. Restrictions were imposed by the Criminal Justice Act of 1982 (Section 1) on the imposition of custodial sentences on offenders under 21 years of age. As has been previously discussed, the new philosophy was most clearly embodied in the Criminal Justice Act of 1991 with its statutory presumption against the imposition of custodial sentences. The examination of 1987 through 1991 sentencing data indicates that a decline in the use of incarceration had already occurred before passage of the 1991 act (Hirschel & Wakefield, 1995, pp. 146–148). Although the percentage of defendants given prison terms remained low in 1992, the downward trend started to reverse in 1993 both in Magistrates' Courts and in Crown Court. The sharpest increases in rates of incarceration were noted for those convicted of property offenses and those with numerous prior convictions (Home Office Research and Statistics Department, 1994b). These prior convictions could now be taken into account in sentencing as a result of passage of the Criminal Justice Act of 1993 (Section 66[6]). From 1992 to 2002 there was a steady increase in the imposition of prison sentences, with 11% of all adult defendants convicted of indictable offenses receiving prison sentences in 1992 and 20% of these offenders being sent to prison in 2002 (Home Office, 2003, p. 95). For summary offenses the rate remained the same at 1% (Home Office, 1993, p. 149; 2003, p. 93). A comparison of international criminal justice statistics for 1998 indicated that England had, with 126 prisoners per 100,000 inhabitants, the second highest per capita imprisonment rate in Western Europe. The English incarceration rate was, however, well below the U.S. incarceration rate of 668 per 100,000 population (Barclay & Tavares, 2000).

Whereas only 20% of offenders convicted of indictable offenses in England were given immediate prison sentences in 2002 (see Table 8-1), in the United States 69% of those convicted of felonies in state courts in 2002, and 83.2% of those convicted in federal district courts in fiscal year 2001, were imprisoned

(Maguire & Pastore, 2003, Tables 5-19, 5-47). Thus, there is further evidence that supports the contention that in general offenders are more likely to be imprisoned in the United States than in England. This conclusion also appears to be supported in general by analyses that take into account the offense for which the offender was convicted. Thus, for example, whereas in England in 2002 76% of adults convicted of robbery and 51% of those convicted of indictable offenses of burglary were given prison sentences (Home Office, 2003, p. 94), 86% of those convicted of robbery and 72% of those convicted of felonious burglary in state courts in the United States received prison sentences (Maguire & Pastore, Table 5-47). The differences in incarceration rates have, however, been narrowing. Whereas in England there have been increases in the percentage of convicted robbers and burglars given prison sentences, in the United States the comparative rates have been decreasing. Thus, whereas in England in 1992 70% of convicted robbers and only 28% of convicted burglars were given prison sentences (Home Office, 2003, p. 94), the comparative figures for the United States for that year were 88% and 75%, respectively (Maguire & Pastore, p. 487).

A further note of caution should be given about these data. They do not take into account possible differences in the prior records of offenders convicted in England and the United States and in the seriousness of the offenses contained in a particular crime category. We have seen, for example, in Chapter 4 that robberies are far more likely in the United States than in England to be committed with a firearm. Thus, one might reasonably expect a larger percentage of robbery convictions in the United States than in England to result in imprisonment. A further issue that will be discussed in more detail later is that both the average maximum sentence and the average time served tend to be far longer in the United States than in England (see, e.g., Farrington & Langan, 1992, pp. 14–17; Langan & Farrington, 1998, pp. 30–35).

■ Miscarriages of Justice

The image of the quality of English justice began to be severely tarnished in the 1970s by a number of well-publicized **miscarriages of justice**. These miscarriages included the cases of the Guildford Four, convicted of bombing the Horse and Groom pub in Guildford in October 1974; the Maguire Seven, convicted of unlawfully possessing the explosive nitroglycerine in December 1974; the Birmingham Six, convicted of bombing two Birmingham pubs in 1974; and Judith Ward, convicted of a coach bombing in 1974.

These cases focused attention on improper police practices, the use of questionable forensic evidence, and the failure of the prosecution to disclose evidence to the defense. They also highlighted how long and difficult the path is to overturning an unsafe conviction. The Guildford Four, for example, convicted in October 1975, only had their convictions overturned in October 1989, while the Maguire Seven, convicted in 1976, had their convictions overturned in 1991.[4]

The process for investigating alleged miscarriages of justice when appeal rights have been exhausted was at that time through the Home Secretary, who had the power to refer such cases to the Court of Appeal (Criminal Appeal Act of 1968, Section 17). Concerns were raised about the orientation of the Home Office in investigating alleged miscarriages of justice, and in particular the presumption that convictions were well founded (Justice, 1993a, p. viii). Based on the consideration that the current process was "incompatible with the constitutional separation of powers as between the courts and the executive," the Runciman Commission recommended the establishment of a new body, the Criminal Cases Review Authority, "to consider alleged miscarriages of justice, to supervise their investigation if further inquiries are needed, and to refer appropriate cases to the Court of Appeal" (1993, p. 182).

As discussed in Chapter 7, a **Criminal Cases Review Commission (CCRC)** was established by the Criminal Appeal Act of 1995. This 11-member commission functions as an independent public body to review suspected miscarriages of justice and decide whether they should be referred to an appellate court because the commission believes that the defendant either has been wrongly convicted or unjustly sentenced (Criminal Cases Review Commission, 2006b). The commission started work in March 1997 and has been highly active, completing as of December 31, 2005, reviews of 7,590 cases. A total of 308 of these 7,590 cases were referred to the Court of Appeal for review of either conviction or sentence. A total of 178 of the 252 cases decided by the Court of Appeal by December 31, 2005, resulted in either the conviction or the sentence being quashed (Criminal Cases Review Commission, 2006a).

■ Conclusion

Although the general features of the initial stages of the court processes in England and the United States are similar, there are some marked differences.

The systems in both countries operate under a presumption of pretrial release, and do in fact release the vast majority of defendants awaiting trial. In England both the police and the courts are actively involved in the pretrial release process. The police may release the accused on police bail, and since 1994 have been allowed to place conditions on release. In England, unlike in the United States, defendants are released by both the police and the courts without requiring them to post money. As a consequence, there are no professional bail bondsmen working in the system. Concerns have long been voiced about the involvement of bail bondsmen in pretrial release in the United States, and the movement for unsecured release should take encouragement from the fact that the failure to appear rates in England compare favorably to those in the United States.

Another significant feature of pretrial release in England has been the establishment of bail hostels. Although limited in number, and subject to the concern that they may be used to house defendants who otherwise would have been released on bail, they constitute a potentially less expensive and more comfortable and more constructive alternative to jail.

The category of offenses triable-either-way adds a certain officially sanctioned flexibility to the court process in England. It is by no means clear how the relevant criteria should be applied in deciding when defendants should have the right to be tried by a jury, and in reclassifying certain triable-either-way offenses as summary offenses, the English can be observed to be paying heed to the factors of cost and time. The problem of defendants who opt for trial by jury and then decide to plead guilty at Crown Court has still to be solved.

England abolished the use of the grand jury in processing serious criminal cases back in 1933, and appears to have suffered no adverse consequences as a result of this action. Indeed, the utility of the surviving committal proceedings (comparable in nature to the U.S. felony/probable cause hearings) came under question, and they were abolished for indictable offenses and reduced to paper proceedings for triable-either-way offenses. With no in-court challenging by opposing counsel of the strength of the prosecution's case, as can take place in the U.S. felony/probable cause hearing, the English have effectively removed committal proceedings as a filter of the strength of the prosecution's case.

The English continue to have a different orientation toward the admissibility of illegally obtained evidence, and many in the United States would in particular applaud the argument that reliable physical evidence should be admissible at trial even if it were obtained as a result of an illegal search. Excluding such evidence allows a patently guilty offender to escape conviction, provides no benefit to the innocent victims of unlawful police searches, and may do nothing to deter police from conducting illegal searches in the future.

As in the United States, the vast majority of criminal cases are resolved by the defendant entering a guilty plea. Though plea-bargaining occurs, and many of the guilty pleas tendered are undoubtedly the result of plea bargains, the process has not been formalized to the same extent as it has in the United States. A fundamental difference between the situation in England and the United States is that in England the prosecutor has not been entitled to recommend a particular sentence and the judge has not been permitted to indicate the specific sentence he or she is inclined to impose. Thus, plea-bargaining in England has generally involved negotiations over the dropping or reducing of charges, activities that do not involve the judiciary.

This suggests that the English have been reluctant to face reality and openly embrace the existence of plea-bargaining. This situation has changed somewhat since passage of the 1994 Criminal Justice and Public Order Act, which codified the previous practice of giving defendants sentence discounts by providing that, in passing sentence on a defendant who has pled guilty, the court should take into account both the stage at which the defendant indicated the intention to plead guilty and the circumstances in which this indication was given. This reduction in sentence is justified on the grounds that, in addition to saving the court time, such a defendant is indicating remorse and consequently deserves a less harsh sentence than a defendant who is found guilty after insisting on pursuing his or her right to trial. Some would suggest, however, that this reasoning simply attempts to mask the reality that a defendant who insists on going to trial is punished more severely for doing so.

Very few criminal cases are resolved through jury trial. The symbolic importance of the existence of the right to trial by jury should not, however, be underestimated. The English place great faith in the ability of random selection of jurors to achieve impartial juries. In addition, since passage of the Criminal Justice

Act of 2003, classes of individuals who had been previously automatically ineligible or excusable from jury duty now have to provide reasons not to serve. Thus, clergymen, doctors, dentists, members of the House of Commons and House of Lords, and even people involved in the administration of justice are eligible for jury service.

Significantly, the rules governing jury trials in England seem to lean more in favor of the prosecution than they do in the United States. There is no detailed questioning (voir dire) of prospective jurors and more limited challenges for cause, with prima facie grounds having to be made before a juror can be questioned to establish such a challenge. Although the prosecution may ask a prospective juror to "stand by for the Crown," and in certain types of cases may engage in jury vetting, the defense no longer enjoys the right to a single peremptory challenge. Finally, a defendant may be convicted by a majority verdict. Many of the differences in the way jury trials are conducted in England and the United States (lack of voir dire, majority verdicts, and absence of alternate jurors, for example) expedite the court process and save the state both time and money. However, it must be questioned whether this is not accomplished at the expense of defendant rights and the overall quality of justice.

The English sentencing schema traditionally has left a great deal of discretion to the sentencer. However, there has been considerable opportunity for appellate review of sentencing. The Court of Appeals provides a broader overview of sentences than occurs in the United States, where sentences can generally only be challenged on the grounds that they infringe the Eighth Amendment's prohibition against the imposition of "cruel and unusual" punishment. Moreover, the court's jurisdiction to hear cases in which "unduly lenient" sentences have been imposed may well be envied by prosecutors in the United States. However, the guidelines and the practice directions that were periodically issued were for the most part directed at sentences imposed at Crown Court.

Concerns about sentencing disparity led to a renewed legislative focus in the early 1990s on proportionality between the offense and the sentence, and to the establishment in 2003 of a Sentencing Guidelines Council. This council, which is aided in its work by a Sentencing Advisory Panel, issues guidelines that must be consulted by all courts when sentencing defendants. Historically, an offender's prior record had not been systematically factored into the sentencing decision, with offenders simply asking the court when passing sentence to take other offenses into consideration. However, the Criminal Justice Act of 2003 provided for previous convictions to be considered aggravating factors justifying imposition of a more severe sentence.

The English are far more likely than their counterparts in the United States to fine a convicted offender, and it may be prudent for officials in the United States to consider using this potential income-generating sentencing option more frequently. Conversely, the English are less likely to use probation, often giving the offender a conditional discharge instead. Greater use of this sentencing option should also be contemplated in the United States, where viable options need to be applied in place of the much overused probation order.

Although the English have a reputation among Europeans for being imprisonment oriented, convicted offenders appear to be less likely to be imprisoned in England than in the United States. If given prison sentences, these are likely to be shorter in England than in the United States. Concerns about burgeoning prison populations have led some of the states in the United States to emulate the current English orientation toward sending fewer nonviolent offenders to prison. The issue of what constitutes an appropriate prison term for conviction of a particular offense is hard to resolve. For the English it would presumably be whatever number of years are needed to "improve outcomes, especially by reducing crime, at justifiable expense" (Home Office, 2001, p. ii).

No system of justice, finally, is immune from miscarriages of justice. The English criminal justice system has seen its share, and as a result established in 1995 a Criminal Cases Review Commission. This independent commission reviews suspected miscarriages of justice and decides whether they should be referred to an appellate court because there is reason to believe that the defendant has been either wrongly convicted or unjustly sentenced. The existence of an independent review authority can help alleviate the public unrest that often accompanies suspected injustice, and the establishment of such an authority should perhaps be considered in the United States.

CHAPTER SPOTLIGHT

- A defendant can be brought before an English criminal court in one of three ways: summons, arrest on warrant, or arrest without warrant.
- A case may be decided at the first hearing if it is a summary offense and the defendant pleads guilty. If the defendant pleads not guilty, a trial date will be set. If the offense carries a sentence of 3 months or less, the defendant can plead guilty by mail. If the case is "triable-either-way," the magistrates will decide whether it can be dealt with summarily or sent to the Crown Court as an indictment.
- The magistrates also decide whether the defendant should be given pretrial bail or be incarcerated.
- England has stringent laws that restrict the reporting of cases to the names of the defendants and the details of the charges, until after the case concludes.
- Unlike U.S. courts where the defendant sits next to his or her attorney in the court, in English courts the defendant sits alone in a separate area known as the dock. The exception to this is in juvenile cases, where the defendant sits with the defense team.
- At all stages from arrest through to trial, there is a presumption of liberty (that a defendant will be given bail) and a presumption of innocence.
- Both the police and magistrates are involved in pretrial release decisions.
- In England, bail does not involve posting any money, although a surety may be requested. Conditions may be placed on bail.
- Bail may be denied if a person accused of murder, manslaughter, or rape has prior convictions for those offenses. If bail is granted over the objections of the police, magistrates must give reasons for this.
- The magistrates normally decide "triable-either-way" cases, unless they decide the case is too serious or the defendant elects for jury trial. Defendants are often motivated to elect for jury trial by the belief that they are more likely to be acquitted. A large number who elect to go to the Crown Court plead guilty before the trial. However, they are likely to receive a more severe sentence in Crown Court.
- There have been various policy and legislative changes to encourage defendants to remain in the Magistrates' Courts and thus lighten the load on the Crown Court. This provides the government with cost savings because it is more expensive to try a case in Crown Court than in a Magistrates' Court.
- When sentencing, the courts may take into account when and where the defendant indicated an intention to plead guilty.
- In England, grand juries were abolished in 1933. There was a process of committal proceedings, which acted as a filter between the Magistrates' Court and the Crown Court. These now exist only for either-way matters, and are conducted on paper.
- England does not have an exclusionary rule, particularly for physical evidence. The rule is that evidence should be admitted provided it is relevant to the matters at issue, has probative value, and will not unfairly prejudice the defendant.

- To have any evidence excluded, the defense must persuade the court that there was a significant breach of the PACE rules or that some other impropriety occurred and this affects the fairness of proceedings so seriously as to warrant exclusion.
- England does not have formal plea-bargaining. However, there is a statutory scheme for reducing sentences based on guilty pleas, which takes into account the stage of the proceedings at which the defendant indicated that he or she would plead guilty.
- All persons between ages 18 and 70 who are registered to vote and have lived in the United Kingdom for at least 5 years are eligible for jury service. Exceptions are limited to those who have been imprisoned or received a suspended jail sentence or community service order within the past 10 years; those placed on probation in the past 5 years, and the mentally disordered. Full-time members of the armed forces may be excused if their jury service would be prejudicial to the efficiency of the military.
- In England, defense and prosecution attorneys have only very limited rights to challenge potential jurors. They may challenge for cause only if and when prima facie grounds for a challenge have been made. The judge may remove any juror who is deemed unfit to serve.
- In very limited circumstances, the prosecution may vet the jury. Such vetting is limited to matters involving national security and terrorism.
- Juries are made up of 12 people. There are no alternates. Convictions are either unanimous or by majority of 10 or 11 of the 12. If jurors are lost for some reason, the majority can be 10 of 11, or 9 of 10.
- A sentence may be imposed on the day of the conviction or subsequent to the issuance of a presentence report.
- Sentencing is based on factors of retribution, rehabilitation, deterrence, and incapacitation.
- A sentence may be appealed to a higher court.
- Fines are far more prevalent in English courts than in the United States. Probation is used less frequently than in the United States. However, England has a system of conditional and absolute discharges for people found guilty. Conditional discharges are typically contingent upon good behavior.
- England uses prison far less than the United States but far more than most of Western Europe. Prison sentences are typically longer in the United States than in England.
- A series of notorious miscarriages of justice in the late twentieth century led to the establishment of the Criminal Cases Review Commission. This commission considers whether potential miscarriages of justice should be referred back to the courts for examination.

KEY TERMS

Committal proceedings: A process by which the prosecution must show the Magistrates' Court that it has the elements of its case prior to the case being sent to the Crown Court. Serves as a filter on either-way cases that are being sent to the Crown Court for trial.

Community rehabilitation: Probation.

Criminal Cases Review Commission (CCRC): An 11-member commission that functions as an independent public body to review suspected miscarriages of justice and decide whether they should be referred to an appellate court because the commission believes that the defendant has been either wrongly convicted or unjustly sentenced.

Criminal Division of the Court of Appeal: Hears appeals from the Crown Court if conviction is unsafe or unsatisfactory, there were an error of law in a ruling, or there were a material irregularity in the conduct of the trial.

Jury vetting: A practice whereby the prosecution conducts detailed background checks on potential jurors.

Miscarriage of justice: Conviction and punishment of a person for a crime he or she did not commit.

Paper committal: A shortened form of committal proceedings where there is no hearing and argument is in the form of written pleadings only.

Plea bargain: An agreement between the prosecution and the defendant, by which the defendant agrees to plead guilty to a particular offense in return for a reduced sentence, a reduced charge, or the dropping of an additional charge or charges.

Plea before venue: In either-way cases, a process by which a defendant is invited to indicate the likely plea before the magistrates consider whether to keep the case or send the matter for trial in the Crown Court.

Pretrial release: Letting a defendant out into the community prior to trial.

Prima facie: A Latin phrase that translates as "on its first appearance," meaning in legal terms that the prosecution can show all the elements of the alleged offense and could prove them if they are not rebutted.

Sentencing Advisory Panel: Established by the Crime and Disorder Act of 1998 to assist the Court of Appeal in developing its sentencing guidelines. Now assists the Sentencing Guidelines Council.

Sentencing disparity: Difference in sentences handed out to different offenders who have similar criminal backgrounds, have been convicted of similar offenses, and would be expected to receive similar sentences.

Sentencing Guidelines Council: An authority mandated to issue guidelines for courts to consult when sentencing defendants.

Triable-either-way: Offenses that can be tried in either the Magistrates' Court or Crown Court, depending upon the severity of the specific case and the sentence likely to be imposed.

PUTTING IT ALL TOGETHER

1. Should the English dispense with the formality and the special dress (wigs and robes) that currently are part of their court process?

2. Does the absence of cameras and sketch artists from the courtroom, and the ability of the press to give only the names of the defendants and the charges in cases being decided by the courts, make it easier for the English to resolve high profile cases fairly?

3. In what circumstances should a defendant be detained prior to trial?

4. Is it helpful to have a category of offenses that is triable-either-way?

5. The abolition of the grand jury, and the more recent limitations of committal proceedings, may have made the processing of criminal cases more efficient, but have these changes undermined fairness to the accused?

6. Should all illegally obtained evidence be excluded at trial?

7. To what extent should greater sentencing discounts be given to those who plead guilty earlier in the course of criminal proceedings?

8. Should people involved in the administration of justice be allowed to serve on juries in criminal cases?

9. Is the jury system in England less fair to the accused than the U.S. system because the English do not use the process of voir dire in jury selection?

10. Is the jury system in England less fair to the accused than the U.S. system because the prosecution can ask a prospective juror to "stand by for the Crown," but the defense does not possess a complementary right to any peremptory challenges?

11. Should unanimity be required for a jury to convict a defendant?

12. Is it desirable to give the prosecution the right to appeal a sentence it considers to be "unduly lenient?"

13. Should either the nature of the current offense or the offender's prior record be paramount in determining the sentence to be imposed?

14. For what criminal offenses should fines be used as the sentencing option?

15. When should probation be used as a sentencing option?

16. What is an appropriate sentence for someone convicted of armed robbery? What factors should influence the sentence?

17. Should each state establish an independent public authority to review suspected miscarriages of justice?

ENDNOTES

1. Eighty-two percent of those convicted (at pp. 11, 23).

2. For example, the Fourth Amendment right against unreasonable search and seizure (*Mapp v. Ohio,* 367 U.S. 643 [1961]), the Fifth Amendment privilege against self-incrimination (*Miranda v. Arizona,* 384 U.S. 436 [1966]), and the Sixth Amendment right to the assistance of counsel (*U.S. v. Wade,* 388 U.S. 218 [1967]—lineup identification obtained in violation of the offender's right to the assistance of counsel).

3. Typically only half of the sentence was actually being served (Home Office, 2001a, p. iv).

4. For a more detailed examination of miscarriages of justice, see *Justice in Error* by Walker and Starmer (1993). The popular film *In the Name of the Father,* directed by Jim Sheridan, provides a gripping account of the case of Gerry Conlon, one of the Guildford Four.

REFERENCES

Ashworth, A., & Redmayne, M. (2005). *The criminal process* (3rd ed.). Oxford, England: Oxford University Press.

Auld, Lord Justice. (2001). *Review of the criminal courts of England and Wales.* London: Home Office.

Baldwin, J. (1985). Pre-trial settlement in the Magistrates' courts. *The Howard Journal, 24*(2), 108–117.

Baldwin, J., & McConville, M. (1977). *Negotiated justice—Pressures on defendants to plead guilty.* London: Martin Robertson.

Baldwin, J., & McConville, M. (1979). *Jury trials.* Oxford, England: Clarendon Press.

Barclay, G. C. (1991, November). *National and local changes in sentencing patterns at courts in England and Wales.* Paper presented at the meeting of the American Society of Criminology, San Francisco, CA.

Barclay, G., & Tavares, C. (2000). *International comparisons of criminal justice statistics 1998.* London: Home Office.

Bottoms, A. E., & McClean, J. D. (1976). *Defendants in the criminal process.* London: Routledge & Kegan Paul.

Criminal Cases Review Commission. (2006a). *Case statistics.* Retrieved January 16, 2006, from http://www.ccrc.gov.uk/cases/case_44.htm

Criminal Cases Review Commission. (2006b). *Our role.* Retrieved January 16, 2006, from http://www.ccrc.gov.uk/canwe/canwe_27.htm

Cross, Sir R. (1981). *The English sentencing system* (3rd ed.). London: Butterworths.

Davies, M., Croall, H., & Tyrer, J. (2005). *Criminal justice: An introduction to the criminal justice system in England and Wales* (3rd ed.). Harlow, England: Pearson.

Department for Constitutional Affairs. (2003). *Consultation paper: Jury summoning guidance.* London.

Doherty, M. (1999). Judicial discretion: Victimizing the villains? *International Journal of Evidence and Proof, 3,* 44–56.

Dyer, C. (1993, May 14). Most JPs reluctant to scrap unit fines. *The Guardian,* p. 2.

Enright, S., & Morton, J. (1990). *Taking liberties: The criminal jury in the 1990s.* London: Weidenfeld and Nicolson.

Farrington, D. P., & Langan, P. A. (1992). Changes in crime and punishment in England and America in the 1980s. *Justice Quarterly, 9,* 5–46.

Hedderman, C., & Moxon, D. (1992). *Magistrates' Court or Crown Court? Mode of trial decisions and sentencing* (Home Office Research Study No. 125). London: H.M.S.O.

Hillsman, S. T., Mahoney, B., Cole, G. F., & Auchter, B. (1987). *National Institute of Justice research in brief: Fines as criminal sanctions.* Washington, D.C.: U.S. Department of Justice.

Hirschel, J. D. (1984). How do the English deal with the problem of illegally seized evidence? *Judicature, 67,* 424–435.

Hirschel, J. D., & Wakefield, W. (1995). *Criminal justice in England and the United States.* Westport, CT: Praeger.

Home Office. (1988). *Criminal statistics England and Wales 1987.* London: H.M.S.O.

Home Office. (1989). *Criminal statistics England and Wales 1988*. London: H.M.S.O.

Home Office. (1990). *Crime, justice and protecting the public: The government's proposals for legislation*. London: H.M.S.O.

Home Office. (1991). *The sentence of the court: A handbook for courts on the treatment of offenders*. London: H.M.S.O.

Home Office. (1992a). *Criminal statistics England and Wales 1990*. London: H.M.S.O.

Home Office. (1992b). *Criminal statistics England and Wales 1991*. London: H.M.S.O.

Home Office. (1993). *Criminal statistics England and Wales 1992*. London: H.M.S.O

Home Office. (2001a). *Making punishments work: Report of a review of the sentencing framework for England and Wales*. London: T.S.O.

Home Office. (2001b). *Criminal statistics England and Wales 2000*. London: T.S.O.

Home Office. (2002). *Criminal statistics England and Wales 2001*. London: T.S.O.

Home Office. (2003). *Criminal statistics England and Wales 2002*. London: T.S.O.

Home Office. (2004). *Criminal statistics England and Wales 2003*. London: T.S.O.

Home Office Research and Statistics Department. (1992). *Effect of reclassification of offences in the 1988 Criminal Justice Act*. London.

Home Office Research and Statistics Department. (1994a). *Criminal appeals, England and Wales, 1992*. London.

Home Office Research and Statistics Department. (1994b). *Monitoring of the Criminal Justice Acts 1991 and 1993—Results from a special data collection exercise*. London.

Hood, R. (1962). *Sentencing in Magistrates' Courts*. London: Stevens.

Hood, R. (1972). *Sentencing the motoring offender*. London: Heineman.

Hood, R. (1992). *Race and sentencing: A study in the Crown Court*. Oxford, England: Clarendon Press.

Jeffrey v. Black. (1977). 1 Q.B. 490.

Johnson, K. et al. (2001). *Cautions, court proceedings and sentencing England and Wales 2000*. London: Home Office.

Jones, P., Tarling, R., & Vennard, J. (1985). The effectiveness of committal proceedings as a filter in the criminal justice system. *Criminal Law Review*, 355–362.

Justice. (1993a). *Annual report 1993*. London.

Justice. (1993b). *Negotiated justice: A closer look at the implications of plea bargains*. London.

Karuma v. R. (1955). A.C. 197.

Langan, P. A., & Farrington, D. P. (1998). *Crime and justice in England and Wales, 1981–96*. Washington, D.C.: U.S. Department of Justice.

Lord Chancellor's Department. (1989). *Committal proceedings: A consultation paper*. London: Home Office.

Maguire, K., & Pastore, A. L. (1994). *Sourcebook of criminal justice statistics 1992*. Washington, D.C.: Superintendent of Documents.

Maguire, K., & Pastore, A. L. (2001). *Sourcebook of criminal justice statistics 2000*. Washington, D.C.: Superintendent of Documents.

Maguire, K., & Pastore, A. L. (2003). *Sourcebook of criminal justice statistics 2002*. Retrieved March 17, 2006, from http://www.albany.edu/sourcebook/pdf.

Matthews, R., Hancock, L., & Briggs, D. (2004). *Jurors' perceptions, understanding, confidence and satisfaction in the jury system: A study in six courts*. London: Home Office Research, Development and Statistics Directorate.

McCabe, S., & Purves, R. (1974). *The shadow jury at work*. Oxford, England: Basil Blackwell.

Chapter Resources

McConville, M., Sanders, A., & Leng, R. (1991). *The case for the prosecution*. London: Routledge.

Morgan, P. M., & Henderson, P. (1998). *Remand decisions and offending on bail: Evaluation of the bail process project*. London: Home Office.

Morgan, R., & Jones, S. (1992). Bail or jail? In E. Stockdale & S. Casale (Eds.). *Criminal justice under stress* (pp. 34–63). London: Blackstone.

National Association for the Care and Resettlement of Offenders (NACRO). (1991). *Imprisonment in Western Europe: Some facts and figures*. London.

People v. Defore. (1926). 150 N.E. 585.

Practice Direction. (2002). 3 All. E. R. 904–955.

Practice Note (Guidelines on Exercise of Right of Standby and Jury Checks). (1988). 3 All E. R. 1086–1088.

Practice Note (Mode of Trial Guidelines). (1990). 1 W.L.R. 1439–1442.

Pratt, J., & Bray, K. (1985). Bail hostels—Alternatives to custody? *British Journal of Criminology, 25*, 160–171.

R. v. Barnes. (1984). Crim. L. R. 119.

R. v. Billam. (1986). 1 All. E. R. 985.

R. v. Ford. (1989). 3 All. E. R. 445.

R. v. Khan. (1996). 3 All. E. R. 289.

R. v. Nathaniel. (1995). 2 Cr. App. Rep. 565.

R. v. Sang. (1979). 3 W. L. R. 263.

R. v. Turner. (1970). 2 W. L. R. 1093.

Raine, J., & Willson, W. (1997). Police bail with conditions. *British Journal of Criminology, 37*, 593–606.

Riley, D., & Vennard, J. (1988). Triable-either-way cases: Crown Court or Magistrates' Court? *Research Bulletin, 25*, 31–36.

Roskill, Lord. (1986). *Fraud trials committee report*. London: H.M.S.O.

Runciman, Viscount. (1993). *The Royal Commission on Criminal Justice: Report*. London: H.M.S.O.

Seabrook, R. (1992). *The efficient disposal of business in the Crown Court: Report of a working party*. London: General Council of the Bar.

Seifman, R. D. (1980). Plea-bargaining in England. In W. F. McDonald & J. A. Cramer (Eds.). *Plea-bargaining* (pp. 179–197). Lexington, MA: D.C. Heath.

Sentencing Guidelines Council. (2004). *Reduction for guilty plea*. London.

Spencer, J. R. (Ed.). (1989). *Jackson's machinery of justice*. Cambridge, England: Cambridge University Press.

Tarling, R., Moxon, D., & Jones, P. (1985). Sentencing of adults and juveniles in magistrates' courts. In D. Moxon (Ed.). *Managing criminal justice: A collection of papers* (pp. 159–174). London: H.M.S.O.

Thomas, D. A. (1979). *The principles of sentencing* (2nd ed.). London: Heinemann.

Thomas, D. A. (1992). *Current sentencing practice*. London: Sweet & Maxwell.

Thomas, D. A. (2002). The sentencing process. In M. McConvillle & G. Wilson (Eds.). *The handbook of the criminal justice process* (pp. 473–486). Oxford, England: Oxford University Press.

Travis, A., & White, M. (1993, May 14). Clarke in retreat on unit fines: Quick fix for Justice Act judges attacked. *The Guardian*, p. 1.

Vennard, J. (1985). The outcome of contested trials. In D. Moxon (Ed.). *Managing criminal justice* (pp. 126–151). London: H.M.S.O.

Chapter Resources

Vigorita, M. S. (2002). Fining practices in felony courts: An analysis of offender, offense and systemic factors. *Corrections Compendium*, 27(11), 1–5, 26–27.

Walker, C., & Starmer, K. (Eds.). (1993). *Justice in error*. London: Blackstone.

Zander, M. (1991). What the annual statistics tell us about pleas and acquittals. *Criminal Law Review*, 252–258.

Zander, M. (1992). *Cases and materials on the English legal system* (6th ed.). London: Weidenfeld and Nicolson.

Zander, M. (2003). *Cases and materials on the English legal system* (9th ed.). London: LexisNexis.

The Correctional
System of England

IV

The Development of Confinement and Corrections in England

9

Chapter Objectives

After completing this chapter, you will be able to:

- Describe the early forms of punishment and how they evolved over time.
- Identify early penal reformers and their impact on the criminal justice system.
- Discuss the history of the death penalty in England and how it came to be abolished.

■ Introduction

As is true of the police and the courts, the development of the correctional system in England has a long and "colorful" history. It is also full of barbaric episodes. In fact, the practice of "correcting" criminals through confinement did not emerge in England until the turn of the nineteenth century (DeFord, 1962; Ives, 1970; Pugh, 1968). Until that time, prisons were bleak places where the primary goal was punishment. There had been efforts at reform, and certain concessions had been made to the concept of corrections; but the road to the present system of corrections in England has been rocky and strewn with the remains of individuals convicted of crimes.

In this chapter, we will look at the development of penal philosophy and practices in England within the context of Europe and in comparison with the emerging United States. A discussion of this nature requires an examination of the chronological development of punishment, confinement, societal reaction to crime, the capital punishment debate, and the emergence of the current system of corrections in England. The first task is a brief examination of the history of punishment as a foundation for the concept of confinement.

■ The Early History of Punishment

Most historians agree that it is nearly impossible to pinpoint the precise beginnings of punishment for criminals. Some believe that it began with the "blood for blood" law in 2347 B.C., established after the Great Flood (Walker, 1973). Others point to references in the Bible concerning punishments for wrongful acts,[1] and write of the many barbaric and cruel methods of inflicting punishment upon wrongdoers (DeFord, 1962; Ives, 1970; Pugh, 1968). Whatever their origin, the early methods of punishment were extremely painful and seemingly unlimited in their variations.

Early forms of punishment included fastening an individual to a tree, which became known as crucifixion. In Carthage, Phoenicia, and Assyria crucifixion was widely used; the Romans crucified only slaves, although this was later changed to include Jews. Out of respect to Jesus of Nazareth, the Christian Emperor Constantine the Great abolished crucifixion in A.D. 337 (*Encyclopedia Americana*, 2004; *Encyclopedia Britannica*, 2005; *New World Encyclopedia*, 2002). One report actually suggested that the practice was abolished in A.D. 325 due to a shortage of wood for the crosses; according to Walker, they had sentenced so many to this fate they could not keep up with demand (1973, p. 21).[2]

In the years before Christ, many of the ancient nations considered burying offenders alive an appropriate penalty for serious crimes. In fact, this method of punishment was widely used on women, in particular women who had broken their vows of chastity. Stoning was also a popular form of penalty for vestal virgins who had transgressed.

The Greeks favored the wheel as their form of punishment; the person was strapped to the wheel and whirled around until dead. In 621 B.C. Draco attempted to deter criminal activity by passage of an extremely harsh code that made even minor offenses such as stealing cabbage punishable by death (Bury, 1937, p. 172; *Encyclopedia Britannica*, 2005; Smith, 1870, p. 1072). The pillory—a device in which an individual stands with head and hands through holes in a large wooden board in a public place for a long period of time—made its debut with the Gauls as a punishment for drunkenness.

Some time before 500 B.C. the Roman king Servius Tullius is said to have made an attempt to separate the types of wrongs an individual could commit: a criminal wrong to be decided by the king, with capital punishment mandated for many of these offenses, and a civil wrong, which would be decided by a panel of citizens. After the fall of the monarchy a formal written code was established: the **Law of the Ten Tables** (two more were added in 451 B.C.). Unfortunately, only remnants of these tables remain, but it is known that they included guidelines for legal procedures, rights, and sanctions (Ives, 1970, p. 3; Lee, 1956, pp. 7–8; Morris & Rothman, 1995, p. 14; Walker, 1973, p. 23).

An interesting feature of these laws was the empowerment of common citizens, who no longer had to rely on powerful rulers to obtain redress for wrongs. For example, a man who had been robbed could search a suspected individual's house, false witnesses could be flung from cliffs, and the borrower of money was placed into the power of the creditor. In effect, these were poor men's laws, designed to set down rules for a mostly illiterate population to bring about order. For instance, under these laws a victim could often dispense his own form of punishment instead of depending on the ruling classes for justice; for example, a slave caught in the act of theft could be killed by his captor. In an elementary way, these rights suggest the early beginnings of the philosophy of fair treatment that would come to characterize society and corrections in England many centuries later. Because they codified rules of behavior for all citizenry, these laws of the Romans became a model for most Western systems of justice, and emphasized the significant role of the state in dealing with offenders (Morris & Rothman, 1995, pp. 14–16; Walker, 1973, p. 25).

Some of the penalties in ancient times were strange indeed. For example, in Sparta fat citizens could be whipped for just being obese; if men did not marry before a certain age, they could be fined and, in some instances, branded (Cartledge, 2003, p. 171; Hodkinson, 2004, p. 113). In England in approximately 450 B.C., a primitive method of ensuring conformity also existed, although there were no formal laws. Those who transgressed society's norms were drowned in quagmires (Andrews, 1980, p. 189). In other European countries offenders were sometimes placed in wooden baskets and burned. Often minor offenders were flogged and/or mutilated—hands, fingers, or ears were cut off, tongues cut out (Andrews, p. 17; Morris & Rothman, 1995, p. 33). Additionally, these punishments were intended to deter offenders and others from a life of crime.[3]

About A.D. 150 the Roman Catholic Church introduced penance as a way to atone for one's sins. This eventually led to a classification of how penitents should carry out their punishment: the weepers, kneelers, hearers, and standers. Three centuries later, in A.D. 410, the Church ordered that all who took refuge in the churches be spared from the reaches of government; this practice continued until recent times (Laurence, 1932, pp. 6–7). As the center of Christianity, Rome began to set the standards for punishment; whipping, ducking, excommunication, and flogging with slender rods all were accepted forms of punishment. By this time, beheading was the chosen form of capital punishment for high-ranking citizens (Walker, 1973).

Burning at the stake was also used. In A.D. 304 Saint Alban, convicted of heresy, became the first person to suffer this fate in England (Andrews, 1980, p. 192; Walker, 1973, p. 27). In A.D. 870 Edmund, King of East Anglia, was bound to a tree, burned, shot with arrows, and finally beheaded—all this for not renouncing the Christian faith (*Encyclopedia Britannica*, 2005). This incident occurred in the village of Bury in Suffolk County. This town is now called Bury St. Edmunds; today a cathedral and shrine commemorate the tragic torture that took place there (*Encyclopedia Britannica*, Walker, p. 28).

It has been suggested that the new and different punishments being introduced in England shortly after the birth of Christ might be viewed as progressive (Walker, 1973, p. 28). At a minimum, this progression demonstrated a concern about the crime problem in England and what could be done about it. One interesting note is that the Anglo-Saxons reduced the penalty for murder to a fine and reserved capital punishment only for crimes of theft. All other crimes, including rape and assault, were punished by fines or corporal punishments (Walker, 1973, p. 28). Currently in England, fines continue to be the most common sentence handed down at Magistrates' Court (Barclay & Tavares, 1999, p. 41). According to some historians, the first set of written English laws may have been those codified by Ethelbert I in Kent in the seventh century. These laws are unique in that Ethelbert I insisted that there be values associated with the harm caused by offenders. For example, every part of the human body was given a monetary value: an eye was valued at £2.50 or approximately $4.99 depending on current exchange rates, a foot was the same, a toenail was sixpence (9 cents). Thus, if an offender committed a physical assault such as hitting someone in the eye or breaking an arm, the offender could pay the fine corresponding to that body part instead of receiving a more severe sentence. Ethelbert I did not believe in capital punishment, because that would deplete the potential fighting force needed to defend his kingdom. Women offenders, however, were considered dispensable and accordingly were drowned (DeFord, 1962; Hibbert, 1968; Walker, 1973).

When the Danes came to England in the eighth century, they initiated much more severe punishments; for example, they preferred to throw offenders from cliffs. The Danes were the first to introduce the concept of slave-value. If a slave was killed, the killer had to pay the "man-bot," or value of that slave, to his master. In fact, every man, from the highest ranking to the lowest, was given a value (*wergild*) that became the appropriate compensation to the victim's family should he be murdered. Exceptions to this were women and criminals; they had no *wergild*, and compensation did not have to be paid in cases of their demise (Fraser, 2003, p. 35; Ives, 1970, p. 3; Laurence, 1932, p. 4).

The roots of the prison system began during the eighth century when the concepts of *infangthief* and *outfangthief* were introduced. *Infangthief* referred to the right of a landowner to bring a thief caught on his land to court. *Outfangthief* was the right of a landowner to bring a thief from outside his land to court. These imply that some sort of prison had to exist in order to hold these men until sentencing and punishment (Ives, 1970; Kenyon, 1983, p. 190; Pugh, 1968).

In the ninth century, after the emergence of the practice of holding thieves for punishment, it was common for most kings to have prisons as holding rooms where offenders would await trial and/or be punished (Pugh, 1968, p. 1). It was at this time, about A.D. 890, that the term *prison (career)* first made an appearance in the codes of laws (DeFord, 1962, p. I; Pugh, 1968, p. 1). "If a man fails in what he has pledged himself to perform he is to be imprisoned, the laws say . . . and while there is to submit himself to punishments of the bishop's devising. Should he escape . . . he is to be recaptured" (Pugh, 1968, p. 2). This is an example of the early use of prison as punishment and not simply a holding facility for offenders awaiting some other form of punishment.

Although the precise date is unknown, it was during the late ninth and early tenth centuries that the first mention of the use of **stocks** is found; they were, in all probability, the earliest form of imprisonment. Later they came to be used not only to hold offenders, but also to punish them through shame and public humiliation (Andrews, 1980, pp. 120–121; Ives, 1970, p. 1; Pugh, 1968, p. 1).

After the death of King Alfred in A.D. 901, a period of harsher physical punishments replaced the system of heavy fines. When Canute ascended to the throne in 1016, he declared that there would be no more capital punishment but rather "gentle punishments decreed for the benefit of the people" (Ives, 1970, p. 8).

However, his idea of "gentle punishments" included such practices as cutting off ears, noses, and lips; gouging out eyes; scalping; removing the tongue; and removing arms, legs, and feet. To his credit, Canute did distinguish between intentional and unintentional acts. After Canute's death in 1035, however, the executioner once again attained a position of prominence (Andrews, 1980, p. 17).

With the Norman conquest of 1066, the use of imprisonment as a punishment appears to have declined. However, the Normans brought with them the **oubliette** (from the French for "forgotten"), a deep hole in the ground where prisoners could be held for a few days before they were punished. In addition, William the Conqueror forbade that "any person be killed or hanged for any cause." However, in the next breath he stated: "Let their eyes be torn out and their testicles cut off"; this generally resulted in death (Andrews, 1980, p. 4; Ives, 1970, p. 8).[4]

Along with these serious penalties, lesser sanctions of compensation for the victim were reserved for minor infractions. Although small confinement facilities existed in castles, manor houses, and similar places throughout the land as appropriate places to punish, the use of imprisonment as a punishment appears to have declined. In fact, the Normans built prisons mainly to hold their enemies rather than criminals (Walker, 1973, p. 31).

William's prohibition of capital punishment did not last. After his death, his son Rufus initiated one capital crime: illegal killing of deer in the royal forests. Rufus himself was killed by a deer hunter and was succeeded by his brother, Henry I, who began to extend the list of capital offenses as a way to combat the severe crime problem he had inherited. He reasoned that because prisons were relatively few and his subjects were poor and could not pay heavy fines, there was only one solution for criminals: execution (Laurence, 1932, p. 4).

With the increase in the power of the king and the waning of the idea that a common man wronged by another should be compensated, offenses were regarded as breaches of the "king's peace." Thus crimes became much more important and serious penalties were decreed. New places were needed to accommodate these transgressors; King Henry II commanded at the Assize of Clarendon in 1166, "In the several counties where there are no gaols, let such be made" (H. M. Prison Service, 1988, p. 4). As the twelfth century came to a close, there were only five counties without prisons. England was beginning to undergo one of many legal reform movements. Some actions originally considered to be civil torts, such as trespassing, were now viewed as criminal because they were perceived to be offenses against the king and were punishable by incarceration (DeFord, 1962, p. 4).

In addition to criminal offenders, people who defaulted on their taxes were detained in these places not only until payment was made, but also as a form of punishment for being delinquent. It was only a small step from this to imprisonment for failure to pay debts. The Statute of Acton Burnell of 1283 and the Statute of Merchants of 1285 provided for creditors to bring action to put debtors in prison until they paid their debts. Some debtors starved to death because they had no family or friends to help pay off their debts. Others were allowed to beg while chained to posts outside the prisons, and many citizens gave alms to the prisoners to help them get out of prison (Greif, 1997).[5]

These early prisons provided the foundation for the penal philosophy and practices in England. The next era in the development of confinement and corrections, although not a pleasant picture, is crucial to an appreciation of the roots of the current correctional systems in England and the United States.

■ The Move Toward Public Displays of Punishment

In the years preceding the medieval period, it appeared as if the spiritual well-being of prisoners became more important than their physical health. Accordingly, going to prison had become an almost fashionable penalty (Walker, 1973, p. 1). By 1405 there was a shortage of prison space; Parliament decreed that all villages should be equipped with stocks to counter the increasing crime problem and lack of facilities for handling prisoners. Stocks were inexpensive and required little or no supervision as compared to prisons. Also, they were quite public and provided the peasants with entertainment, particularly when a nobleman or other member of the upper classes was being punished. Gradually, however, more prisons were built throughout the land and treatment within them became more severe. In 1423 the famous Newgate Prison was built in London. By the end of the fifteenth century, however, hanging became the preferred method of punishment. The right to hang felons was given to every town, abbey, castle, and lord. As early as 1429 a lynching took place; a woman was executed without trial by unauthorized persons, in this case other women. However, the practice of lynching did not seem to gain in popularity until the next century (Walker, 1973, p. 54).

The sixteenth century showed very little progress in matters of humanity. When Henry VIII came to power, he took an extremely hard line toward criminals and punishment. The "common criminals" were harshly treated. This harsh treatment culminated in Henry's Act of 1531—repealed in 1547—which called for offenders to be boiled alive and allowed for executions even to take place on Sundays (Andrews, 1980, pp. 198–199).[6]

When Henry VIII's son, Edward VI, succeeded, he was seen as more compassionate in his treatment of offenders. He abolished the practice of boiling people alive, but introduced mutilation as an appropriate punishment for fighting in church. In about 1552, Edward VI gave the City of London the ancient Bridewell Palace as a place where the poor, beggars, prostitutes, and nightwalkers could be kept and made to work. Eventually, other such places appeared and came to be known as "**bridewells**," or houses of correction (Ives, 1970, p. 15). Although the original burned down in 1666 (in the Great Fire of London), many of the others survived in some form and came to be synonymous with the **gaols** (a place of legal detention) of the time. They spread to every county in England until the middle to latter part of the nineteenth century. As originally intended, the bridewells were well run and gave the poor inhabitants good food, a warm bed, and pay for their work with generous amounts of whipping to ensure that they kept to task. Unfortunately, the bridewells were gradually absorbed into the prison system under the control of the sheriffs. Consequently, they came to resemble most confinement facilities of that time in their brutality and deplorable living conditions.

Examples of Medieval Punishment

After the ascension of Elizabeth I to the throne in 1558, the cruelty that came to characterize this period in the development of English corrections reached unparalleled proportions. A brief look at a few of the innovations and techniques used at this time provides a glimpse of the depths man could sink to in the name of "correcting" another's behavior.

In 1577, gaol fever (typhus) killed all 300 of the officials and citizens attending the Assize (Court) of Oxford within 40 hours (Ives, 1970, p. 17). This served to underscore the need to clean up the conditions of the prisons.

Although hanging was still the most accepted method of capital punishment, because there were large numbers of convicts awaiting execution, other methods had to be substituted. Therefore, new punishments appeared, such as the gibbet, an iron cage in which a criminal was suspended near the site of the crime and left to die. This was thought to deter others passing by as a reminder of the consequences for criminal behavior. Sometimes passersby would shoot the poor wretch to put him out of his misery (Walker, 1973, pp. 75–76).

Hanging, the preferred method of execution, was not much more humane. The convict was tied to the back of a cart and marched to the site of the gallows. The citizens in the procession would flog the criminal and pelt him with objects. One of the more famous gallows was in London at a site called Tyburn near the Marble Arch. The trip to Tyburn was quite an ordeal: the cart would stop at church for prayer, at pubs for drinks, and then on to the gallows as citizens continued to torment the convict. Due to the high volume of executions, a large beam was erected so that multiple hangings could take place complete with a paying audience sitting in bleachers specially built for the occasion. Some reports indicated that as many as 40 prisoners were hung in a day. It was so common that London became known as "the City of the Gallows" (Block & Hostetler, 1997; Walker, 1973, p. 71).

In the seventeenth century, the introduction of transportation was a consequence of the growth of urban crime and the number of criminals. The system of transportation utilized by England grew out of the old practices of banishment and exile, which had been used for many years by ancient countries.[7] Australia and America were frequent recipients of these individuals; this was thought to be as cruel a punishment as any of the others previously mentioned. At first it appeared that England had solved the problem of the disposition of criminals. However, conditions on the convict ships were deplorable and the public began to question this treatment of prisoners. Although transportation continued to be used until the mid-nineteenth century, growing disenchantment during the seventeenth century led to a call for penal reform, first by certain individuals and later by the general public.

Examples of the government's mistreatment abounded. For instance, on November 3, 1726, a woman was condemned to die for murdering her husband. The method called for the executioner to hold the woman by a rope around her neck over flames. After she had strangled to death, the executioner was supposed to lower her body into the flames. However, on this particular occasion the executioner's hands were burned and he dropped the still conscious woman into the fire to die an agonizing death. There was an outcry from the public (Laurence, 1932, p. 9; Parry, 1934, p. 51; Walker, 1973, p. 82). This and other incidents ignited public indignation and with it a call for reform, if only for more efficiency in the manner in which these harsh punishments were carried out.[8]

■ Early Penal Reformers

In the middle of the eighteenth century new penal philosophies began to emerge. In a book published in 1764, a young Italian named Cesare Beccaria challenged the deterrent value of severe punishment. He suggested that life imprisonment was a better deterrent than public executions. Beccaria's book appears to have encouraged the movement for reform. Other criminologists in the eighteenth century stressed "social responsibility" in punishing criminals. Some—for example, Lombroso and his followers—argued that criminals did not carry the full responsibility for their crimes because such behavior could well be a product of hereditary factors over which the criminal had little or no control.

These ideas provided support for the efforts of John Howard, who had begun to agitate for penal reform as early as 1755. He embraced many of the Enlightenment views shared by his counterparts on the Continent. His work on prisons became a life quest as he set out to inform officials of the horrible conditions therein. His knowledge was based upon observation and research both in England and on the Continent, especially in France.

As the number of capital offenses increased throughout England and the Continent, with executions remaining public spectacles, Howard and others began to call attention to the cruelty of these events and the disturbing numbers of young people attending and being influenced by their violence. The execution of Earl Ferrers in 1760 serves to illustrate the nature of these events. A large crowd had gathered to witness this execution for several reasons: the high standing of this individual, the ceremony surrounding this execution, and the fact that it was going to be the first test of a new method of hanging, the drop method. Previously, an individual was hoisted to hang by the neck until strangled to death. However, a new device was designed to drop the floor from underneath the criminal simultaneously with the jerking of the rope. This was to break the victim's neck and bring a swifter death. However, the device malfunctioned and the Earl was left to suffocate and choke to death (Walker, 1973, p. 92).[9]

The rationale for these vicious punishments was that pain and suffering were the ultimate deterrent. It was against this idea that John Howard fought. He supported the notion that man operates of his own free will, is rational, and can be influenced to change his behavior. Howard documented the human wastage within the prisons and in executions in his book, *State of Prisons in England 1777* (1792). This book also addressed the penal theory that there may be a link between structural prison design and the rehabilitative process, and suggested prisoners might have an environment more conducive to behavioral change if they were housed in separate cells.

Howard was not the only activist clamoring for reform. William Eden, Samuel Romilly, and Jeremy Bentham were also vociferous advocates of change (Terrill,

2002, p. 59). Eden was shouted down in Parliament in 1771 when he called for the abolition of the death penalty or, failing that, its use as a last resort. Romilly repeatedly introduced bills in Parliament calling for a reduction in the number of capital offenses and a lessening of the penalties for relatively minor crimes. Although they had only limited success, reformers began to gather momentum; some of John Howard's early ideas were beginning to gain public support. At the end of the eighteenth century, effort was being made to classify prisoners according to the crimes committed and the level of security they required. By 1812 Romilly was successful in seeing the removal of the death penalty as a sentence for sailors; but it was only a beginning. In 1819, shoplifting was still a capital offense, one of more than 200 capital offenses remaining on the books (Ives, 1970, p. 171).

Howard argued that providing clean facilities and nutritious food was a better way to treat prisoners, but also would result in more effective outcomes. Although disagreement exists among historians, Howard is often credited with being either the inventor of the single-cell design or a first ardent proponent (Ives, 1970, p. 172).[10] This design, which is very important in the history of corrections in the United States, has been utilized since the late eighteenth century.

By the turn of the nineteenth century, Parliament, reflecting the public reaction to cruel treatment of criminals, undertook a massive reform of its criminal statutes. Even Sir Robert Peel played a role in this reform. His organization of an established police force in 1829 had ramifications for the penal reforms the public was demanding. He was successful in persuading Parliament to reduce the number of capital offenses and to eliminate executions as a penalty for thefts of trivial amounts.

Eighteenth- and Nineteenth-Century Corrections in England

It is important to note that England and, for the most part, the rest of Europe had shifted from using prisons primarily to hold criminals until their punishment could be implemented to a system where the imprisonment itself became the punishment. Before we examine the many developments in English prisons of this period, two important aspects of penal practice deserve consideration. One is the practice of transportation and the other the extensive use of the hulks.

The Use of Transportation and the Hulks

As mentioned earlier, with the shortage of space in prisons and gaols, officials began to turn to the colonies as an alternative to overcrowding. This seemed an excellent approach in that England could rid herself of unwanted criminals yet have them serve as a source of labor for the voluntary colonists.

After a first attempt at sending prisoners to Sierra Leone, the most natural destination was the mainland colonies of British North America. There are reports of James I sending a hundred criminals to Virginia as early as 1610 (DeFord, 1962, p. 30). By 1662 acts of Parliament began to use the word *transportation* when discussing punishment; this practice called for vagabonds, rogues, and sturdy beggars to be transported to the plantations of the English colonies across the seas.

Transportation became more widely used, and at the end of the seventeenth century a large number of criminals were banished. By 1717 Parliament wanted to solidify the banishment of criminals and authorized the beginning of regular transportation of prisoners across the seas to America and the West Indies (DeFord, p. 23).

As the practice continued, most criminals were sold into slavery upon arrival and served their term in that manner. It is worth noting that at least they were not kept in prisons and had the opportunity to stay after finishing their terms, which many chose to do. Many became leading citizens in their new surroundings and freely intermarried with the voluntary colonists; others, however, became members of the new country's criminal class.

With the revolt of the American colonies in 1776, the whole transportation system came to a halt. This caused an immediate problem of overcrowding in the English prisons, and the government began using old large warships ("**hulks**") to house prisoners. The prisoners aboard these hulks were used for public works projects ashore during the day. Although hulks continued to be used, they were a short-term remedy. Fortunately for England, Captain Cook discovered Australia and the first boatloads of prisoners were sent to Botany Bay in New South Wales in 1777. It is amazing that any survived the journey; it took almost 8 months to travel there. Those who did survive the voyage were faced with a new set of intolerable conditions once they stepped off the boat. The land was undeveloped; they had to build their own quarters; disease was rampant; they were constantly attacked by the natives; and supplies were almost nonexistent. In spite of all this, the transportation of criminals continued in full force for nearly a century.

In 1816, Botany Bay was changed from a penal settlement into a colony, and free colonists (soldiers, families, etc.) were allowed to travel to Australia. The settlement at Botany Bay became a thriving town in which former convicts, who had completed their sentences and were now free citizens, intermingled with the new colonists. In a related development a "**ticket of leave**" system was instituted in 1837 for criminals sent to Australia. This system was a precursor to our parole system in the United States in that prisoners could be rewarded with some remission of their sentence for good behavior. This was a conditional freedom and was subject to close supervision. A large number of ticket of leave prisoners remained in Australia after serving their terms (Bartrip, 1981, pp. 150–181).

With the discovery of gold in 1851 in New South Wales, there was a rush of prospectors to the area. Then the Penal Servitude Act of 1857 abolished transportation. By the end of the nineteenth century the government was supporting only a few convicts remaining in the former colonies (often called "camps"). By this time, Australia had its own penal system and the camps were all but ghost towns, silent reminders of a past fraught with shame and suffering.

As noted earlier, when the revolt in the American colonies ended transportation as a sentence, hulks were used to house prisoners. By 1825, there were 10 hulks in use in England. The first had been moored on the river Thames near London; later, others were moored at various locations throughout the country. The conditions in these hulks were as bad as they were in the prisons. Dark, unclean, disease- and rat-infested, these were certainly not pleasant places to serve out one's sentence. Each day the prisoners were marched ashore in chains under the watchful eyes of guards who would flog and lash them for the slightest of infractions. They would go to their assigned work every day, including Sundays, to put in a full day of labor. On board each night, they had to sleep in hammocks or on mats in "spoon" fashion because they were so overcrowded there was no space to stretch out. The stench was terrible, and though the ships were supposed to be washed down each day, this rarely occurred. There was certainly nothing on board to rehabilitate these criminals; therefore, upon release the same prisoners would return again and again because they were certainly unable to find work back in society. As John Howard and other reformers began to be heard, the hulks came under closer scrutiny and the public did not like what they saw. By 1825 some real reforms had been initiated: a special hulk had been built for boys; the hulks no longer were being used as a dumping ground for the system but rather as an integrated part of the penal system; and by the early nineteenth century some efforts at treatment were undertaken, such as prison schools, libraries, and religious instruction. As transportation came to an end and more prisons were built in England, the hulks were abandoned; the last one was set afire in London in 1857 (Ives, 1970, p. 126).

■ Elizabeth Fry, Reformer

Elizabeth Fry was a Quaker who had taken an interest in the conditions at Newgate Prison in London about 1813 when, as part of her religious mission, she began to visit the prison on a regular basis. The conditions she encountered were intolerable, and Fry began to push for change. Of particular concern to her were the conditions of the women and children. She found them naked, diseased, chained, at the mercy of the male guards, and totally without hope of paying the high fees that would allow them to be released. According to most accounts, against impossible odds and the threat of personal danger to herself, Elizabeth Fry was able to bring about reforms that had been called for one hundred years earlier but had not been implemented. Food, blankets, separation of the sexes, medical treatment, employment of the prisoners, religious instruction, and classification by offense were just some of the changes she was able to initiate (Walker, 1973, p. 105). She brought about many other changes over the years, such as the establishment of several committees similar to her original **Ladies Newgate Committee**, which was committed to raising money and resources for the prisons. She even published a book, *Observations on the Visiting, Superintendence, and Government of Female Prisoners* (1847), concerning the state of prisoner treatment in England based upon her personal observations of prisons and hulks throughout the country (DeFord, 1962, p. 60).

Even as she suffered her last illness in 1843, Elizabeth Fry was working on changing the conditions at the newly opened Pentonville Prison. She died in 1845 at the age of 65, still fighting for the reforms in which she fervently believed. She is perhaps best remembered for her work with female prisoners; however, she fought hard for reforms for both male and female inmates and earned the title of the "Angel of Newgate" (DeFord, 1962, p. 59). Today, in commemoration of her efforts at penal reform, there stands a large statue of Elizabeth Fry in the entrance hall of the Old Bailey criminal court, which was built on the site of Newgate Prison.

■ Prisons of the Nineteenth Century

During the early years of the nineteenth century, England was attempting to pattern her prison practices and structures after the successes the Americans claimed to have in this area (see, e.g., Rothman, 1971, pp. 130–154). The newly appointed inspector of prisons, William Crawford, was sent to the United States in 1835 to observe the designs and treatment methods employed, which focused on the concepts of separation (individual cells and silent work) and reform. Upon his return and report to Parliament, new reforms were undertaken in England. One of the first prisons to be built and function on the American ideas of separation and reform was Pentonville Prison in northern London in 1842. It was designed based on the ideas of Jeremy Bentham along a "panoptican" format that stressed prisoner visibility. The building had single cells arranged in tiers, and the buildings, or cell blocks, resembled the spokes of a wheel, with an observational area at its center allowing the guards surveillance over the prisoners at all times.

Millbank Prison was also constructed during this period, and was intended as a test of the new penitentiary approach taken from the United States, whereby inmates worked in silence and lived in separate cells. It functioned from 1837 to 1844 in this capacity as an alternative for prisoners who were not to be transported. However, as transportation began to wind down, more prisoners were assigned to Millbank and the penitentiary experiment ceased. It became a human warehouse for convicts, eventually holding over 1,000 inmates originally designated for transportation (Tomlinson, 1981, p. 127).

Although Pentonville also had been intended as an experiment in separation, it suffered the same fate as Millbank by becoming a repository for the thousands of prisoners originally scheduled for transportation. Because of their dubious success as warehouses, Millbank and Pentonville became the unintended models for a dramatic upsurge in prison construction in England in the next few years. By 1846, 54 new American-style prisons had been opened across the country (DeFord, 1962, p. 74), and a new director of prisons who valued the concepts of separateness and hard (some would say useless) labor for inmates was chosen to oversee more construction.

With these new prisons being managed by a committee known as the Directors of Prisons, Joshua Jebb, a military man, was given the task of solving the transportation crisis and the problem of overcrowding caused by the abolition of the hulks. Two Penal Servitude Acts passed in 1853 and 1857 provided new regulations for serving time in England. Although the 1853 act allowed the continuation of transportation, only convicts previously sentenced to transportation were actually transported. By 1857 transportation ceased entirely and prison building continued. This encouraged Jebb, appointed the first Surveyor General of prisons in 1844, to proceed with his ideas of separateness and hard labor for the convicts. Many prisons were constructed during this time, and some (e.g., Pentonville and Dartmoor) are still in use today.

One provision of the two Penal Servitude Acts that the public had difficulty with was the concept of remission, or "good time" as it is known in the United

States. As we have described earlier, it had been the practice to release convicts early from transportation depending on their behavior. The same was also true of working on the hulks. Jebb had assumed that this would be true for the new prisons as well; but the public, not concerned if convicts were released early in Australia or transported across the seas, became uneasy when it became clear that this practice would put more convicts back into society at home in England (Tomlinson, 1981, p. 134). The public wanted to be sure that the convicts were adequately punished and that the punishment served as a deterrent not only to the convicts, but also to others.

Eventually, Jebb reconciled this public need with his ideas of separateness and hard labor in what has been called the "progressive stage system." In this program, originally used on the hulks, inmates received rewards for good behavior. Jebb used a sliding scale of payments to convicts with which they could purchase little luxuries, such as tea, butter, and cheese, according to their work performance. He believed that this would provide incentives for the prisoners to reform. As further incentive, the schoolteachers and chaplains would place them in different classes according to their moral conduct. For his efforts, Jebb has been referred to as the "originator and prime representative of the English convict system" (Tomlinson, 1981, p. 137).

◼ Alexander Maconochie and Walter Crofton

Although Jebb's ideas were criticized as being too soft and allowing prisoners to be rewarded for being criminals, others supported and expanded these ideas. Two such pioneers were Alexander Maconochie on Norfolk Island, 800 miles off the coast of Australia, and Sir Walter Crofton and his "Irish system."

In 1840 Alexander Maconochie began using a system of positive rewards to change behavior. The ticket of leave system utilized a series of "<u>marks</u>," or rewards, a convict could earn for good conduct, allowing a prisoner to work his way from a hard labor existence to a relatively pleasant life including visitors, writing letters, and free time. In this process, the individual would pass through five stages of increasing responsibility and freedom: (1) strict imprisonment, (2) labor on the chain gang, (3) limited freedom, (4) a ticket of leave, and (5) unconditional freedom. Maconochie believed that this process would gradually prepare a convict for release into the community, a concept that is still supported today. Although the ticket of leave system was introduced in England in the latter part of the nineteenth century, there was resistance to it and controversy over its benefit to society, and it eventually disappeared (Bartrip, 1981).

Walter Crofton practiced the marks system in Ireland in a modified form, in 1853. Building on Maconochie's ideas, he too believed a convict's behavior in prison should be linked to his eventual release. The "<u>Irish system</u>," as it became known, was a three-stage system culminating in the offender being allowed an early release with supervision and several restrictive conditions. The introduction of this concept of supervised assistance represents Crofton's contribution to the modern system of parole being used in both England and the United States (DeFord, 1962, p. 76).

■ Return to Punishment

By the late nineteenth century, England's new prisons were overflowing and were managed by a group called the Directors of Convict Prisons appointed by the Home Secretary. Despite the aforementioned reforms, the manner in which the prisons were managed was far from progressive. The directors advocated hard labor and used the governors of the prisons to enforce it. They believed that the only reason for a convict to be in prison was for punishment, and that it was appropriate to crush the spirit of the inmate through useless labor.

The Directors of Convict Prisons introduced such mindless activities as the shot drill, which involved convicts passing a 24-pound lead ball around the courtyard for hours at a time to "reduce the energy level of the prisoners." Another activity, "oakum picking," consisted of hand-separating fibers in discarded cable or rope for hours at a time. Any slacking or misbehavior during these exercises was met by a flogging at the hands of the guards (Ives, 1970, p. 205). Finally, the use of the "crank" and the treadmill was quite common. The crank consisted of a box-like apparatus with a handle that the convict continually turned; the handle would move scoops on the end of steel arms to sift sand inside a box. It was a very difficult task, and the prisoners participated in it daily; a quota of revolutions had to be achieved before the prisoner could stop. Again, the sole objective of this activity was to reduce excess energy. With the treadmill prisoners simply walked endlessly in circles to burn off excess energy. At the turn of the twentieth century, 13 treadmills and 5 cranks were still in use in English prisons (Walker, 1973, p. 124).

By the last quarter of the nineteenth century, two types of prisons had been established: central (convict) prisons and local prisons. The **central prisons** were controlled by the central government, with prisoners assigned to them primarily for penal servitude. These were the convicts serving longer sentences who would work on public works projects. The **local prisons**, which were controlled by the local authorities, were for short-term offenders, usually those with less than 2-year sentences. These prisons adopted one of two different systems of treatment. One was the **association system**, wherein prisoners worked alongside each other during the day and returned to their single cells at night; the other was the more popular **separate system** of solitary confinement with prisoners in their cells all of the time (DeFord, 1962, p. 76).

The Prison Act of 1877 ended local control over their prisons. After much study by the government, the 1877 act passed the ownership of every English prison to the Home Secretary, whether it housed convicts or local prisoners awaiting trial. Funding was to be from public funds, and the responsibility for governance of the prisons was to be by the newly formed Board of Prison Commissioners.

The first chairman of the Board of Prison Commissioners, Edmund DuCane, was to influence prison regime well into the early part of the twentieth century with his establishment of a harsh discipline code and attention to rigorous labor. Although Joshua Jebb's system may have been criticized as too soft, DuCane's regime, based on solitary confinement, was criticized as being too harsh. No longer was prison a temporary solution while waiting for transportation; no longer was it subject to public opinion or inmate protest; no longer was it subject to the whims of the Directors of Convict Prisons. DuCane's regime was to become an intricate part of the entire correctional plan of England for the next 30 years (Tomlinson, 1981, p. 144).

As the end of the nineteenth century drew near, criticism of the deplorable conditions of English prisons increased. A series of articles appeared drawing attention to the harsh conditions in the prisons; ultimately, another royal commission was formed to study the situation. After the report of the Departmental Committee of Inquiry into Prisons (the Gladstone Committee) was given to Parliament in 1895, Parliament passed the Prison Act of 1898. One of its provisions abolished excessive corporal punishment; however, its most noteworthy achievement may have been its insistence that the goal of penal philosophy should be reform and that remission should be granted for good behavior (Walker, 1973, p. 128). As examples of the achievement of these goals, England can point to the establishment in the twentieth century of separate institutions for different types of offenders by sex, age, and offense, as well as a progressive system of prerelease treatment centers and preventative detention facilities. However, there is one dilemma with which the English have wrestled throughout the development of their penal philosophy and well into the twentieth century: capital punishment.

■ The Death Penalty in England

Although abolition of the death penalty in England was considered by a select committee of Parliament as early as 1866, it was not achieved; however, *public* executions became an occurrence of the past after 1868. In addition, the efficiency of the executions had improved. Hangings were now swift, sure, and relatively error-free (Walker, 1973, p. 124).

With the abolition of the death penalty in 1908 for offenders under the age of 16 (Children Act of 1908), the focus now shifted toward removing capital punishment for women. For a short period in the early part of the century no women were sentenced to death, and no women were hanged until 1922.

By 1929, few European countries were still using capital punishment. Many states in the United States (e.g., Utah, Nevada, and Washington) had abolished capital punishment, reinstated it, and begun experimenting with techniques other than hanging—the electric chair, cyanide gas, and the firing squad to name a few. England established a Select Committee on Capital Punishment in 1929, which after extensive study and debate concluded that capital punishment served no useful purpose, and recommended a 5-year suspension to test this conclusion. This recommendation was unequivocally rejected.

At the same time, the John Howard League, which had been formed to carry on the work of the great prison reformer, was beginning to gather information from around the world on capital punishment and was making the results known to government officials. Still more countries abolished the death penalty, and England appointed a Royal Commission on Capital Punishment, which sat from 1949 to 1953. During that time all capital sentences were suspended while the commission listened to experts and criminologists from all over the world. Although aware that public opinion still favored capital punishment, the commission concluded that there was no evidence to support the death penalty as a deterrent to murder. With the persistent efforts of a new reformer, Sidney Silverman, the commission presented ample evidence and statistics to support its recommendation for the abolition of capital punishment (Home Office, 1953). Although the work of the commission was highly respected as a valid piece of research, the death penalty was retained as a legal punishment in England (Walker, 1973, p. 140). Concerns about executions in three notorious cases (involving Timothy Evans in 1950, Derek Bently in 1953, and Ruth Ellis in 1955) fueled misgivings about the propriety of state-sanctioned taking of life, and the National Campaign for the Abolition of Capital Punishment was formed. Despite its work, the zealous efforts of Sidney Silverman, and mounting opposition, executions continued, and between 1949 and 1956 more than 100 were carried out.[11] In the period between 1957 and 1964, death sentences had dwindled to 48, and only 29 convicts were actually hanged (Walker, 1973, p. 140).

This was still too many for Sidney Silverman. He continued to push for total abolition based upon his contention that, as a deterrent, a life sentence was comparable to an execution. By 1964, his efforts drew results, and the last two hangings in England took place on August 13, 1964. The Murder (Abolition of Death Penalty) Act was passed in 1965. This act, which originally suspended the death penalty for 5 years, was made permanent by affirmative resolution of both houses of Parliament in December 1969.[12]

Although there was an initial increase in the number of criminal homicides after passage of the 1965 act (from 325 in 1965 to 364 in 1966 and 414 in 1967; Home Office, 1992, p. 72), and a growth in the criminal homicide rate through the end of the twentieth century and beginning of the twenty-first century—809 criminal homicides in 2004–2005 (Coleman, Hird, & Povey, 2006, p. 51; Povey, Walker, & Kershaw, 2005, p. 42)—it is by no means clear that this was in any way occasioned by removal of any deterrent effect of the death penalty. There had been as many as 400 criminal homicides in 1952, and traditionally a majority of English criminal homicides where the victim is known to the offender are composed of killings committed as a result of quarrels, revenge, or loss of temper (see, e.g., Coleman et al., 2006, p. 51), notoriously nondeterrable circumstances. There is still considerable support for the death penalty in England, and there have been legislative attempts to reintroduce it. However, it is unlikely that there will be a death penalty in operation in England in the foreseeable future even with the changing of political control of the government from either Liberal to Conservative or Conservative to Liberal in the early decades of the twenty-first century.[13]

■ Conclusion

Our journey through the early beginnings of punishment and the development of the English correctional system provides insight into the foundations of the systems of corrections in the United States. Early punishment practices involved terrible conditions in medieval (in both senses of the word) English prisons. The courageous efforts of some reformers brought about change. With the onset of transportation and the first people banished to America, we began to see the American heritage as it developed through the stumbles and starts of the English in attempting to deal with an apparently ever-increasing crime problem. In some ways, the United States began to take the lead in penal reform, and the English assessed these practices and adopted many for their own use. In other ways, the United States seemed to be repeating English history, as can be witnessed by the fact that in the later 1980s, New York and other large cities began to use "prison ships" moored in their harbors to alleviate the overcrowding of the jails in their cities (Dinsmore & Jackson, 1994, p. D1; New York City Department of Corrections, 2006; Serrano, 2006, p. 2; *U.S. News and World Report*, 1988, pp. 11–12).

Throughout the developmental process, England and the United States struggled with the capital punishment issue and have not fully resolved it. The juxtaposition of both countries on the issue, with the United States retaining the death penalty and England having abolished it, still stirs fiery debate among the citizens of both nations.

Both England and the United States shared a similar problem over sentencing philosophy as the twentieth century unfolded. Both countries attempted to serve the dual aims of ensuring that society administers the appropriate punishment and at the same time facilitates the successful reintegration of the individual back into society. Chapter 10 will examine how England has chosen to address this problem as she has developed her current correctional system and faces the challenges of the future.

CHAPTER SPOTLIGHT

- It is nearly impossible to pinpoint the time when the punishment of criminals began. The Law of the Ten Tables is one of the earliest recorded examples of a legal code that separated criminal and civil matters with guidelines for procedures, rights, and sanctions.
- These laws also relied heavily on the empowerment of common citizens to provide redress for themselves.
- Ethelbert I may have codified the first set of English laws in the seventh century. This was largely based on a system of fines, something that remains prevalent in English criminal justice to this day.
- Later, systems of prisons for punishment, and capital and other physical punishments, were devised and implemented.
- In the thirteenth century certain civil torts were effectively criminalized, in particular unpaid debts. Debtors were put in prison until they either paid off their creditors or died. These early prisons laid the foundations for modern correctional systems in England and the United States.
- With the development of prisons came the problem of prison overcrowding and the development of alternatives such as public displays of punishment. Stocks were relatively inexpensive and provided public entertainment as well as punishment.
- Another popular alternative and method of reducing prison numbers was hanging.
- Henry VIII expanded capital punishment to permit executions on Sundays and death by boiling alive.
- Henry VIII's successor, Edward VI, abolished boiling and provided the City of London with Bridewell Palace as a place where the poor, beggars, and prostitutes could be detained and made to work. Further "bridewells" subsequently developed across the country. The bridewells were eventually absorbed into the prison system and became gaols.
- In the seventeenth century transportation became popular, and England exported criminals to far-flung parts on the empire. Australia and America were frequent recipients.
- Following the American Declaration of Independence, transportation to those colonies stopped and prisons began to overcrowd. Old large warships known as hulks were employed to house prisoners. The discovery of Australia in 1777 provided another destination for transportation.
- A system of tickets to leave, a precursor to the modern parole system, developed in Australia, where prisoners could be rewarded with early release in return for good behavior.
- The Penal Servitude Act of 1857 abolished transportation.
- Penal reform emerged as an issue in the eighteenth century. John Howard was a leading agitator for such reform who sought to highlight and publicize conditions in jails. He supported the notion that man has free will, is rational, and can change his behavior and promoted the idea that prison systems and structures could be designed to promote rehabilitation.
- Others who called for reform included William Eden, Samuel Romilly, and Jeremy Bentham.

- Reform calls grew in the nineteenth century, and by 1857 transportation had been abolished and the last of the hulks closed and destroyed. Serious attempts at penal reform were instituted and schools, libraries, and religious instruction introduced.
- Elizabeth Fry was a Quaker and prison reformer who was able to bring about reforms in the prisons system that had been called for over 100 years before. These included basics such as food and blankets and the separation of the sexes. She formed the Ladies Newgate Committee and other committees designed to campaign for reform and to raise money and other resources for prisoners.
- England attempted to copy U.S. prison designs in the early nineteenth century, with individual cells and prisoners working silently. When transportation ended these prisons became human warehouses for thousands of inmates, and the experiment failed.
- The situation was exacerbated by the public's hostility to the concept of remission (good time), which had been developed in Australia for persons transported. The public was less supportive of such a scheme when it was introduced in England. In place of remission, a scheme of rewards was introduced whereby prisoners could receive payments to buy additional foods.
- Others, in Australia and Ireland, built on this scheme to develop a scheme of progressive rewards that were intended to lead to, and prepare the prisoner for, freedom. The Irish system included a concept of supervision after release—the precursor to the parole systems in England and the United States.
- By the late nineteenth century the prisons were again less progressive, with prisoners compelled to engage in mindless, purposeless physical activities that were designed to reduce their "excess" energy.
- By the late nineteenth century the prison system was divided into two, with central prisons for serious offenders serving long sentences and local prisons, controlled by the local authority, for lesser offenders with sentences of, typically, less than 2 years. Local prisons transferred to the central government in 1877.
- The Gladstone Committee Report of 1895 and the Prison Act of 1898 led to the abolition of excessive corporal punishment and the institution of remission. There was also the establishment of separate prisons for different offenders, by sex, age, and seriousness of offenses.
- The death penalty in England was abolished in 1965 and the last execution took place in 1964.

KEY TERMS

Association system: Prisoners worked alongside each other during the day and returned to their single cells at night.

Bridewells: Bridewell Palace became a poor house and prison in 1553. The name became synonymous with jails and police stations across England and Ireland.

Central prisons: Controlled by the central government, with prisoners assigned to them primarily for penal servitude.

Gaol: A place of legal detention; now often spelled as "jail."

Hulks: Large, old warships used to house prisoners.

Irish system: A three-stage system culminating in the offender being allowed an early release with supervision and several restrictive conditions.

Ladies Newgate Committee: A committee for raising money and resources for the prisons.

Law of the Ten Tables: Ancient, Roman legal code (later the Law of the Twelve Tables) that provided a formal written code of guidelines for legal procedures, rights, and sanctions.

Local prisons: Controlled by the local authorities, these were for short-term offenders, usually those with less than 2-year sentences.

Marks: A system of rewards a convict could earn for good conduct, allowing a prisoner to work his way from a hard labor existence to a relatively pleasant life including visitors, writing letters, and free time.

Oubliette: From the French for "forgotten"; a deep hole in the ground where prisoners could be held for a few days before they were punished.

Separate system: Solitary confinement with prisoners in their cells all of the time.

Stocks: Early form of imprisonment consisting of a wooden restraint device where an individual would be required to sit for days, facing forward, on a hard, wooden stool with his head, or hands, or feet, and even all three appendages thrust though small holes in a secure wooden board. Often, this took place in public (e.g., in the village square) in order that shame and humiliation would be associated with the restraint of movement.

Ticket of leave: A system instituted in 1837 in which prisoners could be rewarded with some remission of their sentence for good behavior.

PUTTING IT ALL TOGETHER

1. How did early forms of punishment impact the use of punishment in the correctional systems of both the United States and England?

2. How did the early penal reformers shape the present-day criminal justice systems in both the United States and England?

3. Why did England abolish the death penalty?

ENDNOTES

1. For example, Walker speaks of Moses in 1451 B.C. commanding the Jews to build cities as refuges for "manslayers" (1973, p. 19). In 1444 B.C. a similar order was issued by Joshua, and by 1179 B.C. the Law of Moses had made murder a capital crime among the Egyptians and Greeks. In fact, the court of Ephetae ordered the execution of murderers as a deterrent to crime. This might suggest early thinking concerning the power of courts to inflict punishment as a way of combating criminal activity (Walker, 1973, p. 20).

2. Traditional images of crucifixion have been questioned due to the discovery in 1968 of a man who had been crucified during the time of Christ not far from the likely site of Christ's crucifixion. A narrow shelf supported the bulk of the man's body on the upright beam and both his knees were broken and twisted to the side with a nail holding them in place. The outstretched arms were nailed to the crossbar in the traditional manner of crucifixions (Walker, 1973, p. 21).

3. These punishments were separate and distinct from the torture used later on in the Spanish Inquisition as a means to secure confession. Some have even suggested that the more contemporary practice of "police administering a third degree to a suspect" in order to secure a confession is a vestige of the Inquisition (DeFord, 1962, p. 2).

4. Exceptions to the harsh treatments fell under the concept of "benefit of clergy" (which some regard as a primitive form of the present system of probation, discussed in Chapter 10). "Benefit of clergy" came about because the Church believed that only it had jurisdiction over members of the clergy. If the clergy committed crimes, they were not subjected to the criminal courts but were to be handled by the Church courts. Henry II objected to this and insisted that clerics suspected of crimes should be tried by secular courts. A compromise provided that clergy accused of crimes would be tried in secular courts but with the benefit of clergy (i.e., the bishops could claim dispensation for them). The charge would be read, but the state could not present evidence against the accused. The accused would be allowed to give his view of the accusation and bring witnesses. Thus, with the only evidence coming from the witnesses chosen by the accused, most cases resulted in acquittals. Benefit of clergy was later extended to all Church personnel as protection against capital punishment and eventually to all persons who could read. Parliament later declared that certain acts would be felonies without the benefit of clergy and effectively abandoned its use in 1705; however, benefit of clergy was not technically removed from the law until 1968 (Walker, 1962, pp. 74, 146).

5. For a detailed account of the complexities of committal and release from debtors' prisons in the eleventh and twelfth centuries, see Chapter 1 of Pugh (1968). This chapter also described the role of sheriff and the marshall in debt collections, punishment, and the imposition of fines.

6. It has been estimated that some 72,000 persons were executed in England during Henry VIII's reign (Laurence, 1932, p. 8; Walker, 1973, p. 60), including Sir Thomas More.

7. For an account of the practice of banishment, see Ives (1970, ch. 4).

8. While these severe punishments were occurring in England, they had their counterparts on the Continent. At the end of the eighteenth century, Joseph Ignace Guillotin had invented a device to dispense swiftly, efficiently, and "honorably" of individuals by use of a blade dropping from a height to sever a head from the torso. The first official victim was publicly executed on April 25, 1792, by a guillotine with a slanted blade designed by Louis XVI, who was later to die by it himself (Walker, 1973, p. 102).

9. Although there are many accounts of the role of the hangman in the history of English corrections (and the role of the executioner in other countries), most describe the public hanging as a source of catharsis for society. The scene was almost festive, with vendors selling their wares, members of the upper classes renting rooms overlooking the spectacle, and street villains (pickpockets, prostitutes, and others) plying their trade. In later years there were special excursion trains to bring people from all over the country to witness an execution. In fact, in 1899 the travel agent Thomas Cook organized trips across the Channel to witness executions in Paris. It has been suggested that the public execution allowed the populace to absolve themselves of any personal responsibility for the death of another. However, from the first hangman to hold the post until the last, there was always a reluctance for respectable men to perform the hangman's duties. Usually it fell to lower-class individuals appointed by the sheriff to disburse these duties. For an interesting and colorful account of the hangman in the history of English punishment, see Bland (1984).

10. DeFord (1962), Pugh (1968), and Ives (1970) all refer to the earlier uses in Rome, Greece, and France of the system of single cells for imprisoning criminals.

11. Although abolition efforts were gaining momentum, the movement was not assisted by the Archbishop of Canterbury, who suggested in 1956 that the death penalty was not always "unchristian" and wrong. Some felt that because Christians believed in life after death, an executed person would at least find salvation through death and his soul would find "peace in heaven" (Walker, 1973, p. 140).

12. Technically, the crimes of treason and piracy on the high seas remain as capital offenses.

13. In May 2006, Labour Party Prime Minister Tony Blair was facing mounting pressure from both parties to step down from his post prior to the end of his term. Local elections in 2006 had seen the Conservative Party triumphant in regaining posts that had previously been held by Labour Party representatives. In May 2007, Tony Blair announced he was stepping aside from his post as Prime Minister. In June 2007, it appeared as if the most likely successor would be Gordon Brown, Chancellor of the Exchequer. However, even in the face of these changes, the restoration of the death penalty in England would be unlikely.

REFERENCES

Andrews, W. (1980). *Old-time punishments*. Detroit, MI: Singing Tree Press.

Bailey, V. (1981). *Policing and punishment in nineteenth century Britain*. New Brunswick, NJ: Rutgers University Press.

Barclay, G. C., & Tavares, C. (Eds.). (1999). *Digest 4: Information on the criminal justice system in England and Wales*. London: Home Office, Research and Statistics Department.

Barnes, H. E., & Teeters, N. K. (1943). *New horizons in criminology.* New York: Prentice Hall.

Bartrip, P. W. (1981). Public opinion and law enforcement: The ticket-of-leave scares in mid-Victorian Britain. In V. Bailey (Ed.). *Policing and punishment in nineteenth century Britain* (pp. 150–181). New Brunswick, NJ: Rutgers University Press.

Bland, J. (1984). *The common hangman.* Hornchurch, England: Ian Henry.

Block, B. P., & Hostettler, J. (1997). *Hanging in the balance.* Winchester, England: Waterside Press.

Bury, J. B. (1937). *A history of Greece to the death of Alexander the Great.* New York: Random House.

Byrne, R. (1989). *Prisons and punishments of London.* London: Harrap.

Cartledge, P. (2003). *The Spartans.* New York: Overlook Press.

Cole, G. F. (1989). *The American system of criminal justice* (5th ed.). Belmont, CA: Brooks Cole.

Coleman, K., Hird, C., & Povey, D. (2006). *Violent crime overview, homicide and gun crime 2004/2005* (2nd ed.). London: Home Office.

DeFord, M. A. (1962). *Stone walls.* Philadelphia: Ambassador.

Dinsmore, C., & Jackson, J. (1994). Shipyards to build jail barges: Norfolk could be first customer; city officials to tour barge in New York. *The Virginian Pilot*, D1, September 15, 1994, Norfolk, VA.

Encyclopedia Americana. (2004). Danbury, CT: Scholastic Library Publishing.

Encyclopedia Britannica. (2005). Chicago, IL: Encyclopedia Britannica.

Fairchild, E. (1993). *Comparative criminal justice systems.* Belmont, CA: Wadsworth.

Fraser, R. (2003). *The story of Britain.* New York: W.W. Norton and Company.

Greif, A. (1997). *On the social foundations and historical development of institutions that facilitate impersonal exchange: From the community responsibility system to individual legal responsibility in pre-modern Europe.* Paper presented at the Workshop in Political Science, June 12, 1997. Chicago: University of Chicago.

Harding, C., Hines, W., Ireland, R., & Rawlings, P. (1985). *Imprisonment in England and Wales.* London: Croom Helm.

Hibbert, C. (1968). *The roots of evil: A social history of crime and punishment.* Birmingham, AL: Minerva.

H. M. Prison Service. (1988). *Prisons past, prisons future.* London: Home Office.

Hobhouse, S., & Brockway, A. F. (Eds.). (1922). *English prisons today: Being the report of the prison system inquiry committee.* London: Home Office.

Hodkinson, S. (2004). Female property ownership and empowerment in classical Hellenistic Sparta. In T. J. Figueira (Ed.). *Spartan society* (pp. 103–134). Swansea: The Classical Press of Wales.

Home Office. (1953). *Report of the Royal Commission on Capital Punishment.* London: H.M.S.O.

Home Office. (1992). *Criminal statistics England and Wales 1991.* London: H.M.S.O.

Howard, D. L. (1960). *The English prisons: Their past and their future.* London: Methuen.

Howard, J. (1792). *The state of prisons in England and Wales J 1777* (4th ed.). London: Johnson, Dilley & Cadell. (Reprinted by Patterson Smith, Montclair, NJ, 1973.)

Ives, G. (1970). *A history of penal methods: Criminals, witches, lunatics.* Montclair, NJ: Patterson Smith Reprint Series.

Chapter Resources

Chapter Resources

Kenyon, J. P. (Ed.). (1983). *A dictionary of British history*. New York: Stein and Day.

King, R., & Morgan, R. (1976). *A taste of prison: Custodial conditions for trial and remand prisons*. London: Routledge & Kegan Paul.

Laurence, J. (1932). *A history of capital punishment*. Port Washington, NY: Kennikat Press.

Lee, R. W. (1956). *The elements of Roman law*. London: Sweet & Maxwell.

Mayhew, H., & Binny, J. (1862). *The criminal prisons of London and scenes of prison life*. London: Griffin, Bohn.

McConville, S. (1981). *History of English prison administration*. London: Routledge & Kegan Paul.

Morris, N., & Rothman, D. J. (1995). *The Oxford history of the prison: The practice of punishment in Western society*. New York: The Oxford University Press.

New World Encyclopedia. (2002). New York: Funk and Wagnalls.

New York City Department of Corrections. (2006). *Division III: Facilities*. Retrieved January 15, 2006, from www.nyc.gov.

Parry, L. A. (1934). *The history of torture in England*. Montclair, NJ: Patterson Smith.

Peters, E. M. (1995). Prisons before prisons: The ancient and medieval worlds. In N. Morris & D. J. Rothman (Eds.). *The Oxford history of the prison: The practice of punishment in Western society* (pp. 3–47). New York: Oxford University Press.

Povey, N. S., Walker, A., & Kershaw, C. (2005). *Crime in England and Wales 2004/2005*. London: Home Office.

Priestley, P. (1985). *Victorian prison lives*. London: Methuen.

Pugh, R. B. (1968). *Imprisonment in medieval England*. Cambridge: Cambridge University Press.

Rothman, D. J. (1971). *The discovery of the asylum: Social order and disorder in the new republic*. Boston: Little, Brown.

Rumbelow, D. (1982). *The triple tree*. London: Harrap.

Serrano, J. (2006). *Serrano raises concerns over proposed Oak Point Prison*. Press release. Washington, D.C.: Congressman Serrano's Office.

Smith, William (Ed.). (1870). *The dictionary of Greek and Roman biography and mythology* (Vol. 1). Boston: Little, Brown.

Terrill, R. J. (2002). *World criminal justice systems* (5th ed.). Cincinnati, OH: Anderson.

Tomlinson, M. H. (1981). Penal servitude 1846–1865: A system in evolution. In V. Bailey (Ed.). *Policing and punishment in nineteenth century Britain* (pp. 126–149). New Brunswick, NJ: Rutgers University Press.

U.S. News and World Report. (1988, November 14). The far shore of America's bulging prisons, 11–12.

Walker, P. (1973). *Punishment: An illustrated history*. New York: Arco.

Webb, S., & Webb, B. (1922). *English prisons under local government*. London: Longmans, Green.

Wicks, R. J. (1979). *International corrections*. Lexington, MA: Lexington.

Wolff, M. (1967). *Prison: The penal institutions of Britain—Prisons, borstals, detention centers, approved schools and remand homes*. London: Eyre & Spottiswoode.

The Organization and Operations of Corrections

10

Chapter Objectives

After completing this chapter, you will be able to:

- Describe the history of corrections and the philosophical foundation of the present correctional systems in both England and the United States.
- Identify the various types of prisons used in England and the individuals in charge of operating these prisons.
- Discuss the various community sanctions England uses as alternatives to incarceration.

■ Introduction

In Chapter 9 we saw that the continuing concern regarding increasing crime, shifting attitudes about the role of punishment versus corrections, overflowing prisons, dilemmas about proper treatment within the prisons, and public cries for reform have all played a part in the development of a workable system of corrections in England. That corrections system was in a state of crisis for most of the twentieth century. Social conditions have changed in England, contributing to the growing tensions within the correctional system. An influx of minorities into England from Britain's former colonies and the existence of a distinct socioeconomic underclass resulted in disproportionate numbers from the lower classes and minority groups being represented in the total prison population.[1] With regard to prison conditions, England has the dubious distinction of being one of the most frequent defendants in cases brought before the European Court of Human Rights (Berger, 1992; Fairchild, 1993, p. 212; Fairchild & Dammer, 2001, p. 253).

Even in the face of these challenges, the English system of corrections has seen many reforms and positive developments as it moves into the twenty-first century. This chapter will examine this progression and the important features that have characterized the past hundred years in English corrections. We will look at corrections in the first half of the twentieth century, the philosophical foundations of the current system of corrections, the current organization and administration of corrections, institutional corrections, community alternatives to prisons, and the challenges of the future for England's correctional system as it proceeds into the twenty-first century.

■ The Twentieth Century

After being taken over by the Home Secretary as a result of the Prison Act of 1898, prisons did begin to improve, but very slowly. The act banned excessive corporal punishment and, in general, called for a more progressive attitude toward reforms and remission for good behavior. The intention of Parliament was that an individual should leave prison a better person than he or she came in, both mentally and physically. Unfortunately, the officials directly in charge of the prisons, called **governors** in England, were slow to change their ways and often did not interpret the new guidelines the way Parliament had intended. Thus, the reforms passed by Parliament, many of which were heavily influenced by the apparent successes of prisons in the United States, were not fully implemented because of reluctance on the part of the governors to abandon the old practices (Fairchild, 1993, p. 213; Walker, 1973, p. 128).

The Borstal Movement

Early in the twentieth century, another in a long line of reformers, Sir Evelyn Ruggles-Briese, emerged as a man of action. Sent to the United States by the English government to study American systems, Ruggles-Briese developed a sense of appreciation for the American emphasis on reform and education.

In 1904, after studying the state reformatories in the United States, he set up a facility for juvenile males in a village in Kent called Borstal. Although the young men at Borstal were originally scheduled to serve their sentences in adult prisons, Ruggles-Briese managed to have them assigned to his facility and provided accommodations for them much in the fashion of the reformatories in the United States. He was concerned about their mental, emotional, and physical improvement and used the Elmira Reformatory of New York State as his model. Many other such schools were developed and named after the village of Borstal. The **borstal system** was quite structured. Usually for offenders between the ages of 16 and 20, the system was based on a "house" concept that included a daily work schedule, self-discipline, and a system of graduated rewards for performance. All this took place under the supervision of a headmaster who strictly oversaw the rigors of the daily activities. Although considered a leader in individualized treatment for both adult and juvenile correction programs, the borstal system eventually succumbed to the pitfalls of most adult prisons in England: overcrowding, deteriorating conditions, and a never-ending stream of offenders coming into the system (Fairchild, 1993, p. 214). Although the borstal system was recently phased out, its development caused the English judiciary to become aware that children and young persons should not be punished as adult criminals.

There had been unsuccessful efforts for reform of the treatment of children in England in the nineteenth century; by the late nineteenth century magistrates informally began to separate children from adults in their court hearings. This practice was codified by the Children Act of 1908, which revised the entire criminal justice process for juveniles and abolished the death penalty for children under 16 years of age. Quite suddenly, children were the focus of new legislation, and the correctional system began to examine its own practices in the treatment of youthful offenders (Wolff, 1967).

Institutional and Community Alternative Development

The first half of the twentieth century was not nearly as "colorful" in terms of the conditions of the prisons in England as compared to the nineteenth century. The borstal and youth movements had changed some of the thinking about prisons, but there still was continued emphasis on incarceration as punishment and little effort toward progressive treatment of offenders while in prison. English institutions during the first part of the twentieth century continued to emphasize "intentional harshness" (DeFord, 1962, p. 132). Because England had opted for the segregated, separate system, the rigid rule of silence was enforced when the inmates were out of their cells. Originally, this silent segregated system did not even permit staff to converse with inmates in other than a command-like manner, although this policy was not as rigidly enforced. The silence affected both staff and inmates; understandably, insanity rates increased (McConville, 1987, p. 39). This regime, which was in force well into the 1950s, made any type of treatment program difficult to implement.

The physical state of the prisons was also an issue. About 60% of the prisons in use during the first half of the twentieth century had been built in the middle of the nineteenth century, and many are still in use; as of 1980, only 20% of the prisons in England had been built since 1945 (McConville, 1981; Stockdale, 1979, p. 7). This has improved slightly, as more than 20 new prisons have been built since 1980 and nearly 17 of that number have been either built or completely refurbished since 1994 (H.M. Prison Service, 2007).

During the early part of the twentieth century, different types of facilities began to emerge. There were local prisons, much like jails in the United States; regional prisons similar to reformatories; and central prisons corresponding to the penitentiaries of the United States. Among these facilities a classification system developed based upon level of security: minimum, medium, and maximum.

Upon conviction and sentencing, all prisoners would be taken directly to the nearest local prison, where they would be assigned a label of either "**star**" **prisoners** (first-time offenders) or "ordinary" prisoners. Those under the age of 20 were usually assigned to the borstal system. The others were sent to one of the prisons, based on their classification. The same process of classification was used for both sexes, and star prisoners and ordinary prisoners were strictly segregated.[2]

There were certain advantages to the correctional system in place during the first half of the twentieth century. The silent segregation virtually eliminated aggression and rioting. Almost all the riots took place in the public works or agricultural prisons where prisoners worked and congregated together; in contrast, in the segregated prisons the only source of contact and verbal interaction for the inmates was the landing officer, who had the most influence over the daily lives of the inmates. As the rigidity of the silent system eventually relaxed, the inmates often turned to the staff for advice or just minimal social contact. This helped to break down the barriers between staff and inmates, thus reducing the potential risk of violence and rioting after the silent system dissipated (McConville, 1987, p. 39).

Several acts passed during the first half of the twentieth century exerted considerable influence on prisons and penal philosophy in England. The general absence of judicial discretion was altered by the passage in 1907 of the Probation of Offenders Act, which officially established the Probationary Service (Walker, 1973, p. 130). In addition, appellate review of sentences by the Court of Criminal Appeal was introduced by this act, which gave prisoners the right to appeal their convictions and/or sentences (DeFord, 1962, p. 134; Walker, 1973, p. 130). In the following year, 1908, the Prevention of Crime Act was passed, introducing the borstal system for young offenders and preventative detention for habitual offenders. Although subsequent acts carried some measure of corrections reform—for example, requiring judges to consider the appropriateness of the sentence handed down and/or any suitable alternatives now available—the onset of World War II slowed almost to a stop most of the progress that had been made. It was not until passage of the Criminal Justice Act of 1948, when sentencing options were expanded to include more noncustodial alternatives, that new strides in reform were made.

The Criminal Justice Act of 1948 sought to diminish the harsh punitive aspects of sentencing by, for example, abolishing whipping (Section 2) and penal servitude (Section 1). Sir Lionel Fox, former chairman of the Prison Commission, suggested that if people were sent to prison, they should be sent there *as* a punishment, but not *for* punishment (DeFord, 1962, p. 134). As a result of the 1948 act, community-based sentencing options were expanded and all progressive stage systems based on deterrence and/or punishments, such as meaningless labor, were ended. There was more of a focus on rehabilitation, with remission of sentence and a lesser security status depending on the inmate's performance in prison. Further, because remission was now based on the inmate's conduct, the key to early release rested primarily in the inmate's own hands. In addition, after the 1948 act other amenities such as prison libraries and schools began to emerge, and letters began to be considered rights instead of privileges.

By the middle of the twentieth century, English prisons contained many of the features characteristic of institutions in the United States: education, libraries, religious instruction, daily newspapers, legal libraries, common rooms, and the opportunity for input on inmate governance issues. But the most notable change was the emergence of the noncustodial sentence and its development through the extensive use of community alternatives to incarceration, which will be addressed later in this chapter.

Since the days of John Howard and Elizabeth Fry, there have been efforts to improve life for convicted persons outside prison as well as inside. The Penal Reform League was founded in 1907 supplementing the work of the Howard Association, formed in 1896. In 1921, the two organizations merged to form the **Howard League for Penal Reform**, which had as one of its missions the improvement of conditions for offenders both inside and outside institutions. The Howard League continues its efforts in corrections today in both England and the United States, and produces a journal of research findings on corrections.

Before proceeding to the current state of corrections in England, it is important to consider briefly the development of penal philosophy during the middle of the twentieth century, which had an important influence on the English correctional system.

■ Philosophical Foundations of the Present Corrections System

Corrections in the twentieth century were characterized by the dilemma over the use of prison for punishment versus the new individualized treatment ideas of the American Progressive movement (Fairchild, 1993, p. 213; Fairchild & Dammer, 2001). We have also noted earlier the conflict inherent among the four basic rationales for sentencing: retribution (or "just deserts," the punishment focus), incapacitation, deterrence (both specific and general), and rehabilitation.

The problem with incapacitation is that because the vast majority of offenders are eventually released from prison, incapacitation ensures only that an individual is temporarily removed to another society. In the pseudo-society of prison, the inmate adjusts to the rules of that society and often continues to engage in violent or subversive activities, gaining little in the way of treatment and reemerging into society no better, and perhaps even worse, than before.

The role of deterrence continues to be a major source of controversy in both England and the United States because, despite the fact that its effectiveness is questionable, it holds an emotional place in the hearts of our citizens. With regard to specific deterrence, it is generally acknowledged that fear of punishment by the offender in the future may deter some types of property crimes but has little effect on the deterrence of violent and/or impulse crimes. The effectiveness of general deterrence is also questioned by research findings. In our examination of English corrections, we have seen that crime continued to flourish despite public executions, public displays of humiliation (stocks and pillories), and public whippings.

Throughout the twentieth century, rehabilitation was advanced as a rationale for sentencing, particularly by professionals working in the system. The public, however, was not as ready to accept the concept of rehabilitation. Citizens have long been concerned by a lack of focus on punishment, or retribution; it is a commonly held belief that rehabilitation provides inmates with "soft time" in prison and does not work. The controversy in the United States in the 1960s and 1970s questioning the effectiveness of rehabilitation and culminating with

Martinson's report, colloquially known as the "Nothing Works" report (Martinson, 1974), did not go unnoticed by the English. However, in England, as in the United States, efforts at rehabilitation continued as judges and other professionals supported rehabilitation as a consideration for sentencing decisions. This led to increased use of noninstitutional alternatives in both England and the United States, with an increased emphasis on diversion from the courts and prisons (Stockdale, 1979, p. 18).

The English have attempted to balance the competing goals of punishment and deterrence on one hand and rehabilitation on the other by using the former as their primary guideline in sentencing so as to placate society while at the same time pursuing the objective of rehabilitation whenever appropriate, primarily through noninstitutional alternatives. This would appear to be a satisfactory arrangement; however, due to the economic conditions in England, financial support for many of the noninstitutional programs has been largely withdrawn in favor of expenditures for the building of new prisons. Consequently, England resorted to greater use of imprisonment, continuing to feed the already overcrowded institutions that bring back images of prison conditions of past centuries. With these controversies and issues outlined, we will now consider the current organization and administration of corrections in England.

■ Current Organization and Administration of Corrections in England

Since the nineteenth century there has been a shift in the administration of prisons and corrections in England toward the centralized administration in the office of the Home Secretary that exists today. There are four components of the English correctional system: H.M. Prison Service (formerly the Prison Service Agency), the Independent Monitoring Board (formerly the Boards of Prison Visitors), the parole board, and the National Probation Service (formerly the Probation and After-Care Service). All of these components currently fall under the auspices of the recently formed National Offender Management Service (NOMS). In response to criticisms regarding the management of offenders (see Carter, 2003), the Home Office developed a report, *Reducing Crime—Changing Lives* (Blunkett, 2004), that delineated a number of changes to the management of offenders. In 2004 the **National Offender Management Service (NOMS)** was created to oversee the management of offenders, from their entry into the criminal justice system through their exit. This change means that there is a single person (**offender manager**) responsible for a particular offender throughout his or her incarceration and/or community sentence. This service, therefore, is responsible for both the punishment of offenders and the reduction of offending. In creating this service, the Home Office brought together the prison and probation services under one service. However, this organizational change has had little effect on the day-to-day operations of the prison service, which still operates much like it used to (Blunkett, 2004). The importance of the Home Secretary in the English criminal justice system will become apparent as we consider its relative place in the overall hierarchy of public administration. A graphic illustration of this position in the correctional system is presented in **Figure 10-1.**

The Prison Service Agency

The Prison Service Agency came into existence on April 1, 1993, replacing the old Prison Department. Agency status was designed to provide the prison service with "greater autonomy" (H.M. Prison Service, 1992, p. 39) by giving it the responsibility for its own budget. In addition, because of agency status the Home Secretary no longer had to answer questions asked in Parliament about the Prison Service, but could direct them to the Director General of the Prison Agency. However, as a result of the development of the National Offender Management Service, the Prison Service's "formal status as an Agency, in theory semi-independent of the Home Office, will end and it will become an integral arm of NOMS but with continued substantial day to day operational freedom" (Blunkett, 2004, p. 15).

The **Director General**, who is appointed for a renewable fixed period by the Home Secretary with the approval of the Prime Minister (H.M. Prison Service, 1993b, p. 6), is the Home Secretary's principal policy advisor and oversees the day-to-day

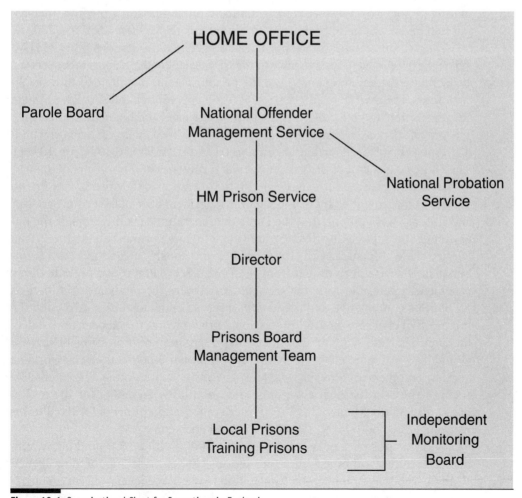

Figure 10-1 Organizational Chart for Corrections in England.

management of the agency. Assigned tasks include ensuring that the members of the Prison Board, which has responsibility for administering the overall state of prisons, are appointed and serve; selecting members for the Independent Monitoring Board; establishing rules for prison personnel; issuing transfer orders for prisoners from one institution to another; granting temporary discharges of prisoners due to ill health; with the Treasury, approving building expenditures and prison closures; and reporting to the Home Secretary (H.M. Prison Service, 1993b, pp. 6–7).

The Prison Service Agency has an annual budget of more than £1.5 billion, and is responsible for administering some 76,000-plus inmates in more than 139 prisons. The agency is accountable to Parliament and must respond to public concerns, inmate concerns, penal reformers, and the official prison inspectors. A number of goals have been set for the Prison Agency, such as maintaining order and control, providing decent conditions and positive regimes, helping prisoners prepare for their return to the community, and delivering services with maximum efficiency (H.M. Prison Service, 1993b, p. 5). In addition, key performance indicators have been established against which progress toward meeting these goals can be measured. Among the priorities enumerated by the first Commissioner of Correctional Services, Martin Narey, were reducing and/or eliminating prisoner escapes, reducing assaults, reducing prison crowding, increasing staff diversity, and improving prison programs (H.M. Prison Service, 2003a).

The administration of the entire prison system is the responsibility of **H.M. Prison Service**, which is responsible to the National Offender Management Service. To facilitate the smooth running of the prison system, it is divided into six directorates, each with specific responsibilities: personnel and services, inmate administration, inmate programs, custody, health care, and finance and planning. The prisons themselves are under the supervision of area managers, each of whom has responsibility for about nine institutions (H.M. Prison Service, 1994f, pp. 10–11). The Director General of the Prison Agency formulates and implements the various policies and procedures necessary for the operation of the prisons. The **Prison Service Management Board** is "the senior management team of the Prison Service," and initially was composed of the Director General, the six directors of the operational and service directorates, and three members from outside the Prison Service (H.M. Prison Service, 1994f, p. 11). Previously the Prison Board members had likewise been drawn from the Prison Service, but had included only two members from outside the service. Criticism of the old board had focused on both its composition and the length of service of its members. Although the inclusion of civil servants and professional prison workers had been applauded, many critics had suggested that greater representation of professional expertise would lead to a more effective board. Judges, probation officers, social workers, and law enforcement officials were all suggested as representing a broader spectrum of the criminal justice system and as suitable members for the prison board. In addition, to avoid political interference, appointments for fixed terms were suggested both for the Director General and senior staff.

With the prison population exceeding 80,000 in June 2007, and many antiquated facilities still in use, there are serious challenges facing the new agency. With the economic problems in England, the Prison Agency has the awesome responsibility for fixing the system. In fact, in June 2007, plans were developed to move prisoners nearing the end of their sentence to open jails (The Times, 2007:2).

Independent Monitoring Boards

Set up as a way to bring outsiders into the prisons and to assist with some of the administrative problems, the Boards of Visitors, or visiting committees, have their origins in the nineteenth century. Just as the Police Lay Visitors Program is utilized as a way of bringing volunteers into law enforcement, the visiting committees bring volunteers to the prisons. The Boards of Prison Visitors were established and given three broad categories of duties: (1) to visit the prison regularly as a board and individually, with each member having the right to visit as often as he or she wishes; (2) to hear complaints about the prison and prison officials and make major disciplinary decisions (rather conflicting roles); and (3) to make an annual report to the Home Office on matters of prison procedure and policy concerning each institution (Borrie, 1976, p. 282; H.M. Prison Service, 1993b, p. 7; Home Office, 1992b, p. 55). After April 2003, Boards of Prison Visitors became known as **Independent Monitoring Boards**, though legislation still refers to them as boards of prison visitors. The board members are appointed by the Home Secretary for renewable 3-year terms and are given training to prepare them for their work. At least half of the members must be magistrates (Borrie, 1976, p. 284; Home Office, 1992b, p. 55).

This is an administrative system unique to England. The English public has long believed that these outside review boards were a necessary structure to guard against the excesses of government not only in corrections, but also in law enforcement. According to some (Fairchild, 1993; Fairchild & Dammer, 2001; Fitzgerald & Sim, 1979, pp. 6–9), it is particularly important to have outside review and access to prisons because of the Official Secrets Act in England. This act, originally passed in 1911 and amended in 1922, makes it illegal for government officials (in this case prison workers) to disclose the internal operations of their departments. It has been suggested that prison officials often used the provisions of this act to keep the public in the dark as to the actual penal policies and operations within the prison walls so as to promote their own interests.

Recently, boards of prison visitors have come under fire from both the public and the prison service. Because of the required composition of these boards, many professionals and prison officials feel that the members are too old, too middle class, and too educated to relate effectively to the prisoners and their problems. In addition, prison officials often resent the fact that "amateurs" are looking over their shoulders as they administer the day-to-day operations of the facilities; they would prefer that individuals who have input into the system have a deeper understanding of the prisons. As previously mentioned, grievances are sometimes brought to the attention of the European Commission on Human Rights rather than reported to the boards of visitors; this can be seen to reflect the mistrust of the board members even by the inmates.

Criticism is not limited to the composition of the boards; it also focuses on the training and preparation the board members receive. Although the 2-day-long training offered at the prison staff college since 1971 has been seen as a way of correcting this deficiency, many believe that it is not adequate. Critics suggest that this is not enough time to learn about the prisoners and their rights, the professional staff, and prison regulations.

Finally, it is interesting to note that some criticism comes from the members of the boards themselves. They often feel frustrated about the inadequate amount of training they receive and their inability to accomplish anything within the system. They voice the opinion that members of the prison staff often do not take the work of the boards seriously and that this does not allow for constructive change in the prisons. The board members frequently point out improvements that should be made in the treatment programs, but their ideas are hampered by the financial condition of the prison department and the ever-increasing crisis of overcrowding.

As Terrill (1992) pointed out, some critics suggested that some of the functions of the boards should be delegated to the courts. Until April 1992, the boards were responsible for adjudicating serious discipline charges against inmates and assisting in parole evaluation. Many independent observers felt that serious discipline charges should be handled by the courts, as has been increasingly the case in the United States, or at the least by professional adjudicators who are assisted by members of the legal profession. As the controversy continued, the Home Office, following the recommendation of the Woolf (1991) report, decided to transfer this function traditionally performed by the boards of prison visitors to the courts.

The Parole Board

Established in 1968, the **parole board** was a rather recent addition to the prison system. As in the United States, parole in England is based on the idea that an inmate who has earned the privilege may be released from the prison before his or her sentence has been completed, serving the remainder of the sentence under supervision within the community. (England refers to this as "on license.") In England this was viewed as a connecting link between probation, after-care, and the entire penal system (Bochel, 1976, p. 225). There is perhaps a greater distinction between the words *parole* and *probation* in the United States, whereas in England they are often used interchangeably.

The parole system was established after the issuance of a report in 1965 entitled *The Adult Offender* in which the government suggested that prisoners who earn the privilege should be released early and allowed to carry out the remainder of the sentence outside prison under the supervision of a probation officer (Home Office, 1965). The report indicated that a large number of inmates could benefit more from supervised treatment within the community than from being kept in prison for a longer period. The prison population was expanding rapidly at this time, and this may have influenced the recommendations.

The parole proposal was introduced in the wide-ranging Criminal Justice Bill of 1966. The lengthy discussion in Parliament focused not so much on the concept of parole as on the appropriate machinery to implement it. The probation service was thought to be the logical agency to supervise parolees, but it was already overburdened. In addition, the probation service was concerned about its role in parole. It preferred a treatment-oriented approach and did not want to become custodial in its relations with parolees (Bochel, 1976, p. 226).

The Parole Unit was set up in the Probation and After-Care Service on April 1, 1968. Practically, this meant extra duties for probation workers and prison welfare officers, because more reports and other paperwork were required for each. This was seen as more demanding and perhaps less rewarding work than probation supervision; however, those eligible for parole are probably a lower-risk group of individuals than the prison population in general, because they have had to demonstrate exemplary behavior in order to be considered for parole (Fitzgerald & Sim, 1979).

Until 1992, there were four stages to the parole process. The first stage involved the examination of the prisoner's case by a local review committee made up of the prison governor; a member of the boards of visitors, who was usually also a magistrate; a senior prison officer; and a private citizen. As in the United States, the inmate was allowed to present evidence on his or her own behalf, and usually an interview was granted with a member of the local review committee. If approved, the case was then sent to the Home Office, which was the second stage of the process. For less serious cases (those generally involving a maximum sentence of 4 years), this was the final stage of the process. More serious cases had to be considered by the parole board. The third stage consisted of obtaining the parole board's recommendation. The 20 members of the board were divided into smaller panels to hear individual cases. The board was drawn from the community and included judges, medical professionals, police officers, and others who had some experience in the social services (Fitzgerald & Sim, 1979, p. 95). Once the recommendation was made, the case was returned to the Home Office because only the Home Secretary can grant the parole. This fourth stage (along with the second stage) could take a very long time, often 4 months or more, and the parole system was frequently criticized for this delay because it naturally caused anxiety and tension in the system, particularly among the candidates for parole. The Home Secretary, it should be noted, usually accepted the recommendation of the parole board panel (Fitzgerald & Sim, 1979, p. 95).

Upon entry into prison, convicts used to receive remission of one third of their sentence. Part or all of this remission could subsequently be lost for misbehavior. Until passage of the Criminal Justice Act of 1991, an inmate was typically considered for parole after serving approximately one third of the sentence. With remission, those serving determinate sentences were released free from restriction after serving two thirds of the sentence. Originally, a minimum of one year had to be served in prison before eligibility for parole could be considered. This was changed in 1984 so that those serving less than one year could also be paroled. This had the immediate effect of nearly tripling the number of inmates eligible to apply for parole; consequently, the work of the parole division was substantially increased. This system was changed, again, following the Criminal Justice Act of 2003. Currently, for prisoners serving less than 4 years, release at the halfway point is automatic. Prisoners serving 4 years or more are eligible for parole at the halfway point in their sentence, but release is not automatic. If the parole board does not recommend release for these prisoners at the halfway point, release is automatic after two thirds of the sentence has been completed.

The rate of those successful in having their parole granted jumped from 55% in 1981 to 61% in 1986. By 2002, this rate was approximately 53%, which is the highest release rate since 1992. In 2002, parole was granted to 3,180 of the 6,010 inmates considered for parole. Also in 2002, 420 paroled inmates were returned to prison for violating the conditions of their parole; this represents about 13.1% of those released on parole that year. This figure does not take into account new offenses (Home Office, 2002).[3]

The length of time a parolee is under supervision is generally short, but has increased over time. In the early 1990s, about 90% of those on parole served less than a year and 45% less than 6 months (Barclay, 1993, p. 67). However, by 2002, the average amount of time spent on parole was approximately 18 months. Approximately 17% served less than a year, and 83% served more than a year. For those inmates serving sentences of less than 15 years, the average time spent on parole was 17.9 months. Inmates serving sentences of 15 years or more spent an average of 39.9 months on parole.

Although parole was a welcome addition to the options available to the prison system to try to counter the influx of prisoners, the parole process presented many problems of its own. In addition to creating tension, the slow-moving machinery of the parole process could lengthen the amount of time an individual had to wait for parole; in some instances, prisoners misbehaved while awaiting the decision of the board, thereby making themselves ineligible for parole. One author suggested that had parole decisions come through earlier, inmates might have been paroled before the misconduct and might not have faced the prospect of more prison time (Fitzgerald & Sim, 1979, p. 98).

The Criminal Justice Act of 1991 significantly changed parole eligibility requirements and the parole process, thus reducing the work of the parole board. It abolished remission, with the result that prison disciplinary offenses are now dealt with by the award of extra days in custody (Section 42). The act provided for the automatic release of all prisoners serving less than 4 years after serving half of their sentence unless additional time is imposed for misbehavior in prison (Section 33). Those serving 4 years or more may be released on license (supervision with conditions) by the Home Secretary on the recommendation of the parole board, after serving half their sentence (Section 35) and are normally to be released after serving two thirds of their sentences (Section 33). Except for those sentenced for less than 12 months who are released without supervision, supervision lasts until three quarters of the sentence has been served (Section 33).

In addition to reducing the number of cases considered for parole by providing for the automatic release of all prisoners serving less than 4 years, the Criminal Justice Act of 1991 streamlined the parole process. The local review committees were abolished, and stages one and two of the old process were eliminated. As authorized by the act, the Home Secretary designated the parole board as decision-maker for those serving between 4 and 7 years. The Home Secretary, on the basis of the parole board's recommendation, decides on those serving 7 years or more (Gibson, Cavadino, Rutherford, Ashworth, & Harding, 1994, p. 200). The Parole Board Transfer of Functions Order of 1998 later authorized the parole board to decide parole applications from prisoners serving less than 15 years. Decisions regarding prisoners serving 15 years or more continue to be the responsibility of the Home Secretary.

■ The Prisons of Today

As mentioned previously, imprisonment is the most severe sentence handed down by the English courts. In 2004, approximately 42% of offenders sentenced to immediate custody were sentenced by the Crown Court. The length of prison sentences imposed by the Crown Court in England increased through the 1990s, but has recently begun to decrease. Thus, whereas in 1981 the average sentence length for all indictable offenses other than those carrying life sentences was 16.7 months for males and 10.9 months for females; in 1991 the figures were 20.5 and 18.1, respectively (Home Office, 1992a, pp. 158–160). By 2004, the average sentences for men and women were 16.6 and 11.6 months, respectively (Home Office, 2004a).

The length of sentences imposed in England is relatively short in comparison with those given in the United States. A 2002 survey of felony sentences imposed by state courts in the United States revealed that the average sentence was 57 months. For those convicted of robbery the average sentence was 85 months for males and 55 months for females; for those convicted of burglary it was 41 months for males and 24 months for females (Durose & Langan, 2002). This compares with 2004 English figures of 40 months for males and 26 months for females for robbery, and 17.5 months for males and 16 months for females for burglary (Home Office, 1992a, pp. 158, 160). Part of this differential may be accounted for by the more serious nature of offenses in the United States (e.g., a far larger percentage of offenses committed with firearms). In addition, some research noted that there may be less of a difference in sentence length when actual time served, rather than the sentence imposed, is made the focus of examination (Lynch, 1993).

After rising gradually until it reached a peak of 49,900 in 1988, the prison population in England started to decline (Home Office, Research and Statistics Department, 1994, p. 1, Table 2). Following implementation of the Criminal Justice Act of 1991, with its statutory presumption against imposition of custodial sentences, there was a further sharp decline in the population from 45,835 at the end of September 1992 to 40,606 at the end of December 1992. However, upon reaching that low in December 1992 the prison population began to rise again at a fast rate, reaching 48,788 in March 1994 (National Association for the Care and Resettlement of Offenders [NACRO], 1994) and 70,860 by 2002 (Councell, 2003), and 80,977 in 2007 (The Times, 2007). In fact, the Home Office has recently had to adjust its population projections because of a growing divergence between the actual population and the projected population. The anticipated population is considerably lower than the actual population. The Home Office suggests that this may, partially, be the result of an increase in the remand population of untried prisoners (deSilva, Cowell, & Chow, 2005). Other factors that have contributed to this large increase in the prison population over the last 10 years include the custody rate in the courts, the average sentence length imposed, and the number of cases that pass through the court system (Councell, 2003). It is interesting to note that data on the prison population in the United States indicate that the number of inmates in U.S. prisons rose from 330,000 in 1980 to nearly 949,000 in 1994 (Gilliard & Beck, 1994), and to almost 1.5 million in 2004 (Harrison & Beck, 2005), with an increase from 139 sentenced prisoners per 100,0000 population in 1980 to 344 per 100,000 in 1993 (Gilliard & Beck, 1994, pp. 1, 9) and 486 per 100,000 in 2004 (Harrison & Beck, 2005).

As a result of the skyrocketing increase in the number of prison inmates in England, new and somewhat controversial measures were considered, including the building of new facilities, privatization, and the use of police cells as prison cells. Much of the controversy was spurred by the report of an inquiry committee headed by Lord Justice Woolf concerning the riot in April and May 1990 in Manchester's Strangeways Prison. The report on this prison, one of the oldest in the system, was released on February 25, 1991, and among many factors pointed to overcrowding in the facility as the root cause of the riot (Woolf, 1991). This report came at a time when the Home Office Research and Statistics Department released figures to indicate that in order to bring about a 1% decrease in crime, there would need to be a corresponding increase of 25% in the jail population (Tarling, 1993).

Of the total prison population of 70,860 in 2002, approximately 18% were untried or unsentenced prisoners. The majority of sentenced females were incarcerated for violence and/or drug offenses. The majority of sentenced males were incarcerated for violence, burglary, and/or drug offenses (Councell, 2003). As is true in the United States, the prison population had a disproportionately high number of lower class, unskilled, ethnic minority, and young inmates. In 2002, ethnic minority groups made up 22% of the male prison population and 29% of the female prison population. The ethnic minority prison population has increased 124% since 1992 (Councell, 2003). Operating at about 1,000 prisoners over operational capacity in 1994 (H.M. Prison Service, 1994a, pp. 7, 23), by October 2005 the prison system was operating at around 2,000 below operational capacity (H.M. Prison Service, 2005b). However, by May 2006, the prison population had increased to 110% of designed capacity to a record 77,677 persons in custody (National Offender Management Service [NOMS], 2006, p. 1).

Types of Prisons

The primary classification of English prisons is into local and training prisons. **Local prisons** serve the equivalent function of jails in the United States and house both offenders awaiting trial and those serving short sentences, typically less than 6 months. However, local prisons are increasingly being used to house inmates serving sentences of more than 1 year (Terrill, 2002). In addition to the local prisons there are remand centers, which house only offenders awaiting trial. A major difference between the English local prisons and the U.S. jails is that the former are part of a unified prison service, whereas the latter are typically administered by locally elected sheriffs who have no correctional background.

Training prisons are the institutions to which convicted offenders serving longer terms are sent. They can be either open or closed; closed training prisons are characterized by the presence of walls or fences to prevent escapes. About one third of British prisons are open, with no fences or walls. The specific prison to which an offender is sent is determined by both the offender's security classification and the length of the prison sentence (H.M. Prison Service, 1994a, p. 3). Male prisons are further classified into four security levels, A–D, with Category A prisoners being the most dangerous. Those in Categories A–C are in closed prisons; Category D prisoners are in open prisons. Women and juvenile prisoners can be classified as Category A prisoners, but are not categorized otherwise. Depending on the risk posed by female and juvenile prisoners, they may be held in open or closed prisons. In 2005, there were 40 local prisons, 17 open training prisons, 83 closed training prisons, 2 semi-open prisons, 14 remand centers, 34 young adult centers, and 18 female institutions in the English prison system (H.M. Prison Service, 2005a).

The Criminal Justice Act of 1991 expanded the jurisdiction of juvenile courts to 17-year-olds, raised the minimum age at which an offender can be sentenced to detention in a young offender institution from 14 to 15, and set a new minimum of 2 months for the detention of young offenders as opposed to the minimum of 21 days for males and 4 months for females that had existed prior to passage of the act (Criminal Justice Act of 1991, Section 63). Currently, there are a number of sentencing options available for juvenile offenders: discharges, fines, referral orders (ages 10–17), reparation orders, community orders (ages 16–17), community rehabilitation orders (ages 16–17), curfew orders, action plan orders, supervision orders (can last up to 3 years), parenting orders, intensive supervision and surveillance programs, detention and training orders (ages 12–17 and lasting for 4 months to 2 years), and long-term custody (Home Office, 2005a).

There are three types of custody centers for juvenile offenders: secure training centers, secure children's homes, and young offender institutions (YOIs). **Secure training centers (STCs)** house juveniles through age 17 and are primarily for education and rehabilitation. **Secure children's homes** are typically reserved for juvenile offenders between the ages of 12 and 14. However, females may be housed in secure children's homes up to age 16 and male offenders who are deemed to be particularly vulnerable may be housed in these facilities up to age 16. Secure children's homes focus on the physical, emotional, and behavioral needs of the juveniles that they house. Finally, **young offender institutions (YOIs)** house 15–21-year-old offenders.

As mentioned above, the local prisons serve the same purpose as jails in the United States—they house short-term prisoners either convicted or awaiting trial. In an effort to keep only dangerous prisoners in long-term training prisons, nonviolent offenders with sentences of more than 4 years are often held in local prisons as well; this was not the original intent of local prisons. Thus, the aforementioned controversy concerning overcrowding and what to do about it may, in reality, be a concern about local prisons and not long-term training prisons. In March 1994, local prisons were operating at 20% over capacity, and 8,488 prisoners were living two in a single cell (NACRO, 1994, pp. 2–3).

Although recently discontinued, there were also several "dispersal" prisons, which were maintained primarily to house the highest-security prisoners and the most dangerous inmates. These dispersal prisons were characterized by extreme control measures with constant inmate checks, surveillance cameras, and counting and recounting of inmates (Fitzgerald & Sim, 1979, p. 33).

As has been pointed out, deterrence through punishment is still a major objective of England's prison system. This appears to be the primary goal of prisons to carry through the twenty-first century. However, the other objective of rehabilitation has not been totally forgotten. As in the United States, vocational, educational, and therapeutic programs are present in most correctional facilities in England, and group counseling techniques are utilized in many of them. In fact, H.M. Prison Service lists its current goals as holding prisoners securely; reducing reoffending; maintaining safe, effective, and efficient establishments; and treating prisoners humanely (H.M. Prison Service, 2004f). This, along with the implementation of the NOMS in 2004 suggests that England is attempting to incorporate a philosophy of reintegration/rehabilitation into its punishment system. However, along with this move toward a reintegration/rehabilitation philosophy is also a move towards fiscal efficiency, because many of the provisions in NOMS allow for increased contestability and/or privatization of both prison and probation services (Bawden, 2006; National Association of Probation Officers [NAPO], 2006).

One continuing feature of the prisons in England is the attempt to ensure that all prisoners work during their stay in prison. This goal has been an integral part of every rehabilitative program in England and the United States: give the prisoner an opportunity to leave prison with some marketable skills and experience that can be utilized on the outside. As laudable as this goal may be, it is not easily attainable. Due to overcrowding and lack of resources, many prisoners find that the "skills" they are learning consist mainly of cleaning, laundry, cooking, and similar activities. This is comparable to the situation in the United States, except that in the United States there sometimes are not even enough of these activities to occupy the inmates' time; this inevitably results in increased inactivity and boredom.

As stated earlier, it appears that funds in England will continue to be more readily available for security measures than for meaningful treatment programs or work-related activities. This continues to be a challenge that England will carry through the twenty-first century.

Prison Staff

Over the past few decades, the prison staff in England has changed significantly, both in number and function. In 1978, there were 14,365 prison officers in posts throughout England and Wales; in April 2004, this number had increased to 46,310 permanent staff, including 2,580 part-timers treated as half positions (Civil Service Statistics, 2004a). Although the prison population between 1987 and 1992 decreased by 3% and the number of permanent staff increased at the same time, the male prison population between 1992 and 2002 increased by 57% and the female prison population increased 184%. However, the number of permanent staff did not experience the same increase, increasing only by 27% (Councell, 2003, p. 2). The prison staff continues to be a predominantly white male force. According to H.M. Chief Inspector of Prisons, race relations have improved in some prisons, but not all (Owers, 2005, p. 21). As of March 2005, approximately 5.7% of the prison staff were ethnic minorities (H.M. Inspectorate of Prisons, 2005, p. 117). As of October 2003, approximately 31.4% of the prison staff were female (Civil Service Statistics, 2004b, p. 1). Between April 2004 and March 2005, 49% of new staff members recruited were female and 8% were ethnic minorities (H.M. Inspectorate of Prisons, p. 125).

The Prison Service projected that the average cost per prisoner-place in 2004–2005 (total prison expenditure divided by the number of places available whether filled or not) would be more than $48,000 per year (H.M. Inspectorate of Prisons, 2005, p. 103). However, because of prison overcrowding, the government spends less than that for each inmate incarcerated. In 2004–2005, the cost per inmate was about $46,000, a 2.7% increase from the previous year

(H.M. Prison Service, 2004a, p. 80; 2005c, p. 103). Given the continuing overcrowding problems, this figure is expected to remain approximately the same or slightly increase given normal inflationary costs (NOMS, 2006). These figures are somewhat larger than those in the United States, where the actual cost for care of prisoners per year in 2001 averaged about $23,000 (Stephan, 2004, p. 1). Salaries continue to be the single most significant cost associated with operating the prisons, accounting for 66% of the total cost. Another major cost factor is prison construction. Twenty-one new facilities have been constructed since 1980 (H.M. Prison Service, 1994b, p. 8; 2005a).

The prison service staff consists primarily of three categories: the governor grade, which is similar to the wardens and other administrators in the United States; the uniformed staff, including prison officers, senior officers, and principal officers; and the professional and technical staff, which provide medical, vocational, educational, and other support services. The chief administrator of the prison is the governor, who has total responsibility for the operation of the prison. He (virtually all are male) is assisted by assistant governors, of which there are five grades. These individuals are assigned to the various wings of the prisons and to specific operations within the wings; in addition, they have responsibility for the rehabilitative care of the inmates (H.M. Prison Service, 2004c). Traditionally, the prison service throughout England has not been unified, which is in contrast to law enforcement, where there is a national police system with 43 forces.

In general, the uniformed prison officer begins his or her career in uniform and will finish it at the top of the uniformed grade as principal officer. Governor-grade officers generally enter the system laterally, depending on qualifications and testing; thus, they may move up the career ladder never having served in a uniformed officer capacity.

Selected from outside the prison service, governors attend the Prison Staff College at Wakefield in northern England, or Newbold Revel in Warwickshire, before being assigned to an institution. At the governor level, university education is valued and encouraged before and after coming into the Prison Service. Social work, criminal justice, psychology, and counseling are areas in which governors are expected to pursue further education.

The **Intensive Development Scheme** was recently implemented as a program developed to allow college graduates to enter H.M. Prison Service as prison officers and potentially advance to the level of Trainee Operational Manager, a middle management governor position. The Prison Service suggests that participants in this program could be leading a unit within a prison 3 years after joining the service (H.M. Prison Service, 2004b).

Although there has recently been an intensified effort to recruit women and minorities both locally and nationally, this has not been as successful as had been desired. Efforts have been more successful in recruiting women than minorities, but in general this still presents a major challenge for the Prison Service. This situation, of course, is not unique to England.

There are no formal educational requirements for prison officer applicants. However, successful applicants are required to take the **Prison Officer Selection Test (POST)**, which tests the applicants' mathematical, listening, reading, and writing skills (H.M. Prison Service, 2004c, 2004d). In addition, the applicant must be between 18 and 57 years of age and physically and mentally sound. Although having been convicted of an offense is not an automatic disqualification, the potential prison employee must be able to explain any criminal record.

The training process for recruits typically begins with a month at a local prison before attending the Prison Staff College for the residential portion of the training. The first 8 weeks at the college involve learning custody techniques and methods of supervising recreation, work, and leisure activities. Subsequently, a large portion of time is devoted to the rehabilitative needs of inmates. Upon completion of the training, the new staff member is assigned to a facility for a 1-year probationary period. Early in the training, the recruit is asked about a preference for the first posting. If possible, this request is honored; however, given the current crisis in the prisons, this often is not possible.[4] Prison staff are expected to continue their training through a series of in-service programs each year.

The third category of staff, technical and professional, is composed of both full- and part-time employees. The educational programs are required by law, and it is the responsibility of the local authority in which a prison is located to see that educational programs are provided for inmates. This is similar to institutions in the United States, where most states provide education toward a general education degree as a minimum in their prisons (Adler, Mueller, & Laufer, 1994, p. 420; Timmins, 1989, pp. 61–76). For the most part, civilian instructors carry out the educational and vocational programs. The funds are provided by the Home Office.

Traditionally, the majority of prison officers were drawn from the ranks of the armed services. This has created by default a heavily paramilitary atmosphere in most of the institutions. In 1970, 80% of the staff at Strangeways Prison in Manchester had military backgrounds, and over 90% of that number had more than 5 years of service with the military. In Albany Prison, 95% of senior and 65% of basic-grade officers had military experience. Although this has changed since the abolition of compulsory military service in England, the military influence is still quite prevalent in the Prison Service. One author suggests many new recruits are often given the impression that they have indeed joined a military service (Fitzgerald & Sim, 1979, p. 121).[5]

Prison Programs

In attempting to meet the prisons' dual objectives of punishment and rehabilitation, a prime requirement is that all prisoners work while serving their sentences. However, as mentioned earlier, the current crisis of overcrowding in prisons makes it almost impossible to enforce the goal that "employment must be provided for all people over school age, who are fit to work" (Fitzgerald & Sim, 1979, p. 55). As in the United States, wages for the work inmates perform are extremely low, with the minimum weekly payment allowed set at £4 in 2002. For those prisoners who are willing to work, but are unemployed because of lack of available opportunities, the minimum payment is £2.50 per week (H.M. Prison Service, 2002, p. 117; 2003b, p. 116).

Prison work can be classified by type: industry, farming, maintenance, domestic duties, outside work, and full-time education, including vocational and trade training. Most of the work performed by inmates is industry work, which includes laundry, tailoring, textiles, and woodwork. As newer facilities have been built, workshops have been added, but due to overcrowding and lack of resources, there have also been closures. Often the facilities are in place for these work activities, but the funds to pay for the instructors and work supervisors are not available. In general, the bulk of the work products have been for the use of the prison department, with the remainder sold to other government agencies and a small portion to outside consumers (Fitzgerald & Sim, 1979, p. 56). It was estimated in 1994 that the approximately 6,500 prisoners working in prison industries and farms produced £47 million worth of goods and services a year (H.M. Prison Service, 1994a, p. 5).

Of the various types of work, the farming industries have been the most profitable. The prisons offering farming are mostly open prisons, and the farming inmates are the "best risks" and require little security supervision. The overhead is small enough for the profits to be high.

The works department is responsible for taking care of the maintenance of the institutions. In addition, these inmates are used for building new facilities and improving existing ones. This makes economic sense because the cost of inmate labor has been estimated to be 25% less than that of outside contract labor (Fitzgerald & Sim, 1979, p. 57; H.M. Prison Service, 2004a). The prisoners must be transported to the site and closely supervised; however, it remains a more economic method of taking care of the buildings and facilities in the system.

The domestic assignment is usually considered boring drudgery. It includes cooking, cleaning, hospital orderly work, and the reception work of checking new inmates into the prison. Many of the facilities are antiquated, so the equipment and tools are old. Thus, most of the work in this area is dull and monotonous and can destroy morale among the prisoners. The one bright spot, as in institutions in the United States, is quite often the kitchen detail. It is a prized assignment because it allows prisoners access to extra rations.

Educational programs have been required in prisons by statute since 1823; however, participation has always been voluntary and usually is a low priority of prison officials. Education expenditures account for a small amount of the overall prison budget, and the programs are usually carried out by local education authorities with part-time instructors.

Many institutions have educational programs that allow inmates to complete courses of study from the elementary levels through a university education (H.M. Prison Service, 1994a, p. 6). As in the United States, few inmates entering the prison system are educationally prepared to meet the challenge of pursuing advanced education. In addition, inmates may be frequently transferred from one institution to another and often do not have the opportunity to complete a course of study started at one prison before being transferred to another.

Educational programs and opportunities are given a great deal of publicity outside the prison walls, but in actuality the situation is rarely as positive as portrayed. Vocational education programs through the work assignments are usually more beneficial than academic education in that an inmate may be able to learn a skill that could translate into gainful employment on the outside. According to the Chief Inspector of the Adult Learning Inspectorate, education and training opportunities in at least 60% of the prisons are grossly inadequate (H.M. Inspectorate of Prisons, 2005, p. 30). One educational program that has high participation in prison is physical education and recreation. With its roots in the exercise programs of the past century, English prisoners, like their counterparts in the United States (Telander, 1988, p. 82), have equipment and facilities devoted to their physical fitness. They are popular programs with the inmates, particularly in the men's prisons, and resources often are strained to accommodate all the requests for use of facilities. In the newer prisons, these physical fitness facilities often rival those of private health facilities on the outside. Weights, aerobics, and games such as ping-pong and soccer are the most popular activities; however, teachers attempt to include instruction in many aspects of fitness, including diet and overall wellness (Asmussen, 1992).

Finally, all prisons make provisions for the medical and religious needs of the inmates. The medical service has the responsibility for the general medical and mental well-being of the inmates. All institutions have a hospital, or at least a hospital wing, and there are full surgical units at some of the larger institutions. For example, H.M.P. Pentonville in north London opened a new, modern, and fully equipped hospital wing in 2004. The medical staff consists of full- and part-time employees, most of whom belong to the **Health Care Service for Prisoners** (until May 1992, the Prison Medical Service), an organization separate from the National Health Service of England. Hospital officers, who are prison officers with nursing training, assist the medical officers. Inmates who cannot be adequately treated within prison may be taken to outside medical facilities. Over the years, the treatment received by inmates has been a source of controversy; prisoners often complain about the quality of their medical care. The medical officer is often in a delicate position because inmates frequently complain of illness to get out of work assignments and to interrupt the routine of their daily prison lives. It is the responsibility of the medical officer or the hospital officer to determine the veracity of these complaints and sort out the malingerers from those actually requiring treatment (Fitzgerald & Sim, 1979, p. 79). By April 2006, the responsibility for health care in the prisons transferred to Primary Care Trusts (PCTs) (BBC, 2006; NOMS, 2005, p. 26). PCTs are local organizations, but are also a central part of the National Health Service and receive 75% of the National Health Service's budget (National Health Service, 2004).

All prisons have religious worship facilities and provide for the diversity in religious orientations of the inmates. The Church of England and the Catholic Church traditionally have been the most prominent in the prisons. However, in recent decades with the influx of minorities into England and the overrepresentation of minorities and other ethnic groups in the prisons, the range of religious facilities has increased. Religious instruction, counseling, and guidance are provided as resources allow.

Women in Prison

As of December 31, 2004, women made up approximately 6% of the prison population in England (H.M. Prison Service, 2004e), which is similar to the percentage of female prisoners in the United States, where 7% of the prison population is female (Harrison & Beck, 2005, p. 1). More than 4,000 female inmates (H.M. Prison Service, 2004e) are housed in the 18 institutions in England and Wales designed for women (H.M. Prison Service, 2005a). Although the female prison population has never been large, the needs of women prisoners are different and treatment varies accordingly. The central aim of imprisonment for women in England has been individualized treatment.

Until the late 1970s, women were sent to one of four regional prisons upon conviction. Holloway Prison in northern London, which opened in 1853, was considered the main women's facility. It began to be used exclusively for women in 1902 and has undergone many structural changes over the years. By 1970, Holloway had been torn down and rebuilt; today, it is primarily used as a re-mand center and local prison for women in the south of England. It resembles a hospital in appearance and contains extensive medical and surgical facilities as well as advanced treatment for women with psychiatric needs.

At first glance, women's facilities give the impression of being more relaxed than men's institutions, perhaps reflecting the belief that women pose less of a security risk than men. However, the conditions and problems in the institutions are comparable. The daily regime in the women's prisons is similar to that in men's prisons in that women are expected to work; a great deal of the work is domestic, and it can be boring and monotonous.

The main differences between men's and women's facilities exist in living accommodations. Women are generally assigned to rooms with toilets rather than cells, and these are more aesthetically pleasing. Also, inmates are allowed to wear their own clothes rather than a prison uniform (H.M. Prison Service, 2003b; Home Office, 1977, para. 178).

One difference that distinguishes women's prisons in the United States from those in England is that in four English facilities there is a fully outfitted maternity unit with space for 68 mothers and their children. As recently as 2003, there were plans for additional units to be added (Ash, 2003, p. 16). In the United States, only the state of New York has facilities for mothers and their children, and those are housed in just three institutions (LaRosa, 1992, p. E1). The English units are for women who have babies while in prison or immediately before entering the institution. In two of the units, mothers are allowed to care for their own infants until the babies are approximately 9 months old. In the other two units, mothers can care for their children up to the age of 18 months (Ash, p. 16). Mothers feed, bathe, clothe, and play with their infants until suitable placement can be found on the outside. This approach reflects the traditional concept of women's role in English society and the government's appreciation for that position: early nurturing and bonding between mother and child should be preserved if at all possible. While recognizing that the prison environment is not the optimal atmosphere in which to raise a child, the English feel it is still preferable to separating the mother from the baby at this important stage of a child's development (Ash, 2003; Home Office, 1977, para. 178).

■ Alternatives to Incarceration

Although the Prison Service is the unit within the Home Office with responsibility for institutional care, the **National Probation Service** is the unit responsible for the overall administration of noninstitutional programs. In 2001, as a result of the Criminal Justice and Court Services Act, the Probation and After-Care Service became the National Probation Service. The National Probation Service has recently been subsumed under NOMS; however, the National Probation Service continues to be responsible for supervising offenders sentenced to community sentences. In this section we will briefly examine the development, organization, administration, and operation of noninstitutional dispositions.[6]

Noninstitutional dispositions can be divided into two types: supervised programs, such as probation and community service; and unsupervised noncustodial dispositions, such as absolute discharges, deferred sentences, fines, and suspended sentences. The majority of noninstitutional sentences are imposed on offenders who have committed property offenses, particularly burglary and theft. However, between 1994 and 2004 there was a 43% increase in community sentences for crimes of violence (Home Office, 2005b, p. 50).

There are two general types of offenders subject to noninstitutional supervision: those subject to care by court order, such as probation and community service, and those in both voluntary and statutory after-care. It is important to remember that although in the United States probation and parole are ordinarily kept separate, in England the National Probation Service handles both types of cases. In 2004, 35% of offenders sentenced for indictable offenses received a community sentence (Home Office, 2005b, p. 43). The total number of individuals under community supervision in 2004 was 202,000—179,000 from Magistrates' Courts and 23,000 from Crown Courts (Home Office, 2005b, p. 2).

Probation

From the inception of probation in England, its purpose has been to remold offenders into good, honest, law-abiding citizens. Probation was the earliest form of a supervised noninstitutional disposition in England, and its aims and development parallel to those of the United States. The period following World War II in England was characterized by an emphasis in corrections on social casework, and the underlying assumptions that offenders could be rehabilitated and that the best place to do this was in the community (Raynor, 1988, p. 4). It was believed that the offender needed supervision while going through this process and that probation would provide the necessary guidance while the offender remained free in the community.

Although casework was the accepted method until the 1970s to bring about rehabilitation, probation workers began to employ other types of treatment strategies such as group work, community work, task-centered work, contracts, transactional analysis, family therapy, and behavior modification. This is similar to what was occurring in the United States in the same period.

The National Probation Service is located administratively within the National Offender Management Services (NOMS), which, in turn, is located administratively within the Home Office (see **Figure 10-1** earlier in the chapter). However, control is mostly centered in the local probation committees. Composed of magistrates and citizens, these local committees oversee the operation and resources required by the probation offices. If hostels or day-care units are needed, the committee is responsible for securing them, along with any other funds, which are provided by the local authority for such items as salaries and contracted services.

Originally, probation officers were frequently volunteers and did not have to meet any standard criteria in terms of education, background, or experience. When the Labour Party came to power in England in 1997, training for the probation service was taken out of social work education. Currently, there are nine different training consortia responsible for recruitment and training of probation officers. A Diploma in Probation Studies has been developed and is delivered through these consortia, while working with academic institutions. A Diploma in Probation Studies is the required qualification for trainees to become probation officers. There are three career options within the National Probation Service. Trainee probation officers (TPOs) must be at least 20 years old at the time of appointment and must complete a 2-year training program that will result in a Diploma in Probation Studies. Following completion of their training, TPOs are eligible for employment as probation officers. Probation Service officers work under probation officers and receive in-house training. Probation Service officers must have five General Certificates of Secondary Education (GCSEs) or equivalent work experience. Finally, supervisors are used to manage offenders' sentences to compulsory unpaid work. There are no age or educational requirements for supervisors (National Probation Service, 2005, pp. 6, 14, 17).

Between 1981 and 1991 the percentage of female probation officers increased from 36% to 47% (Barclay, 1993, p. 71). As of 1993, about 5% of probation officers were from minority ethnic groups (Home Office, Research and Statistics Department, 1993). This has not changed dramatically in this century.

Probation orders are normally issued for 1 year but can be issued for as short as 6 months and as long as 3 years. Originally issued as an alternative to sentencing, probation was made a sentence of the court by the Criminal Justice Act of 1991 (Section 8) with the result that probation orders can now be combined for a single offense with other sentences such as fines and community service orders. As in the United States, they may require one or more conditions related to behavior, work, activities, schedules, or supervision. Often, group counseling is required. There may be an order to attend a day training center for skill improvement in order to acquire the necessary basic skills to help secure employment. Or there may be a requirement that the offender reside at a probation hostel, where activities can be more closely monitored.

If an offender violates a condition of probation, he or she may be subjected to a revocation proceeding and the possibility of incarceration; however, community service or a fine is often given for breaching a condition of probation. In 2003, 35% of offenders on probation breached their orders (Home Office, 2004b, p. 5.7). The percentage of offenders who breached their probation orders that were sentenced to prison as a result varied by type of probation order breached. Offenders breaching drug treatment and testing orders were most likely to be sentenced to prison (50%) whereas offenders breaching community punishment orders were least likely to be sentenced to prison (12%) (Home Office, 2004b, p. 5.8).

Community Service

Another form of noninstitutional sentence in England is a **community order**, a form of unpaid community service.[7] Community orders cover a wide range of possible requirements, which can be combined depending on the severity of the offense and the criminal history of the offender. Community orders allow judges to individualize both the punishment and treatment requirements for each offender. Offenders sentenced to a community order remain in the community but must follow the requirements of the order. These requirements may include, but are not limited to, compulsory unpaid work, participation in specified activities, programs designed to change the offender's behavior, prohibition from certain activities and/or geographic areas, curfews, residence requirements, mental health treatment, drug rehabilitation, alcohol treatment, supervision, and attendance center requirements (Home Office, 2005a, pp. 11–12). Empowered by the Criminal Justice Act of 1972 (Section 15), English courts began to issue these orders in the early 1970s. They were intended as a sentencing disposition for any adult offender who had committed an offense for which he or she could be sent to prison; however, young persons between the ages of 17 and 21 who are not first-time offenders are currently the most common recipients of this supervised alternative to incarceration. These orders often require offenders to work between 40 and 240 hours under the direction of a community service organizer in the probation service. They may work in hospitals or nursing homes, help with community-based programs, provide physical labor for community work projects, or perform work for youth organizations. If the offenders do not carry out their work assignments, they can be fined and/or subjected to another, more severe, sentence.

Although hailed as an innovative technique for working offenders, community orders have received some criticism from both the public and probation officers. Probation officers cite concern over a lack of one-on-one contact with community service order clients, due primarily to the large number of cases and the impersonal settings in which the orders are carried out, whereas traditional probation work has usually involved close interpersonal contact. A general concern is that the aim of a community service order is somewhat ambiguous. It could be viewed as punitive, as rehabilitative, or as redress for the community. This conflict in objectives may diminish its benefit to the offender (NOMS, 2005; Raynor, 1988, Ch. 1; Young, 1979).

Financial Penalties

Fines are the most frequent penal sanction for criminal offenses, accounting for about 70% of all sentences in English courts and, more specifically, some 73% of summary sentences in Magistrates' Courts (Home Office, 2005b, Table 1-2 and Figure 1-1). **Victim compensation**, which is the payment of money to the victim, is also a frequent order.[8] In 2004, 16% of offenders sentenced in Magistrates' Court for indictable offenses and 7% of those sentenced in Crown Court for indictable offenses were ordered to pay compensation to their victims (Home Office, 2005b, Table 4-10).[9] The amount of the payment to the victim is based not only on the loss or injury suffered, but also on the offender's ability to pay, which may include the offender's own financial obligations as well as income. Fines are graded on levels between 1 and 5. Level 1 fines have a maximum of £200 (approximately $395) and Level 5 fines have a maximum of £5,000 (approximately $9,900). A sentence may include both a compensation order and a fine; when the offender has insufficient means to pay both, the courts have been directed to give preference to imposing a compensation order (Powers of Criminal Courts Act of 1973, Section 35). The Magistrates' Courts sentencing guidelines provide suggested tariffs for injuries—for example, up to £100 for a bruise, £125 for a black eye, £500–1,000 for the loss of a non-front tooth, and £1,500 for the loss of a front tooth (Judicial Studies Board, 2005, Section 5-32).

The unit fine system, established by the Criminal Justice Act of 1991 (Section 18) and abolished by the Criminal Justice Act of 1993 (Section 65) set the amount of the fine on a strict formula based on both the seriousness of the offense and the offender's financial circumstances. The formula was perhaps discarded prematurely. Nevertheless, the courts still take both the seriousness of the offense and the offender's financial circumstances into consideration when setting fines, and can order that there be an attachment to the offender's earnings. Offenders often can—and do—pay on the installment plan. Though the courts may modify the amount as a result of changes in the offender's circumstances, the success of financial penalties may depend on the state of the economy, and they can pose a dilemma for the offender. Either the offender risks violating the court order by not paying or risks being caught engaging in another illegal activity in an attempt to raise the required money (a situation not uncommon to most criminal justice systems). In the past, this dilemma has often been cited as a leading cause of prison overcrowding because the court can order imprisonment for non-payment of fines (Stanley & Baginsky, 1984, p. 204). The number of prisoners in custody for defaulting on a fine has decreased over time. In 1993, 30% of prisoners under sentence were there for fine defaulting. By 2004, this had decreased to 1.3% (Home Office, 2004b, Table 7-1). The maximum amount of time the court can order the offender to serve for failing to pay a fine or satisfy a compensation order depends on the amount owed.

Suspended Sentences

The 1967 Criminal Justice Act provided for another alternative called the **suspended sentence**. With a court-ordered sentence of confinement for up to 2 years, imprisonment can be suspended for a maximum of 2 years and the offender will not be imprisoned unless he or she is arrested for another offense punishable by incarceration. A suspended sentence order may carry additional requirements similar to probation and parole orders. When a suspended sentence of more than 6 months is imposed, the offender may be placed under the supervision of a probation officer, though this is unusual (Wasik, 1993, p. 164). If the individual is convicted while on a suspended sentence, he or she may be ordered to serve the remainder of the suspended sentence in prison, along with additional time imposed for the new offense. In 1991 such imprisonment was imposed in 70% of these cases (Wasik, p. 165); this rate has remained constant (NOMS, 2006).

As a result of the Criminal Justice Act of 1991, the court must first conclude that imprisonment is appropriate, and then can suspend imprisonment only if there are "exceptional circumstances" (Section 5).[10] It was expected that these provisions would result in decreased use of the suspended sentence and a consequent decrease in imprisonment for breach of the suspended sentence. As noted above, this has not been the case (NOMS, 2004, 2006).

An intermittent custody order, though not an alternative to custody, does allow offenders to spend only part of the week in prison and part in the community under the supervision of the National Probation Service (Home Office, 2005a, p. 15). Intermittent custody orders can only be imposed if a fine or community service is not sufficient, but incarceration is too severe (Judicial Studies Board, 2005, Section 3-68).

Nonpunitive Dispositions

The English use a number of dispositions that may be considered nonpunitive and do not entail any type of supervision. These include the absolute discharge, the conditional discharge, and the deferred sentence.

The **absolute discharge** is imposed when the court "considers that no further action is required on its part beyond the finding of guilt" (Home Office, 1986, p. 18). However, an offender receiving an absolute discharge may be required to pay compensation. With the **conditional discharge**, the offender remains liable for resentencing for the offense if he or she commits another offense within a stipulated period, which can be no longer than 3 years. This condition that the offender not commit another offense is the only one that the court can impose. If the offender commits another offense, he or she may be resentenced for the original offense. In some ways the absolute discharge may be compared to unsupervised probation in the United States, but without the involvement of the probation office.

With the **deferred sentence**, the actual sentencing of an offender can be postponed for up to 6 months after conviction (Powers of Criminal Courts Act of 1973, Section 1); this gives the court time to determine whether the offender requires supervision. This grace period is intended to provide an opportunity for the offender to change; if rearrested, the offender would normally face a sanction for the original offense as well as for the new one (Stanley & Baginski, 1984, p. 82). However, the offender must agree to the deferment of his or her sentence (Judicial Studies Board, 2005, Section 3-98).

■ The National Association for the Care and Resettlement of Offenders (NACRO)

The emergence of noninstitutional alternatives and care for offenders has not been due solely to the efforts of the government. Volunteer organizations have contributed greatly to this effort. The **National Association for the Care and Resettlement of Offenders (NACRO)** is one such organization.

Founded in 1966 by volunteers, ex-prisoners, and other citizens, its goal has been to train hostel workers and to develop hostels, short-term shelters, and accommodations for persistent but less serious offenders. NACRO has been the main nongovernmental agency working to improve and initiate new techniques for existing rehabilitation programs (Pointing, 1986; Raynor, 1988; Stanley & Baginsky, 1984). It has been primarily concerned with three areas of community-based corrections: the use of volunteers, increased use of diversion, and recommending to Parliament reforms for treating offenders in the community. NACRO regularly publishes information packets, briefings, and position statements on a wide variety of correctional issues. It has supported more active and effective use of volunteers by the probation department, the development of a wide variety of diversionary projects such as hostels, and the reduction of maximum sentences. In addition, it has proposed that penalties for so-called public order offenses such as drunkenness, marijuana use, vagrancy, begging, and soliciting be decriminalized to the point where incarceration is not required. (National Association for the Care and Resettlement of Offenders, 2007)

■ Conclusion: The Challenges of the Future for Corrections in England

There are critical issues facing England as she moves through the twenty-first century in every area of the criminal justice system, and particularly in corrections. These concerns can be classified into four basic challenges: increased crime and prison overcrowding, the antiquated physical state of prisons, the current status of prisoners' rights,[11] and the administrative structure of noninstitutional programs.

In the United States, the increase in crime is not a new phenomenon, and the correctional system has been expanding in response for a long period of time. This is not to suggest that the crisis has been solved. In England, however, the increase in prisoners, particularly those serving time for violent and drug-related offenses, is relatively recent; they constituted 34% of the prison population in 1981 but jumped to 57% by 1992. This has resulted in pressure on government officials to come up with solutions (Barclay, 1993, p. 60).

As a result of the large increase in reported crime in recent years, there have been immediate calls for action from the citizenry, but there is disagreement about what action should be taken. The government advocated the building of new facilities and the continued incarceration of criminals (Rose, 1993a). If continued, this type of action will have a strong impact on the general state of the prisons and may thwart attempts at increased use of noninstitutional alternatives. On the other hand, professionals in the correctional field point to the results of research that indicate that "prison is not the place to rehabilitate" offenders, and have advocated the increased use of community and noninstitutional corrections (Tuck, 1988, p. 88). After peaking in 1988, the prison population decreased to some extent. This was undoubtedly aided by the Criminal Justice Act of 1991 with its statutory presumption against the imposition of custodial sentences. However, the prison population is again on the rise (NOMS, 2006).

Along with the United States, England has begun to experience the challenge of balancing the punishment and deterrent goals of prison with prisoners' rights. The United States has dealt with this issue for a relatively long period of time in comparison to England. As noted earlier, as a result of a lack of constitutional rights such as those provided in the United States by the Eighth Amendment provision against cruel and unusual punishment, a disproportionate number of cases have come to the attention of the European Court of Human Rights regarding the treatment of inmates in English prisons (Fairchild & Dammer, 2001, p. 253).

Specific prisoners' rights questions have been raised concerning control, the classification system, the use of drugs to modify inmates' behavior in prison, and whether a prison is the proper place for treatment of offenders guilty of drug-related crimes. Perhaps, as some critics suggest, a hospital or other center designed for that specific purpose would be better suited for treatment than a maximum- or medium-security institution. From a practical standpoint, however, the cost and security risk of such treatment facilities may be prohibitive.

The general state of prisons in England will continue to challenge the government in the future. Even if England's economic problems were ameliorated in the near future, the country would still be left with several antiquated and outmoded facilities. Although there has been an increase in the construction of new prisons, given the aforementioned increase in reported crime and the prison overcrowding crisis, this may not be sufficient.

Experiments implementing new ideas, such as privatization of the prison system or the **Custody Plus** program where offenders spend half of their sentence in the community on probation (NAPO, 2006), have met with mixed results. In the early 1990s, the government executed a contract with an organization called Group 4, which agreed to run private prisons. Although a report on the first inspection of a private prison in the Wolds in Yorkshire reflected problems, such as drug abuse among prisoners (Mgadzah, 1993), it appeared that some of the initial difficulties, such as loose security and inadequate personnel training, were alleviated. According to the director of the Prison Agency, privatization will continue to be used commensurate with the level of performance by the organizations providing services. Such innovations have also been implemented in the United States and appear to have been received with limited enthusiasm from both the corrections community and the citizenry.[12]

Finally, in 2001 the movement to centralize the probation services in England occurred and became the National Probation Service. Although administratively linked with the Home Office within the NOMS, the Probation and After-Care Service is funded and partially controlled by local authorities. Not surprisingly, this leads to a lessening of differences in service and a reduction in the lack of standardized procedures for handling offenders. Merging into a single national probation service reporting directly to the Home Office has been seen as a way to standardize procedures, remove duplication of services, clarify lines of communication, and facilitate positive interagency cooperation. Cost effectiveness has improved and accountability issues are clearer. Detractors predicted that this type of change would lower morale and result in a "loss of identity" among the probation staff (Terrill, 1992, p. 77). Only time will support or refute this.

In England as well as in the United States, change comes slowly and with great resistance, especially in major institutions of the government. For England to resolve any crises in corrections, there must be an attempt to balance the needs of the public to reduce crime with the treatment needs of the offenders. The establishment of goals and key performance indicators for the prison service may be regarded as a welcome development and one that should be emulated more widely in both England and the United States. As one observer notes, (Bawden, 2006): "Collection and monitoring of reoffending rates is still very basic . . . it surely ought to be possible to collect data on reoffending, noting which combinations of prison and community sentences the offender had completed." Although there may be disagreement about both the goals and the manner in which progress toward meeting them is measured, there is some merit in clearly articulating public policy and measuring the degree to which it is being achieved.

CHAPTER SPOTLIGHT

- The Home Office took over the management and control of prisons in 1898. However, reform was slow due to opposition from prison governors. Many of the reforms proposed were heavily influenced by apparent successes in the United States.

- In the early twentieth century, the British government sent Sir Evelyn Ruggles-Briese to the United States to study the U.S. prison system. He was impressed by the American emphasis on education and reform.

- In 1904, Ruggles-Briese established a new facility for young men modeled upon the Elmira Reformatory in New York. It was located in the Kent village of Borstal, and this gave the borstal system its name.

- The borstal system was for youths between 16 and 20 years old; it was highly structured and included a daily schedule of work with self-discipline and a system of graduated rewards.

- Ultimately, borstals succumbed to the pitfalls of other prisons, with overcrowding and deteriorating conditions. They recently have been phased out. However, they succeeded in causing the courts to recognize that young persons should not be punished as adults.

- The Children's Act of 1908 revised the criminal justice system for juveniles and abolished the death penalty for children.

- English prisons continued to emphasize a policy of intentional harshness with segregation and a rigid rule of silence. Prisoners were not allowed to converse with one another and were restricted to communications with staff. This system lasted into the 1950s. It increased insanity with little, if any, scope for treatment. On the upside, it all but eliminated prison riots.

- In the early twentieth century different types of prisons developed. These included local prisons, regional prisons, and central prisons. These were akin to the different levels of security in the United States. After trial, prisoners were taken to the local prison for assignment. Those under 20 years of age were usually assigned to a borstal.

- The 1948 Criminal Justice Act sought to diminish the harshness of the system and abolished whipping and penal servitude. The policy changed to one whereby people were sent to prison *as* punishment not *for* punishment. Community-based sentencing options were also expanded and there was more of a focus on remission of sentences and rehabilitation.

- There are four basic rationales for sentencing: retribution, incapacitation, deterrence (both specific and general), and rehabilitation. One of the basic problems with incapacitation is that eventually the vast majority of prisoners are released.

- The role of deterrence is controversial; specific deterrence has little effect on impulse crimes, and research has also led to questions about the effectiveness of general deterrence.

- The English have attempted to balance deterrence and rehabilitation. The former placates the general public and the latter is primarily pursued through non-institutional alternatives.

- Since the nineteenth century there has been a shift in England toward the centralization of corrections. All of the various elements now fall under the auspices of the National Offender Management Service (NOMS), which manages prisoners from their entry into the criminal justice system to their exit.

- The Prison Service Agency was created in 1993 to provide greater autonomy to the management of prisons. The service managed its own budget and freed the Home Secretary from the responsibility of answering questions in the House of Commons regarding the prison service. Such questions could be directed to the Director General of the Prison Service. This approach ended with the creation of NOMS.

- The Prison Service Agency is accountable to Parliament and must respond to public concerns, inmate concerns, penal reformers, and official prison inspectors. It has set a number of goals, including maintaining decent conditions and preparing prisoners for their return to the community.

- Prisons are subject to the Independent Monitoring Boards, who are laypersons, formerly known as the Boards of Prison Visitors. They have three broad duties: to visit the prison regularly, to hear complaints and make major disciplinary decisions, and to present an annual report to the Home Office on matters of prison procedure and policy. The Home Secretary appoints board members for renewable 3-year terms.

- The parole board, as in the United States, is based on the principal that an inmate who has earned the privilege may be released from jail before his or her sentence is concluded. In England, the remainder of the sentence is served under supervision on license.

- By 2006, prisoners serving less than 4 years were routinely released at the halfway point in their sentence. Prisoners serving more than 4 years are eligible for release at the halfway point, but such release is not automatic. It is automatic at the two-thirds point.

- The Criminal Justice Act of 1991 abolished remission and replaced it with a system of adding days to custody as punishment for misbehavior in prison.

- A serious riot occurred in Strangeways Prison in 1990, leading to a major inquiry by a senior judge, Lord Justice Woolf. The Woolf Report found that overcrowding was the root cause of the riot. At the time, Home Office statistics suggested that to bring about a 1% reduction in crime there would need to be a 25% increase in the prison population.

- As in the United States, the prison population in England has a disproportionate representation of unskilled, lower class, and ethnic minority groups.

- In England, the primary classification of prisons is local prisons and training prisons. The former are primarily for people awaiting trial or serving sentences of less than 6 months. There are also remand (detention) centers for persons awaiting trial. Training prisons are for persons who have already been convicted and who have been sentenced to longer terms of imprisonment. These can be either open or closed. Closed prisons have fences and walls; open prisons have more relaxed security and no walls or fences.

- The prison a person is sent to is determined by the person's security classification and length of sentence.
- Male prisoners are classified from A to D. Category A prisoners are the most dangerous. Category A to C prisoners serve sentences in closed prisons. Category D prisoners serve their sentences in open prisons.
- Women and juveniles may be categorized as Category A prisoners but are otherwise not classified.
- The Criminal Justice Act of 1991 expanded the jurisdiction of the juvenile courts to 17-years-olds and increased the minimum age for detention from 14 to 15.
- There are three types of custody centers for juveniles: secure training centers, secure children's homes, and young offender institutions.
- There are attempts to ensure that all prisoners work during their stay in prison. This is regarded as part of the rehabilitative process and is designed to give prisoners marketable skills for when they are released.
- Prison work can be classified by type, including farming, industry, maintenance and domestic duties, outside work, or full-time education.
- Prisoners are also used to build new facilities and to maintain older ones. The combination of profits from work and savings benefits the Prison Service.
- Education programs have been required since 1823, but participation is voluntary and it has usually been a low priority for prison officials. Most prisons allow inmates to participate in education from the elementary level through to university degrees.
- There are also vocational education schemes that allow prisoners to train and learn skills that can be useful upon release. There are also physical education and recreational opportunities.
- All prisons provide for the medical needs of prisoners. All institutions have a hospital or hospital wing, which is the responsibility of the Health Care Service for Prisoners. Prisoners who cannot be adequately treated inside the prison may be taken to an outside facility.
- All prisons have worship facilities and provide for the diversity of religious orientations of prisoners.
- Over the past three decades the number of prison officers (guards) has increased significantly, even during times when the prison population declined. The cost per inmate of the prison service has also increased.
- The prison service staff consists primarily of three grades: the governor grades (the administrative staff), the uniformed staff (officers), and the professional and technical staff.
- Governors and officers are generally on separate career tracks, although an intensive development scheme enables college graduates to enter as uniformed officers and then move up the ranks to a middle management governor position.
- There are no formal educational requirements to become a prison officer beyond passing the Prison Officer Selection Test.
- As of December 31, 2004, women made up approximately 6% of the English prison population. They are housed in 18 specifically designated facilities

in England. The central aim of women's institutions is individualized treatment.

- Women's prisons are perceived as more relaxed; women wear their own clothes and are assigned to rooms with toilets, rather than cells. However, like their male counterparts, they are expected to work. Four female facilities are fully outfitted with maternity units, with space for 68 mothers and their children.

- Non-institutional care is the responsibility of the National Probation Service, also a part of the Home Office. The National Probation Service is responsible for supervising offenders assigned community sentences.

- Non-institutional dispositions fall into two types: supervised and unsupervised. Supervised includes probation and community service. Unsupervised includes deferred and suspended sentences and fines.

- Probation was the earliest form of non-institutional disposition in England, and it developed parallel to the system in the United States.

- A Diploma in Probation Studies is the required qualification to become a probation officer. Trainee probation officers must be at least 20 years of age at the time of appointment and must complete 2 years of training.

- Probation orders are typically for 1 year but can be for any time between 6 months and 3 years. They may be combined with fines and community service orders.

- If an offender violates a probation order, he or she may be subject to revocation proceedings and could be incarcerated, although a fine or community service is more likely.

- A community service order is a form of unpaid service to the community, and can consist of a wide range of possible requirements. It is normally tailored to the needs of the individual. If offenders do not carry out their work assignments they can be fined or subject to a more severe sentence.

- Fines are the most common sentence, accounting for 70% of sentences. Victim compensation (restitution) is also a frequent order.

- Fines are graded on levels between 1 and 5. Level 1 has a minimum of £500 and Level 5 has a maximum of £5,000.

- Another alternative is suspended sentences. A sentence of up to 2 years may be suspended for up to 2 years. This is contingent on the offender not being convicted of another offense that is punishable by a period of incarceration. Additionally, there may be other contingencies.

- Under the 1991 Criminal Justice Act, the court must first find that imprisonment is appropriate and then, if there are exceptional circumstances, the court may suspend the sentence.

- Another alternative is the intermittent custody order whereby a person spends part of the week in prison and part in the community under the supervision of the National Probation Service.

- There are nonpunitive dispositions: the absolute discharge, the conditional discharge, and the deferred sentence. The absolute discharge is imposed when the court decides that a finding of guilt is sufficient. A conditional discharge effectively continues the sentencing for a given period of no more than 3 years, to be discharged contingent on good behavior. If the offender violates a condition, the offender may be resentenced for the original offense as well as being sentenced for the new offense. Under a deferred sentence, the actual sentencing of an offender may be deferred for up to 6 months from conviction.

- The National Association for the Care and Resettlement of Offenders (NACRO) is a voluntary sector organization that has campaigned for and provided practical support for non-institutional alternatives to jail. This organization funds and runs hostels and shelters for offenders, administers diversion schemes, and lobbies Parliament and the government regarding prisoner rights and the criminal justice system.
- In England, as in the United States, the development of the correctional system has been marked by attempts to balance punishment and deterrence with prisoners' rights. In the United States this has been a long-term process, due to the effect of the Eighth Amendment. In England it has been more recent, with referrals to the European Court of Human Rights and the Human Rights Act of 1998.
- England has also moved to a more centralized system and, in the past decade, for the first time experimented with privatization of parts of the corrections system.

KEY TERMS

Absolute discharge: Imposed when the court "considers that no further action is required on its part beyond the finding of guilt" (Home Office, 1986, p. 18). However, an offender receiving an absolute discharge may be required to pay compensation.

Borstal system: Housing for offenders between the ages of 16 and 20; the system was based on a "house" concept that included a daily work schedule, self-discipline, and a system of graduated rewards for performance.

Community order: A form of unpaid community service.

Conditional discharge: The offender remains liable for resentencing for the offense if he or she commits another offense within a stipulated period, which can be no longer than 3 years.

Custody Plus: Privatization of prison programs where offenders spend half of their sentence in the community on probation.

Deferred sentence: The actual sentencing of an offender can be postponed for up to 6 months after conviction.

Director General: The Home Secretary's principal policy advisor who is appointed for a renewable fixed period by the Home Secretary with the approval of the Prime Minister. He or she oversees the day-to-day management of the Prison Service Agency, and also formulates and implements policies and procedures necessary for the operation of the prisons.

Governors: Officials directly in charge of the prisons.

Health Care Service for Prisoners: An organization separate from the National Health Service of England that was developed to operate as a medical support and welfare service dedicated to the health care needs of prisoners. (Until May 1992 this was known as the Prison Medical Service.)

H.M. Prison Service: (Also known as the Prison Service Agency.) One of four components of the English correctional system; it falls under NOMS. The administration of the entire prison system is the responsibility of this agency.

Howard League for Penal Reform: One of this organization's missions is the improvement of conditions for offenders both inside and outside institutions (in England and in the United States).

Independent Monitoring Board: Has three broad categories of duties: (1) to visit the prison regularly as a board and individually, with each member having the right to visit as often as he or she wishes; (2) to hear complaints about the prison and prison officials and make major disciplinary decisions; and (3) to make an annual report to the Home Office on matters of prison procedure and policy concerning each institution. This is an administrative system unique to England.

Intensive Development Scheme: A new program developed to allow college graduates to enter H.M. Prison Service as prison officers and potentially advance to the level of trainee operational manager, a middle management governor position.

Local prisons: Serve the equivalent function of jails in the United States and house both offenders awaiting trial and those serving short sentences, typically less than 6 months or up to 1 year.

National Association for the Care and Resettlement of Offenders (NACRO): Its goal has been to train hostel workers and to develop hostels, short-term shelters, and accommodations for persistent but less serious offenders.

National Offender Management Service (NOMS): Created to oversee the management of offenders from their entry into the criminal justice system through to their exit from the system.

National Probation Service: Responsible for supervising offenders sentenced to community sentences.

Offender manager: A single person responsible for a particular offender throughout the offender's incarceration and/or community sentence.

Parole board: Established in 1968, the board is made up of 20 members drawn from the community and includes judges, medical professionals, police officers, and others who have some experience in the social services. The parole board reviews parole applications from prisoners serving less than 15 years.

Prison Officer Selection Test (POST): Tests the applicant's mathematical, listening, reading, and writing skills.

Prison Service Management Board: The senior management team of the Prison Service; initially was composed of the Director General, the six directors of the operational and service directorates, and three members from outside the Prison Service.

Secure children's homes: Focus on the physical, emotional, and behavioral needs of the juveniles that they house.

Secure training centers (STCs): Typically reserved for juvenile offenders between the ages of 12 and 14 and are primarily for education and rehabilitation. However, females may be housed in secure children's homes up to age 16 and male offenders who are deemed to be particularly vulnerable may be housed in these facilities up to age 16.

Star prisoners: First-time offenders.

Suspended sentence: Imprisonment can be suspended for a maximum of 2 years, and the offender will not be imprisoned unless he or she is arrested for another offense punishable by incarceration.

Training prisons: The institutions to which convicted offenders serving longer terms are sent. They can be either open or closed; closed training prisons are characterized by the presence of walls or fences to prevent escapes.

Victim compensation: The payment of money to the victim.

Young offender institutions (YOIs): House 15–21-year-old offenders.

PUTTING IT ALL TOGETHER

1. How did England and the United States learn from one another when developing their individual correctional systems?

2. What are the goals of the new National Offender Management System, and how does it differ from the former system?

3. How do you think England could handle the present challenges it faces within the correctional system?

ENDNOTES

1. In 1991, for example, about 15% of the male and 30% of the female sentenced prison population was known to be from ethnic minority communities. At that time, these high rates were attributed to drug involvement (Barclay, 1993, p. 61). By 2003, 20% of the male and 15% of the female sentenced population was known to be from ethnic minority communities (Home Office, 2004a). It was suggested this difference for females could be attributed to changes in the categorization process utilized between 1991 and 2003.

2. England continued to imprison people for debt. As time went by, many in English society began to use the installment plans of purchase that characterized American culture. Easy credit led to a significant increase in debtors, witnessed by the fact that in 1946 there were only 243 persons in England and Wales sent to prison for failure to pay their bills; in 1960, this figure had risen to 5,675, and a new prison for debtors was opened in 1962 (DeFord, 1962, p. 133).

3. The reconviction rate of inmates for indictable offenses within 2 years of release from prison is about 55% percent for both adult males and adult females, according to figures released in 2002 (Home Office, 2002).

4. As of 1989, opposite-sex-related posting was made possible; men could serve in women's facilities and vice versa. This appears to have met less resistance from both inmates and staff than had been anticipated. In fact, it has made for more career opportunities for prison staff.

5. For a discussion of the influence of the military and its role in rehabilitative efforts, see Fitzgerald and Sim (1979).

6. There are a number of detailed works on the earlier organization and operations of the Probation and After-Care Service and the non-institutional

programs available in England. See, e.g., Bochel (1976), Pointing (1986), Raynor (1988), Stanley & Baginsky (1984), and Whitehead (1990).

7. For a detailed description of the history, development, implementation, and operation of community service orders, see Young (1979).

8. In England, the term *restitution* is used to refer to the restoration of property to the victim. A restitution order may be imposed in conjunction with any other sentence (see the Theft Act of 1968, Section 28).

9. There is a Criminal Injuries Compensation Board, which is empowered to award compensation to the victims of crimes of violence regardless of whether the offender has been convicted or even apprehended (see, e.g., Hirschel, 1973).

10. A variation of the suspended sentence, the partly suspended sentence, was introduced by the Criminal Law Act of 1977 (Section 47). This provision allowed offenders sentenced to between 6 months and 2 years to serve part of the sentence (at least 28 days and not more than three quarters of the whole term) in prison and have the rest of the sentence suspended. The partly suspended sentence was abolished by the Criminal Justice Act of 1991.

11. For a more thorough discussion of the development of prisoners' rights, see Livingston and Owen (1993).

12. For an early analysis of the U.S. experience with prison privatization and its potential for success in England, see Ryan and Ward (1989).

REFERENCES

Adler, E., Mueller, G., & Laufer, W. (1994). *Criminal justice*. New York: McGraw-Hill.

Ash, B. (2003). Working with women prisoners. London: H.M. Prison Service. Retrieved January 7, 2006, from http://www.hmprisonservice.gov.uk/assets/documents/10000339WorkingwithWomenPrisoners.doc.

Asmussen, K. (1992). *A comparison of English and American correctional physical education and recreation programs*. Unpublished Research. University of Nebraska at Omaha.

Barclay, G. (1993). *Digest 2: Information on the criminal justice system in England and Wales*. London: Home Office.

Bawden, A. (2006, April 4). The jury is still out on reform of the system. *Guardian Unlimited*. Retrieved June 6, 2007 from http://www.guardian.co.uk/public/features/story/0,,1746526,00.html.

BBC News (2006, April 9). The challenge providing NHS care in prison. Retrieved June 6, 2007 from, http://news.bbc.co.uk/1/hi/health/4876874.stm.

Berger, V. (1992). *Case law of the European Court of Human Rights*. Dublin: Round Hall.

Biles, D. (1988). *Current international trends in corrections*. Sydney, Australia: Federation.

Blunkett, D. (2004). Reducing crime—Changing lives: The government's plans for transforming the management of offenders. London: Home Office.

Bochel, D. (1976). *Probation and after-care: Its development in England and Wales.* Edinburgh: Scottish Academic.

Borrie, G. (1976, May). The membership of boards of visitors of penal establishments. *Criminal Law Review,* 281–298.

Camp, G., & Camp, C. (1991). *The corrections yearbook, 1991: Adult prisons and jails.* South Salem, NY: Criminal Justice Institute.

Carter, P. (2003). *Managing offenders, reducing crime.* London: Home Office.

Cavadino, M., & Dignan, J. (1992). *The penal system: An introduction.* London: Sage.

Civil Service Statistics. (2004a). Civil service staffing: April 2004. Retrieved January 6, 2006, from http://www.civilservice.gov.uk/management/statistics/publications/xls/staff/staff_apr04_4nov04.xls.

Civil Service Statistics. (2004b). Civil service staff in post at 1 October 2003 for all staff and senior civil service level, by gender. Retrieved June 6, 2007 from http://www.civilservice.gov.uk/management/statistics/publications/pdf/gender_oct03.pdf.

Councell, R. (2003). *The prison population in 2002: A statistical review.* London: Research, Development, and Statistics Directorate.

Curtis, S. (1989). *Juvenile prevention through intermediate treatment.* London: B. T. Batsford.

Davis, G., Boucherat J., & Watson, D. (1989). Pre-court decision making in juvenile justice. *British Journal of Criminology, 29,* 219–235.

DeFord, M. A. (1962). *Stone walls.* Philadelphia: Chilton.

deSilva, N., Cowell, P., & Chow, T. (2005). *Home Office statistical bulletin: Updated and revised prison population projections—2005–2011.* London: Research, Development, and Statistics Directorate.

Durose, M. R., & Langan, P. A. (2002). *State court sentencing of convicted felons, 2002.* Washington, DC: U.S. Department of Justice.

Evans, P. (1980). *Prison crisis.* London: George Allen & Unwin.

Fairchild, E. (1993). *Comparative criminal justice systems.* Belmont, CA: Wadsworth.

Fairchild, E., & Dammer, H. R. (2001). *Comparative criminal justice systems.* (2nd ed.). Belmont, CA: Wadsworth.

Fitzgerald, M., & Sim, J. (1979). *British prisons.* Oxford: Basil Blackwell.

Garland, D. (1985). *Punishment and welfare.* London: Gower.

Gibson, B., Cavadino, P., Rutherford, A., Ashworth, A., & Harding, J. (1994). *Criminal justice in transition.* Winchester, England: Waterside.

Gilliard, D. K., & Beck, A. J. (1994). *Prisoners in 1993.* Washington, D.C.: U.S. Department of Justice.

Gosling, T. (2006, April 26). Reform must sell the rough with the smooth. *Guardian Unlimited.* Retrieved June 6, 2007, from http://society.guardian.co.uk/crimeandpunishment/comment/0,,1761089,00.html.

Hall-Williams, J. (1970). *The English prison system in transition.* London: Butterworths.

Hall-Williams, J. (1987). *New kinds of non-custodial measures—The British experience.* Resource Materials No. 32. Tokyo: UNAFEL.

Harrison, P. M., & Beck, A. J. (2005). *Bureau of Justice Statistics bulletin: Prisoners in 2002*. Washington, D.C.: U.S. Department of Justice.

Hayes, M. (1981, May). Where now the right to bail? *Criminal Law Review*, 20–24.

Hirschel, J. D. (1973). The criminal injuries compensation board. *Current Legal Problems, 26*, 40–65.

H.M. Inspectorate of Prisons. (2005). *Annual report and accounts: April 2004–March 2005*. London: Home Office.

H.M. Prison Service. (1993a). *Business plan, 1993–94 (April)*. London: H.M.S.O.

H.M. Prison Service. (1993b). *Framework document (April)*. London: H.M.S.O.

H.M. Prison Service. (1992). *Perspectives on prison: A collection of views on prison life and running prisons*. London: H.M.S.O

H.M. Prison Service. (1994a). *Background brief H. M. Prison Service in England and Wales (revised Feb. 3)*. London: H.M.S.O.

H.M. Prison Service. (1994b). *Corporate plan, 1994–97 (April)*. London: H.M.S.O.

H.M. Prison Service. (2002). *Prisoners= information book: Male prisoners and young offenders*. London: Prison Reform Trust.

H.M. Prison Service. (2003a). *Corporate plan, 2003/2004 to 2005/2006* and *Business plan 2003–2004*. London: Home Office.

H.M. Prison Service. (2003b). *Prisoners= information book: Women prisoners and young offenders*. London: Prison Reform Trust.

H.M. Prison Service. (2004a). *Annual report and accounts: 2003–2004*. London: Home Office.

H.M. Prison Service. (2004b). Intensive development scheme. Retrieved January 6, 2006, from http://www.hmprisonservice.gov.uk/careersandjobs/typeswork/intensivedevelopmentscheme/.

H.M. Prison Service. (2004c). Prison officer. Retrieved January 6, 2006, from http://www.hmprisonservice.gov.uk/careersandjobs/typeswork/prisonofficer/.

H.M. Prison Service. (2004d). Prison officer selection test (POST). Retrieved January 6, 2006, from http://www.hmprisonservice.gov.uk/careersandjobs/post/.

H.M. Prison Service. (2004e). *Prison population & accommodation briefing for 31st December 2004*. Retrieved January 7, 2006, from http://www.hmprisonservice.gov.uk/assets/documents/1000089B31_12_04_pop_report.doc.

H.M. Prison Service. (2004f). Statement of purpose. Retrieved January 5, 2006, from http://www.hmprisonservice.gov.uk/abouttheservice/statementofpurpose/.

H.M. Prison Service. (2005a). List of prisons. Retrieved January 4, 2006, from http://www.hmprisonservice.gov.uk/assets/documents/1000105Fprisonlist.doc.

H.M. Prison Service. (2005b). Monthly bulletin—October 2005. Retrieved January 4, 2006, from http://www.hmprisonservice.gov.uk/assets/documents/10001484 pop_bull_oct_05.doc.

H.M. Prison Service. (2005c). Annual report and accounts: April 2004–March 2005. London: Home Office.

H.M. Prison Service. (2007). Locate a prison. Retrieved June 6, 2007, from http://www.hmprisonservice.gov.uk/prisoninformation/locateaprison/.

Home Office. (1965). *The adult offender*. London: H.M.S.O.

Home Office. (1977). *Prison and the prisoner*. London: H.M.S.O.

Home Office. (1986). *The sentence of the court*. 4th ed. London: H.M.S.O.

Home Office. (1990). *The inside story*. London: H.M. Prison Service.

Home Office. (1991). *Custody, care and justice: The way ahead for the prison service in England and Wales.* London: H.M.S.O.

Home Office. (1992a). *Criminal statistics England and Wales 1991.* London: H.M.S.O.

Home Office. (1992b). *Report on the work of the Prison Service, April 1991–March 1992.* London: H.M.S.O.

Home Office. (2002). *Prison statistics—England and Wales: 2002.* London: TSO.

Home Office. (2004a). *Criminal statistics—England and Wales: 2003.* London: Research, Development, and Statistics Directorate.

Home Office. (2004b). *Offender management caseload statistics, 2003.* London. Retrieved January 4, 2007, from http://www.homeoffice.gov.uk/rds/pdfs04/hosb1504.pdf.

Home Office. (2005a). *Crime, sentencing, and your community: Sentencing explained.* London: Home Office Communication Directorate.

Home Office. (2005b). *Sentencing statistics: 2004.* London: Research, Development, and Statistics Directorate.

Home Office, Research and Statistics Department. (1993). *Ethnic origins of probation staff 1992.* London.

Home Office, Research and Statistics Department. (1994). *The prison population in 1993 and long term projections to 2001.* London.

Howard's 27 ways. (1993, October 15). *New Statesman & Society*, p. 5.

Judicial Studies Board. (2005). Magistrates= adult court bench book. London: Judicial Studies Board. Retrieved January 3, 2006, from http://www.jsboard.co.uk/magistrates/adult_court/complete.pdf.

King Independent Monitoring Board. *Frequently asked questions.* Retrieved January 3, 2006, from http://www.imb.gov.uk/faqs/.

King, R., & Morgan, R. (1979). *Crisis in the prisons: The way out.* Bath: University of Bath Press.

LaRosa, P. (1992, May 12). Babies behind bars: In 3 NY prisons inmates who give birth may keep their babies with them. Dr. Spock endorsed the idea, but critics are queasy. *Los Angeles Times*, p. E1.

Lewis, L. (1993, October 24). Tories fall foul of judges and church leaders. *Guardian Weekly*, p. 8.

Livingston, S., & Owen, T. (1993). *Prison law.* London: Clarendon.

Lynch, J. P. (1993). A cross-national comparison of the length of custodial sentences for serious crimes. *Justice Quarterly, 10* (4), 639–660.

Maguire, K., Pastore, A. L., & Flanagan, T. J. (1993). *Sourcebook of criminal justice statistics 1992.* Washington, D.C.: U.S. Government Printing Office.

Maguire, M., Vagg, J., & Morgan, R. (1985). *Accountability and prisons: Opening up a closed world.* London: Tavistock.

Martinson, R. (1974). What works? Questions and answers about prison reform. *The Public Interest, 35*, 22–54.

McConville, S. (1981). *A history of English prison administration.* Vol. I. London: Routledge & Kegan Paul.

McConville, S. (1987). Some observations on English prison management. *American Correctional Association Monographs, 11*.

Mgadzah, R. (1993, August 26). Time to review the key question. *The Independent*, p. 22.

Mills, H. (1993, October 16). Howard takes off kid gloves. *The Independent*, p. 1.

National Association for the Care and Resettlement of Offenders (NACRO). (1994). *Prison overcrowding—recent developments*. London: NACRO, in-house publication.

National Association for the Care and Resettlement of Offenders. (2007). Nacro services. Retrieved June 28, 2007, from http://www.nacro.org.uk/ services/.

National Association of Probation Officers. (2006). *Future of the Probation Service— New legislation: from H. Fletcher*. London: National Association of Probation Officers, in-house publication.

National Health Service. (2004). Primary care trusts. Retrieved January 6, 2006, from http://www.nhs.uk/England/AuthoritiesTrusts/Pct/Default.aspx.

National Offender Management Service. (2004). Population in custody: Monthly tables. London: Home Office.

National Offender Management Service. (2005). *The national reducing re-offending delivery plan*. London: Home Office.

National Offender Management Service. (2006). *Population in custody: Monthly tables*. London: Home Office.

National Probation Service. (2005). *Careers in probation: An informative guide featuring careers, information, interviews with probation staff, and an up-to-date look at probation work*. London: National Probation Directorate.

Offenders moved to open jails (2007, June 18). *The Times*, p. 2.

Owers, Anne. (2005). *Annual report of H.M. chief inspector of prisons for England and Wales*. London: TSO.

Pointing, J. (1986). *Alternatives to custody*. Oxford: Basil Blackwell.

Pratt, J. (1986). A comparative analysis of two different systems of juvenile justice: Some implications for England and Wales. *The Howard Journal, 25*, 33–51.

Prison Service News. (2006, May). First prisons transfer health care commissioning. Retrieved June 6, 2007, from http://www.hmprisonservice.gov.uk/ prisoninformation/prisonservicemagazine/index.asp?id=1241,18,3,18,0,0.

Raynor, P. (1988). *Probation as an alternative to custody: A case study*. Aldershot, England: Avebury.

Richardson, G. (1984). Time to take prisoners' rights seriously. *Journal of Law and Society, 11*, 1–32.

Richardson, G. (1985). The case for prisoners' rights. In M. Maguire, J. Vagg, & R. Morgan (Eds.). *Accountability and prisons: Opening up a closed world* (pp. 19-28). London: Tavistock.

Rose, D. (1993a, August 22). Howard plans harsher prisons. *The Observer*, p. 1.

Rose, D. (1993b, October 17). Top judges: Jail not the answer. *The Observer*, p. 1.

Ryan, M., & Ward, T. (1989). *Privatization and the penal system: The American experience and the debate in Britain*. Stratford, England: Open University Press.

Smith, G. (1985). The community corrections option. *Resource Materials,* No. 28. Tokyo: UNAFEL, pp. 135–148.

Stanley, S., & Baginsky, M. (1984). *Alternatives to prison: An examination of noncustodial sentencing of offenders*. London: Peter Owen.

Stephan, J. J. (2004). *State prison expenditures, 2001*. Washington, D.C.: U.S. Department of Justice.

Stern, V. (1987). *Bricks of shame: Britain's prisons*. Harmondsworth, England: Penguin.

Chapter Resources

Stewert, G., & Smith, D. (1987). Help for children in custody. *British Journal of Criminology, 27*, 302–310.

Stockdale, E. (1979). The correctional system in England and Wales. In R. Wicks & H. Cooper (Eds.). *International corrections* (pp. 3–21). Lexington, MA: D. C. Heath.

Tarling, R. (1993). *Report from Research and Statistics Department. Home Office.* London: H.M.S.O.

Taylor, I. (1993, October 19). You cannot cage them forever. *The Independent*, p. 2.

Telander, R. (1988, October 17). Sports behind the walls. *Sports Illustrated*, p. 82.

Terrill, R. (1992). *World criminal justice systems: A survey.* (2nd ed.). Cincinnati, OH: Anderson.

Terrill, R. (2002). *World criminal justice systems: A survey.* (5th ed.). Cincinnati, OH: Anderson.

Timmins, W. (1989). Prison education in Utah: From "The Peawiper to South Park Academy." *Journal of Offender Counseling, Services, and Rehabilitation, 14*, 61–76.

Tough talk won't stop criminals. (1993, October 17). *The Observer*, p. 22.

Travis, A. (1993, October 24). Research flaws jail policy. *The Observer*, p. 8.

Tuck, M. (1988). The future of corrections research. In D. Biles (Ed.). *Current Australian trends in corrections* (pp. 71–89). Sydney, Australia: Federation.

Walker, P. (1973). *Punishment: An illustrated history.* New York: Arco.

Wasik, M. (1993). *Emmins on sentencing.* (2nd ed.). London: Blackstone.

Whitehead, P. (1990). *Community supervision for offenders: A new model of probation.* Aldershot, England: Avebury.

Wicks, R., & Cooper, H. (1979). *International corrections.* Lexington, MA: D. C. Heath.

Willis, A. (1986). Alternatives to imprisonment: An elusive paradise? In J. Pointing (Ed.). *Alternatives to custody* (pp. 18–38). Oxford: Basil Blackwell.

Wolff, M. (1967). *Prison.* London: Spottiswoode.

Woolf, Lord Justice. (1991). *Prison disturbances April 1990: Report of an inquiry.* London: H.M.S.O.

Young, W. (1979). *Community service orders: The development and use of a new penal measure.* London: Heinemann.

Younger, K. (1977). Sentencing. *The Howard Journal, 11*, 17–21.

Ancillary Issues

V

Terrorism

11

Chapter Objectives

After completing this chapter, you will be able to:

- Discuss the history and evolution of terrorism and terrorist legislation in England.
- Explain the history and reasons behind the animosity between northern and southern Ireland and England.
- Explain the background and reasons behind why Al-Qaeda and Osama bin Laden are at war with Western society.
- Discuss the reasoning and practical applications of why terrorist legislation is difficult to define and apply.
- Explain how and why various terrorist-related legislative acts may impinge on civil liberties.
- Discuss the methods used to prevent future acts of terrorism.

■ Introduction

On the morning of September 11, 2001, four passenger aircraft were hijacked by terrorists from the group <u>Al-Qaeda</u>. Two of these aircraft were crashed into the World Trade Center towers in New York City, killing over 2,800 people. The third aircraft crashed into the Pentagon in Washington, D.C., killing 184, while a fourth crashed into a field in Shanksville, Pennsylvania, after passengers fought with the terrorists. Forty passengers and crew died.

On Thursday, July 7, 2005, three bombs exploded within 50 seconds of each other on three London Underground trains and one, an hour later, on a bus near Tavistock Square in central London (7 July 2005 London bombings, Wikipedia:

2007:1). Fifty-six people were killed including the four bombers. Approximately 700 people were injured by the blasts. Al-Qaeda eventually claimed responsibility for the attack, although the extent of their involvement remains unclear. These bombings were the deadliest attacks in London since World War II and the second deadliest terrorist attacks in England since the bombing of Pan Am Flight 103 over Lockerbie, Scotland in 1988. On July 21, 2005, four more explosions occurred on three London Underground trains and the fourth on a bus in Shoreditch East London. However, due to faulty bomb construction, only the detonators exploded. A fifth bomb was found abandoned in West London two days later. There were no fatalities and no one was injured by these attacks. The police arrested all suspected bombers, and although it was not definitively proven, the bombers may have had ties with Al-Qaeda. They claimed to have been motivated by opposition to the Iraq war and images of U.S. and British troops killing civilians (7 July 2005 London bombings Wikipedia: 2007:1).

Although the United Kingdom is no stranger to terrorism on its own soil due to organizations such as the Irish Republican Army (IRA), these three sets of attacks nearly four years apart have placed the United States and England in the unfamiliar territory of the suicide bomber and the large-scale international terrorist attack devastatingly and successfully perpetrated by a few. The consequences to the world at large have been far-reaching. Pressure has been brought to bear on many nations to bring all of those responsible for the September 11, July 7, and July 21 attacks to justice, and each country has since had to constantly reassess its ability to prevent future attacks. Each country continually determines its readiness by way of intelligence-gathering capabilities within and across nations, enacts legislation that endeavors to balance effective investigative tactics with civil liberties, and prepares to cope efficiently and effectively with the aftermath of attacks if and when they happen.

In this chapter, we examine the changing nature of the terrorist threats that England has faced and describe the roles played by the Irish Republican Army and Al-Qaeda. In addition, we describe and assess the legislation that England has enacted in an attempt to meet the new challenges posed by international terrorism, particularly since September 11, 2001, and July 2005. Emphasized in this analysis is the tension that inevitably exists between attempting to maximize public safety and security and maintaining civil liberties.

■ Legislating Against Terrorism

In the creation of any type of criminal legislation, there are four basic philosophies or objectives. Ideally, legislation should:

1. Protect the public from criminal acts
2. Not encroach upon the liberties of the individual
3. Be fair to the accused
4. Have a deterrent effect on future criminal acts

Legislation should emphasize and strive to balance these objectives if the law is to be just and fair and to do what it is designed to accomplish. The challenges that have always existed are clearly defining the act that is to be considered illegal and applying the law in a uniform and fair manner. These challenges have been particularly problematic for those entrusted with the creation of terrorist legislation. Defining a law to combat terrorism would appear to be straightforward because the acts inherent in terrorism (murder, kidnapping, inflicting injuries, destruction of property, etc.) have always been easily understood and consistently acknowledged in nearly all societies as being wrong, and they have been legislated as such. Historically, the orientation of terrorism, particularly in the United States, has been that the criminal law could adequately deal with "conventional" terrorist acts without the need for more specific legislation. This worked well in the United States because most terrorist acts were domestic in nature and not often carried out by highly organized groups such as the IRA. Consequently, because of the IRA, English legislation was more specific towards terrorism.

With the 9/11 and 7/7 attacks, the legislative "balance" in creating just and effective laws has been altered considerably in the United Kingdom and the United States. We are no longer dealing with the "conventional" terrorist acts of kidnapping, bombs in packages, political assassinations, and so forth, so we must devise punishments for terrorists that go beyond those meted out for conventional criminal acts. Terrorist acts now perpetrated are so threatening and may be so devastating and so easily completed that some would argue the recent legislation passed in the United States and United Kingdom has necessarily shifted the emphasis to the prevention of future terrorist acts. If this is the case, it is at the expense of the other two basic philosophies, fairness to the accused and the encroachment upon individual civil liberties. It appears the goal of deterrence has also gone by the wayside because it is apparent that no punishment created through legislation will deter idealistic terrorists or suicide bombers. Prevention is now the primary emphasis, and the recent legislation and the ensuing civil liberties debates have reflected this.

With this approach, the Patriot Act in the United States and the Prevention of Terrorism Acts in England now encourage the policing of day-to-day acts to determine who the terrorists are and to find out what they may be doing. Whom we associate with, the owning of a cell phone, our Internet access, our free speech, and individual curfews targeting certain people now fall under the auspices of the prevention of terrorism. Through this new terrorist legislation, it appears the net is widening and is snagging not only those people whose activities may be perceived as recruiting, organizing, and planning to carry out terrorist activities, but also those people whose actions are routine, basic, and what most people fiercely regard as private. Thus, there is much gray area as to what activities may be considered under the umbrella of "terrorism."

Given the added pressure and urgency to create effective legislation in these more threatening times, public perception is that defining terrorism for legislative purposes should now be simpler because we are dealing with terrorism on a seemingly larger scale with greater threats and with greater potential for loss of life. However, in the past, as it still is very much the situation today, legislating against terrorism was very difficult due to the politics, the problems inherent in defining what constitutes a "terrorist act," and the fact that the ideologies surrounding terrorism change with history, time, geography, and the perpetrators. Perspectives have also varied depending on whether there was a national policy on engaging in terrorism, if it was a form of state repression, and if the perpetrators were foreign or domestic. The result of this was that although legislators were pressured to generate legal definitions broad enough to encompass most potential terrorist acts, each act was assessed on a case-by-case basis analyzing who was the target, what the possibility of success was, and realistically, what action could be taken against the perpetrators (White, 2002, p. 9). Further complicating the matter was the problem of defining who a terrorist might be. The term *terrorist* may be synonymous with terms such as *freedom fighter* and *patriot*. Even Osama bin Laden was seen in a more positive light in the 1980s when he was fighting the Soviets in Afghanistan and the United States was supplying him with funds and weapons. History is rife with these definitional "interpretations."

Further, today representatives of Al-Qaeda and other groups have stated on numerous occasions that the British and the Americans are the terrorists because of their past political policies towards Arab nations and, more recently, the invasions and occupations of Afghanistan and Iraq. Thus, the questions become who are the "terrorists," and who is right? Perspective weighs heavily in this analysis. As a result, and for the reasons discussed above, legislators, governments, and organizations have needed to cover all eventualities, perspectives, motivations, actions, and conflicts of ideologies by defining "terrorism" very generally.

Therefore, the challenge for us is that, although England and the United States are strong allies and are politically and somewhat culturally compatible, a comparison of terrorism legislation that encapsulates the intentions, histories, politics, and legislative evolution of two countries such as England and the United States is difficult to conduct. This is especially the case before and after the terrorist attacks of 9/11 because of the rapid development in terrorism in recent times and the legislators' desire to keep pace, particularly in England after the July 2005 attacks.

What follows is a brief history of terrorism in England, a review of the pertinent legislation of the last 30 years or so to note the struggle with legislating against terrorism, and a discussion of the parallels in both nations concerning the need for the protection of the people and the balancing of their civil liberties.

■ Terrorism in England: A Brief History

Although not the only source for terrorist activity, the varying factions of the **Irish Republican Army (IRA)** provided most of the terrorist threats faced by England in the twentieth century. Generally, the goal of the IRA has been the reunification of Ireland as a single and independent unit from Great Britain. The roots of the modern day Irish "troubles" are, however, deep.

Resistance of some type against British occupation of Ireland has existed since the 1790s, but the seeds of the conflict were sown much earlier during the Protestant reformation of the 1500s. This occurred when Henry VIII, wanting to be annulled from Catherine of Aragon, broke from the Catholic Church in Rome; he created the independent Church of England, placing himself as the supreme leader of the church, and legislated a death sentence for those who did not recognize his authority. Henry attempted to extend his new church into Ireland, declaring himself head of the Church of Ireland in 1536, but Irish Catholics would neither tolerate nor accept the new church and rebelled. From a more secular perspective, Henry's successor, Elizabeth I, further exacerbated problems when she seized the best agrarian sections of Ireland, displacing many Irish, and gave the land to her subjects to colonize Ireland (White, 2002, p. 80). These tracts of land were known as the **Plantation of Ulster**, and in the 1600s the Plantation of Ulster was expanded, further displacing more Irish. These patterns of the imposition of English will and the displacement of Irish by the English were to repeat many more times.

Also contributing to English/Irish animosity were Oliver Cromwell's campaigns in Ireland. In 1641, in revenge for Catholic attacks on Protestant settlements, he killed thousands, thus crushing this Irish rebellion. In subsequent battles, particularly the **siege of Drogheda** in 1649, Cromwell clearly demonstrated his hatred of the Irish by not only killing all soldiers after their surrender, but also killing the civilians. By 1650, the situation for the Irish was dire in that more than one third of the Irish population had either been killed or exiled with their lands taken by the Act of the Settlement of Ireland in 1652 and given to British settlers or soldiers (Oliver Cromwell).

Another source of conflict, the outcome of which is still celebrated today by Protestants in Ireland, occurred between 1689 and 1691 between James II, a Catholic, and William of Orange, King of England, who was a Protestant. James II used Ireland as a base from which to revolt against King William (White, 2002, p. 81). During the siege of the Protestant town of Derry, 13 apprentice boys shut the gates of the city and replied "no surrender" when the demand came from James to surrender. After the siege, in the course of which many starved and succumbed to disease, William of Orange broke through the naval blockade and brought relief to the starving citizens of Derry. James was later defeated at the Battle of Boyne, forever entrenching Protestants in Ireland, and to this day Protestants have celebrated the Battle of Boyne and the apprentice boys through parades and other festivities (White, p. 81).

Since the 1700s, a pattern of revolt, starvation, and defeat by English rule was repeated many times in Ireland; in 1801 problems were galvanized by the British Parliament when it passed the Act of Union, which incorporated Ireland into the United Kingdom. Those in the north, primarily Protestants, were in favor of the act whereas those in the south, Catholics, argued for a constitutional government and an Ireland independent of the British. Further aggravating the situation were economic factors. The north's industry and manufacturing prospered while in the south, the distribution of land, most of which still was owned by Protestants, set the stage for a very poor standard of living for southern Catholics for many years.

In the nineteenth century much occurred to strengthen the resolve of those Irish who desired independence. Numerous uprisings against the British took place, and in the first half of the century between uprisings, the Irish people mobilized politically as opposed to militarily. By the middle of the century the vast majority of the Irish people were demanding legislative independence from Britain, which consistently denied that independence. Also around that time the Great

Hunger of 1845 to 1862, a famine that starved a million people and brought about the emigration of more than a million more Irish, further reduced the people's abilities to stage any more effective uprisings. Despite these events, the Irish Party strove for independence from Britain on three more occasions in 1886, 1893, and 1911 with scattered support from a few in Britain's Parliament. These efforts were met with fierce opposition by the northern Protestants who thought the only way their limited power would remain and not be claimed or usurped by the southern Catholics was to maintain contact with Britain. Further clearing the "battle lines" was that by the latter part of the nineteenth century, nearly all northern Protestants were Irish, with little to no British blood or direct ties to Britain. The result of this was that the conflict was an Irish one, with the northern Protestant Irish able to call on British police and army for help, giving them a great advantage when conflicts arose. Consequently, the southern Catholic Irish, feeling the need for a level playing field, sought help and solutions elsewhere. In response to this void, various factions such as the **Irish Republican Brotherhood (IRB)** gradually formed, organized, and fought for Irish independence from Britain. After home rule was again defeated in the British Parliament, the conflict came to a head in the 1916 Easter Uprising.

The **Easter Uprising** was the result of the instability created by the Government of Ireland Act of 1914, which dictated that Ireland would gain regional self-government after World War I. The act was never implemented for fear that civil war might be sparked if the act came to fruition. In reality, both the north and the south had been secretly arming for the civil war that was thought to be inevitable after the British left Ireland. Both sides wanted to dominate politically and militarily after the British were out of Ireland. On Easter 1916, leaders of the IRB, Patrick Pearse and James Connelly, thought it best to strike out militarily at both the British rule and the northern Irish. Fighting started in Dublin. The timing was advantageous because the British were preoccupied in the war against Germany. The British, who arrived in force in Dublin to quell the uprising, experienced a week of heavy fighting. In the end, the IRB negotiated terms of surrender with the better-trained and equipped British Army. In the negotiations, Pearse addressed himself in the correspondence with the British as the "Commanding General of the Irish Republican Army." A new faction, the IRA, was created.

Irish support for the Easter Rebellion was not great by any means. However, this was to change as a result of a serious misjudgment by the British. In retaliation for the treachery of this uprising (the Irish rebelling while Britain was fighting Germany), the British army rounded up dozens of "conspirators"; executed Pearse, Connelly, and many others; and imprisoned thousands more (White, 2002, p. 84). This martyred Pearse and Connelly, and the goals and ideology of the IRA were embraced in favor of revolution against Britain.

As we have seen, direct clashes between the British army and the Irish citizens were the most common mode of conflict up to the 1920s. After that, the IRA, under the leadership of Michael Collins, perceived its causes would move farther and faster through a campaign of terror and guerilla warfare. The British responded by creating a semi-autonomous government in Northern Ireland designed to combat the IRA. Thus, the British were entrenched in Northern Ireland and the conflicts continued.

In spite of all the conflicts and death, an Irish Free State did eventually evolve. As discussed in Chapter 2, in 1921 the country was split into two sections. The Irish Free State was composed of 26 of 32 counties and came into being due to the Irish Free State agreement signed in December 1921. The remaining six counties, which were mostly Protestant, retained their ties with England. The new Irish government, although still very much under the influence of the Parliament in England, did manage some autonomy in that they had their own judicial system and their own consultive assembly, which replaced their parliament. Despite these advances, the fact that the six counties still have ties with England has remained a source of controversy to this day due to the remaining long-standing Catholic/Protestant disagreements.

From the 1920s until the 1960s, support for the IRA waxed and waned because of the organization's extensive use of violence, internal strife, a pro-German stance during World War II, and the declaration of both northern and southern Irish governments that the organization was outlawed, thus making the organization secret. In addition, given that the undercurrents in the organizations were religious and political, reaching consensus within and across factions was difficult at best.

In 1969, the IRA split into two groups. The majority, known as the **Officials**, advocated a Marxist ideology that would be accomplished without the use of violence, whereas the **Provisionals** advocated terrorism as a necessary evil that would bring about Irish reunification. Refueling the IRA's and northern Irish Catholics' animosity towards the government in the previous 40-year time period was the reduction of the civil rights of northern Catholics by the Northern Ireland government, further repressing them. Now all that was needed was a proper impetus or spark to place the IRA into the forefront of politics and power. The northern Catholic civil rights campaign, which was the response to the reduction in civil rights by the Protestant government, was the key.

In 1969, when it was clear this campaign was ineffective, battle lines were drawn yet again between Irish Protestants and Catholics. On August 15, 1969, Protestants organized marches in Londonderry and Belfast to commemorate the traditional Apprentice Boys celebration, which, as was discussed above, noted a defeated Catholic uprising by James II against the English king William of Orange in 1689. During the march in 1969, Protestants armed themselves with rocks, petrol bombs, and clubs and attacked Catholic neighborhoods. Not surprisingly, retaliatory violence broke out and 3 days later the British sent the army to keep the peace. Now with a more clearly delineated enemy (i.e., the Protestant government with the British army), these sparks became the opportunities for the IRA to rally support and advance their causes in spite of the fact that the British army was initially welcomed by the Catholic community; it offered protection from the Protestants, the Northern Irish government, and its security forces. Violence then shifted to Great Britain proper, with targets being public, military, and political in nature. As it is today with Al-Qaeda with its targeting of government entities and other "soft targets," the IRA hoped that by attacking such targets, they would frighten the citizens, damage the British economy, and change the political will of the British government so that the IRA's goals would be realized (Simonsen & Spindlove, 2000, p. 79).

Between the late 1960s and 1994, the IRA was particularly active in kidnappings, assassinations, drug dealing, extortions, robberies, and bombings. Of note were the attacks of July 22, 1972, known as the Bloody Friday bombing in Northern Ireland, in which 22 bombs killed 9 people and injured 130; the 1979 bombing and murder of the Earl of Mountbatten of Burma and members of his family and a local child off the coast of the Irish Republic; and the 1984 Brighton hotel bombing, which was a failed attempt to kill members of the British Cabinet and Prime Minister Margaret Thatcher. These were all examples of high profile attacks, known in IRA parlance as "spectaculars." From September 1, 1994, to February 1996, and again from July 1997 to the time of this book's writing, the IRA has observed negotiated cease-fires commonly referred to as the "peace process." However, dissident republicans and splinter groups continue to pursue violent activities in England, the Irish Republic, and Northern Ireland in support of a campaign to expel the British from Northern Ireland. The 1994 cease-fire did reduce the overall levels of violence perpetrated by the IRA in England; in return, the IRA gained the representation of Sinn Fein, which is a left wing/socialist political movement of the IRA, in any negotiations with the British and Irish governments.

Given the cease-fires in the last 10 years and the waning of activities of the IRA, terrorism trends in England have been changing, and crime statistics have reflected this. For example, in 1998 for the first time detentions of international terrorists under the English anti-terrorist statutes outnumbered those of domestic terrorists (Home Office, 2001, p. 1). However, U.K. agencies are still vigilant where the IRA is concerned. The number of examinations for more than one hour by police agencies with Irish suspects is still greater than international suspects, but these numbers are reversing (Home Office, 2001, p. 7). The IRA was also dealt with in legislation. Thus, international groups, fueled by the fallout of 9/11, have moved to the forefront of the U.K. government's attention, and by the year 2000, the Terrorist Act had outlawed 25 international terrorist groups in the United Kingdom. It is not unreasonable to assume that the IRA's decrease in terrorist activities may be not only from the cease fires and a desire to maintain legitimacy, but also most recently from the desire to avoid being likened to "extreme" terrorist groups such as Al-Qaeda. Prior to 9/11, the bombing in Omagh, Ireland on August 15, 1998, placed the IRA and its various factions on uncertain footing, and illustrates this line of thought further. This car bombing, it is surmised, missed the original target. However, it destroyed the peace process. Twenty-nine people were killed and over 200 injured. None of the people in this total was linked with security forces or the target, and they would have been classified by the IRA as innocents. Dingley (1998) proposed that this was a major mistake for several reasons. First, the level of violence used was too great, and consequently it did not evoke the desired fear but instead elicited revulsion from many in and out of Ireland. Second, this level of violence was too high for the IRA's supporters. And third, it drew international condemnation. Quite simply, the bombing elicited too much and the wrong type of attention, just as the Taliban experienced in Afghanistan and Al-Qaeda experienced after the 9/11 attacks. The political, emotional, and moral environments, particularly since 9/11, are not very conducive to terrorist violence of any sort, and as of July 28, 2005, the IRA agreed to lay down its arms and planned to pursue peaceful and democratic means to pursue its objectives. At the time of this writing, the IRA had committed no further acts of terrorism on English soil, and has been doing what has been dubbed by the English press as "soul searching." Thus, the focus has shifted to Al-Qaeda.

Al-Qaeda is a fundamentalist Islamic organization with cells worldwide. It has been widely accepted as being responsible for the 9/11 attacks on the World Trade Center in New York and the Pentagon in Arlington, Virginia. Established in 1988 by Osama bin Laden after the Soviet occupation of Afghanistan, there are ironic roots to the group's inception in U.S. policy. In the Carter administration, a scheme was put forth by the CIA to place the Soviet Union in a situation whereby the pro-Soviet government in Afghanistan would be destabilized by radical Islamic groups trained and funded by the CIA. With that partially accomplished in the late 1970s, the Soviet Union invaded the country and Arab-Muslim groups, mainly the Mujahidin including bin Laden, joined the fight and finally ousted the Soviets in 1989. Thus, the early version of Al-Qaeda was provided weapons by, was funded by, and was loosely allied with the United States against a common enemy. This shaky alliance dissolved after the Soviets withdrew from Afghanistan. After the Soviet occupation, radical Islamics, numbering between 10,000 and 20,000 (Katzman, 2005), without a direct enemy to fight, turned to terrorism to advance their goals. They formed training camps in Afghanistan and recruited militant Muslims from around the world.

In 1990, the United States and Great Britain sent troops into Saudi Arabia in preparation to invade Kuwait to drive out the Iraqi invaders. After the war had started, Iraq threatened to invade Saudi Arabia during the Gulf War, and bin Laden volunteered his soldiers and services to the Saudi royal family. His offer was rejected in favor of utilizing American and British forces. Bin Laden interpreted this as a major insult for two reasons. One was the rejection. The second was simply the harboring of U.S. troops in Saudi Arabia, which was seen as a violation of religious sacred soil. From there, bin Laden went to Sudan in 1991 where he bought property, and housed and trained Al-Qaeda members. Under pressure from the U.S. and Egyptian governments, bin Laden was expelled from Sudan in 1996, after which he went to Afghanistan where he aided the Taliban in gaining power and controlling Afghanistan.

More recently, bin Laden and Al-Qaeda's hostility towards the United Kingdom and the United States stems from their interpretation that both countries are openly hostile to Muslims, citing as their reasons their respective foreign policies towards Muslim countries, the invasion and occupation of Iraq, support for Israel, and the U.K. and U.S. military bases in Islamic countries. Ultimately, Al-Qaeda's goals are to unite Muslims against the United States, defeat Israel, overthrow regimes seen as "non-Islamic," and expel Westerners from Muslim countries (Patterns of Global Terrorism 2001. [U.S. State Department] 2001: 105).

■ Terrorism Legislation

Because the United Kingdom has such a long history with domestic terrorism, and because the threat has shifted to an arguably more dangerous international enemy, the English have altered and refined their laws several times to reflect today's threats. The United Kingdom has been very proactive in the creation and evolution of legislation that is designed to combat and deter terrorist activities. Most of the modern era legislation has been entitled the "Prevention of Terrorism." The first of these was passed in 1974 to counteract the activities of the IRA and its various factions. To keep the legislation current, it was updated in 1976, 1979, 1984, and 1989. These acts were repealed in 2000 by the Terrorism Act of 2000, which was itself updated in response to 9/11 by the Anti-Terrorism, Crime and Security Act of 2001, and most recently by the controversial Prevention of Terrorism Act of 2005. The latter was introduced early in 2005 not only to update the 2001 act, but also to combat civil liberties problems resulting from the earlier acts and to address fears stemming from the July 2005 attacks.

In the following sections, we present an overview of the substantive highlights of these anti-terrorism statutes. Of particular concern is the difficulty of applying and enforcing these anti-terrorism laws while maintaining civil liberties in a modern society. With the passage of these acts, one can observe an increasing struggle with civil liberties issues and greater centralization of power in the government. Also, note a dramatic shift in the orientation of the anti-terrorism legislation after the July 2005 attacks, which potentially jeopardizes civil liberties and human rights. Further, even after years of experience, legislative definitions in England remained general, though they are still more focused than the definitions contained in the parallel U.S. legislation.

Prevention of Terrorism Acts

The **Prevention of Terrorism Acts** were a series of progressive acts started November 29, 1974, that bestowed emergency powers upon the police when terrorism was suspected. The acts built upon each other, with legislators retaining the facets of previous acts that had been effective and rejecting those that had not. The acts were initially created in response to IRA bombing campaigns in the early 1970s after increases in IRA activities. The 1974 act, which was created in response to the Birmingham pub bombings in November 1974, contained three parts with three schedules. The three parts were concerned with proscribed organizations, exclusion orders, and general/miscellaneous information. Specifically, the act designated the IRA as a prohibited organization, made any support for the IRA illegal, made it possible to exclude from the United Kingdom persons involved in terrorism, and gave the police powers to arrest suspected terrorists and detain them for 48 hours on their own authority (double the time they could detain nonterrorists), with the possibility of an extension of detention for a further 5 days authorized by the Secretary of State. The act also gave the police the power to examine travelers entering or leaving the United Kingdom or Northern Ireland for the purpose of determining whether they were involved in terrorism.

The 1974 act was replaced on March 25, 1976, by the Prevention of Terrorism (Temporary Provisions) Act of 1976. This act made it an offense to contribute to or solicit contributions towards acts of terrorism, or to withhold information relating to acts of terrorism or persons committing them. In addition, specific groups, such as the Irish National Liberation Army, were labeled as terrorists and placed on the list of prohibited organizations. It also allowed the subjects of exclusion orders that had been in force for 3 years or more to have their exclusion orders reviewed.

The Prevention of Terrorism (Temporary Provisions) Act of 1984

The Prevention of Terrorism (Temporary Provisions) Act of 1984 replaced the 1976 act and came into operation on March 22, 1984. The Prevention of Terrorism (Supplemental Temporary Provisions) Order of 1984 came into operation on March 27, 1984. Three important additions to the previous act were: First, the act enabled the Secretary of State to authorize more than one extension of detention, provided that the total period did not exceed 5 days (Sections 12[4]–[5]). Second, the act limited exclusion orders (whereby an individual may not enter the United Kingdom) to 3 years' duration. At the end of the 3 years, exclusion orders expired, although they might be renewed for an additional 3-year period by the Secretary of State. Exclusion orders of indefinite duration made under previous legislation would, if not previously revoked, expire on March 21, 1987 (Section 3[4]). Third, the required period of ordinary residence in the United Kingdom, which exempted a British citizen from exclusion, was reduced from 20 years to 3 years.

The Prevention of Terrorism Act of 1989

The Prevention of Terrorism Act of 1989 had seven parts, four substantive and three technical. Substantively, the act added the following: First, in addition to the IRA and the Irish National Liberation Army, which had been designated as prohibited organizations under the 1974 and 1979 acts, other organizations were made illegal. Being a member of, associating with, soliciting funds for, or showing support for such an organization was now an arrestable offense (Part III, Section 9). Also, contributing, receiving, holding, or soliciting financial support for terrorism was an offense (Part III, Section 9). Second, exclusion orders against individuals could be issued to keep undesirables out of Northern Ireland and the United Kingdom if the Secretary of State was satisfied that the specific individual had been involved in the commission, preparation, or instigation of acts of terrorism or was attempting or attempted to enter the United Kingdom with a view to being concerned in the commission, preparation, or instigation of such acts of terrorism (Schedule II, Section 4[4]). The maximum sentence was 5 years with unlimited fines.

Also espoused by the act was that individuals could be arrested without a warrant and on reasonable suspicion that they were guilty of offenses or "concerned with the commission, preparation or instigation of acts of terrorism" (Part IV, Section 14). Individuals could be detained up to 48 hours, and this could be extended to 5 days by the Home Secretary. The legal rights of those under arrest were also restricted under this provision.

The Terrorist Act of 2000

On July 20, 2000, Parliament passed the Terrorist Act of 2000 (TACT), which came into force on February 19, 2001. It superseded and repealed the Prevention of Terrorism (Temporary Provisions) Act of 1989 and the Northern Ireland (Emergency Provisions) Act of 1996, and was in response to the perceived increased threat of international terrorism as well as the terrorism problems originating from Northern Ireland, although it preceded the events of September 11, 2001, by several months. To date, this act has been arguably the most comprehensive of the terrorism acts, and the intent of its creation was to streamline past legislation. Thus, this new act reformed and broadened the previous legislation and placed it in service on a more permanent basis. As of early 2004, approximately 544 people had been arrested under this act. Ninety-eight people had been officially charged (Home Office, 2004).

The most noteworthy aspect of the act was that it outlawed 25 international terrorist groups in the United Kingdom and identified 14 Northern Irish groups that were to be treated in the same way (see Schedule II). The Secretary of State could proscribe organizations that he or she believes to be terrorist organizations, thus giving the secretary a great deal of power (Part II, Section 3). Those groups could apply to be removed from this list. If a person was found to be a member of, supported, or professed membership in such an organization, it was an offense. The maximum penalty for this offense was 10 years in prison.

Further, in reference to ordinary people, if a person discovered information during the normal course of their employment or business duties that another person was involved in terrorist activities, it was the former person's legal obligation to notify authorities (Part III, Section 19). An individual was also liable to prosecution where such a situation was apparent but was not noticed by said individual.[1] Police were given increased powers to investigate terrorism, such as arrest powers whereby a constable may arrest without a warrant a person who is reasonably suspected of being a terrorist (Part V, Section 41[i]). A person may then be held for up to 48 hours, allowing for firmer establishment of the facts (Part V, Section 41[iii]). Police were also given broader stop and search powers in that a constable may stop and search a person who he or she reasonably suspects to be a terrorist to discover whether that person is in possession of anything that may constitute evidence that he or she is a terrorist (Part V, Section 43[i]). A constable may also apply for a search warrant of premises where a suspected terrorist may be.

TACT 2000 created new terrorist offenses including inciting terrorist acts, seeking or providing training for terrorist purposes in the United Kingdom or overseas, and providing instruction or training in terrorist activities such as firearms use, the use of explosives, or chemical, biological, or nuclear weapons

Notably, exclusion orders designed to keep "undesirables" out of the United Kingdom were not included in this act. In addition, the offense for withholding information concerning terrorist activities was removed. (This differs from the 1989 act in that there was no stipulation concerning employment. Anyone not reporting potential terrorist information could be charged with this offense.) And as noted by Walker (2002b), there were no provisions for victims as may be found in the Patriot Act and other past legislation in the United States and in legislation found in France (p. 3). Internment without trial was also not present in the act.

The Terrorist Act of 2000 was also the best attempt to define "terrorism," although the term was defined very broadly. The definition also added the aspect of international and non-Irish domestic terrorism. Terrorism was defined as:

> . . . *the use or threat of action where the use or threat is designed to influence the government or to intimidate the public or a section of the public, and the use or threat is made for the purpose of advanc-*

ing a political, religious or ideological cause. "Action" involves seri-
ous violence against a person, involves serious damage to property, en-
dangers a person's life, other than that of the person committing the
action, creates a serious risk to the health or safety of the public or a
section of the public, or is designed seriously to interfere with or seri-
ously to disrupt an electronic system. (Terrorist Act of 2000, p. 1)

Therefore, there was also a new international aspect meaning that "action," "government," and "public" as used above applied beyond the U.K. borders and to anyone, anywhere in the world.[2]

By comparison, the USA Patriot Act, also recognizing the international threat, was also very general and further delineated domestic and international terrorism. The definition of "international terrorism," as may be found in 18 U.S.C. §2331 and as referenced in the USA Patriot Act, refers to activities that:

> *(A) involve violent acts or acts dangerous to human life that are a*
> *violation of the criminal laws of the United States or of any State,*
> *or that would be a criminal violation if committed within the*
> *jurisdiction of the United States or of any State;*
>
> *(B) appear to be intended—*
>
> > *(i) to intimidate or coerce a civilian population;*
> >
> > *(ii) to influence the policy of a government by intimidation or*
> > *coercion; or*
> >
> > *(iii) to affect the conduct of a government by mass destruction,*
> > *assassination, or kidnapping; and*
>
> *(C) occur primarily outside the territorial jurisdiction of the United*
> *States, or transcend national boundaries in terms of the means*
> *by which they are accomplished, the persons they appear intended*
> *to intimidate or coerce, or the locale in which their perpetrators*
> *operate or seek asylum (18 U.S.C. §2331).*

Domestic terrorism in the act is the same, with the exception that it includes the requirement that the terrorist acts in question must "(C) occur primarily within the territorial jurisdiction of the United States" (USA Patriot Act, Section 802).

As we will discuss in detail later in the chapter, both the English and the American definitions of terrorism in both sets of legislation were criticized early and often for being too broad in coverage, potentially including groups and activities for which the legislation was never intended, and thereby encroaching on civil liberties. Accordingly, by both countries' definitions, work strikes by nurses, public transport workers, or airline workers could be interpreted as terrorism depending on how the laws are interpreted and applied. Further, in England, any Internet site, billposting, or speech that may advance a terrorist point of view or political ideology may be interpreted as inciting terrorist activity and seen as a serious violation; in contrast, in the United States such activities are normally protected under the First Amendment unless violence is provoked. Other distressing aspects of TACT 2000, as noted by Walker (2002a), were that there was no democratic accountability for the operation of this type of law and that it was neither successful nor comprehensive enough to avert the creation of the Anti-Terrorism, Crime and Security Act of 2001. Thus, many point to these two legislative acts as the watershed marks whereby concern for civil liberties was balanced against security needs, the latter not only encroaching on the former, but also encroaching on common citizens' rights, not just terrorists and their interests.

Anti-terrorism, Crime and Security Act of 2001

Like the Patriot Act in the United States, the Anti-terrorism, Crime and Security Act of 2001 (ATCSA) was passed quickly in response to the 9/11 attacks. Its design was not to replace TACT 2000, but to add to it in order to deal with new terrorist tactics and potential large-scale terrorist threats. The act contained 129 sections organized into 14 parts. The first three parts dealt with terrorist property including forfeiture of cash and property, the process of the forfeiture, and the details as to what the powers of disclosure for such information were. This new legislation overlapped a great deal with previous sections of TACT 2000, with parts either simply replaced or amended to fit the new threats. The only new addition to these first three parts was the allowance of public institutions to disclose information concerning criminal investigations or proceedings in the United Kingdom or overseas. Between January 2002 and September 2003, 796 disclosures were made with 169 (21%) related to terrorism (Anti-terrorism, Crime and Security Act 2001 Review: Report, 2003, p. 28).

Part 4, which was the most controversial, dealt with the handling of immigration and asylum issues. The controversy was that resident foreign nationals who were suspected of terrorist activities and who were denied asylum in Britain might be held indefinitely without trial because of national security concerns, because they are deemed an international terrorist, or because they had links to terrorists. As of December 2003, 16 foreign nationals had been detained. Two left the United Kingdom after having their certification by the Home Secretary as a terrorist reviewed (Anti-terrorism, Crime and Security Act 2001 Review: Report, 2003, 51). In the past, if there had not been sufficient evidence to prosecute those suspected of terrorism, those suspected terrorists were deported. However, in the case of *Chahal v. United Kingdom* (1996), which was tried in the European Court of Human Rights, it was ruled that no deportation may occur to another country where there was substantial evidence that the person in question would be subjected to torture or inhuman or degrading treatment.

Also controversial was Part 5 concerning crimes motivated by race and religion. This part included an offense whereby it was illegal to incite religious hatred that motivated another person to commit a criminal offense. As was found in the creation of U.S. hate crimes legislation, the language in the legislation often criminalized too wide a range of behavior while including religious behavior that was not necessarily illegal but a matter of personal preference. Thus, much was open to interpretation. In reference to race, it was thought that the legislation had the effect of protecting some religious groups but not others (Anti-terrorism, Crime and Security Act 2001 Review: Report, 2003, p. 69).

Parts 6–14, which were updated and broadened from TACT 2000 due to 9/11, dealt with matters of dangerous materials such as weapons of mass destruction, pathogens, and toxins. The balance of the parts concerned the security of the nuclear industry, aviation, police powers, retention of communications data, and bribery and corruption. The latter two parts concerned miscellaneous and supplemental information. For policing, the TACT provisions guided how police investigated and arrested terrorists in day-to-day operations.

Other additions beyond what was in TACT 2000 were designs to prevent immigrants from abusing immigration procedures and to streamline immigration procedures beyond the European Union's efforts to manage migration flows, particularly from Eastern Europe. The Home Secretary had the power to detain foreign nationals suspected of terrorism who could not be immediately removed from the United Kingdom until deportation. The new act also strengthened security around viable terrorist targets such as airports, nuclear sites, and dangerous substances held in labs and universities. Further, the act increased the penalties for hate crimes, particularly those that were religion or race-based. Finally, exclusion orders were reinstituted and the offense for withholding information concerning terrorist activities was reactivated.

As previously mentioned, controversy and court cases arose as to the legality of the detainment of 16 foreign nationals under the ATCSA, Section 4. The contention was based on criteria set by the European Convention on Human Rights under Article 5 (1a, 1c, 1f). The article noted the circumstances whereby an individual may be deprived of his or her liberty; these were that the person who was being detained must be brought to court to face charges for having committed a criminal offense, or the detained person should be brought to court to face a deportation hearing. In other words, a person may not be held based on a suspicion without his or her day in court.

The ATCSA 2001 criteria for detaining individuals was not based on reasonable suspicion, nor was it the intent of the legislation to go forward with any type of court or deportation proceeding. There were provisions whereby the detained may appeal the Home Secretary's detainment decision. However, the argument made by the government was that due to the nature of these types of cases (i.e., terrorism), the evidence that may be presented in court might be of such a sensitive nature that national security and those who protect it might be affected even if detainees' representatives did have the proper security clearance. Thus, a detainee actually receiving their day in court was a tenuous, somewhat secretive, and a red tape–laden process.

The argument from the United Kingdom was that Article 5 of the European Convention did not apply in these situations because the provisions that allowed for holding detainees under the ATCSA 2001 were allowable due to the extreme threat and circumstances that terrorism and its perpetrators create. Therefore, under these circumstances, the United Kingdom may opt out of its obligations under Article 5 of the European Convention. The argument was that this was allowable under Article 15 because during extreme times such as war and public emergencies that threaten the life of the nation, said derogation is needed and should be limited to the extent required by the urgent need of the situation (Council of Europe, 1966, Article 15[1]). To clarify, the United Kingdom could opt out of the European Convention on Human Rights articles and inter the suspects only when there was an emergency that was immediate and threatening on a national scale.

Accordingly, the legal challenges went forward in October 2004, and on December 16, 2004, the House of Lords Judicial Committee ruled that Part 4 of the ATCSA was in violation of Article 15 of the European Convention on Human Rights. The reasoning was that the threats to English national security were not demonstrably imminent enough to warrant the detainments, and that the act was only applicable to resident foreign nationals. Substantively, the result was that although the European Convention did not have the power to force the United Kingdom to change the ATCSA legislation, the ruling placed political pressure on the U.K. government to make amendments. To remedy these problems, the Prevention of Terrorism Bill of 2005 was proposed on February 22, 2005, which repealed the Part 4 powers, replaced them with a system of control orders, and, yet again, modified British terrorism legislation.

Prevention of Terrorism Act of 2005

The Prevention of Terrorism Act of 2005 was proposed to work through the House of Lords' December 2004 rulings that declared Part 4 of the Anti-Terrorism, Crime and Security Act of 2001 illegal and in conflict with Article 5(1) of the European Convention. On February 22, 2005, the day the bill was introduced, a briefing paper was issued to the libraries of the House of Lords and the House of Commons to aid the respective members in their deliberations on the act. In these papers, the government went to great lengths to justify its strategies in combating terrorism through this latest piece of legislation. The legislators of both houses were under great pressure to find a legislative solution because the previous legislation that held the detained individuals was to expire on March 14, 2005. Time was of the essence.

In the briefing papers, the government made its case that there was still a very serious and a very different threat to the United Kingdom, and explained why the shift toward prevention was paramount and, therefore, so was the need for control orders (Home Office, 2005b). What follows is a summary of the briefing papers to note the justifications for a more preventive strategy to terrorism, and for the Prevention of Terrorism Act of 2005.

Paper One: International Terrorism: The Threat

Paper One, entitled, "International Terrorism: The Threat," noted that many British nationals had been involved in and trained for terrorist activities on British soil. The threat was perceived as more international now than in the past in the areas of recruiting, funding, false documentation, raising material, and using technology to reach and send terrorist messages. These abilities were also now more global, and terrorists were better trained, highly ambitious, and used less restraint, and although conventional weapons had been used thus far, their desire to use nuclear, biological, or chemical weapons was well known (Home Office, 2005b: pp. 2–5). It was also emphasized that conventional tactics against terrorism will not always work against terrorists who are willing to commit suicide to further their causes.

Paper Two: The Government's Strategy to Counter the Threat

The government's basic strategy to combat terrorism was to challenge the ideological motivation of Al-Qaeda and other groups and emphasize a society of inclusion for people of "all faiths, backgrounds and convictions" (Home Office, 2005b: p. 6). In other words, if hearts and minds were changed by this more inclusive society, the result would be to reduce the attractiveness of these terrorist groups, which would hinder recruitment. British society needed to recognize the sophistication, ambition, and resilience of these terrorists. The case was made that these groups have been beaten down and their infrastructures have fallen apart. However, they did recover and utilized local grievances for recruiting purposes. Another strategy cited was that because the terrorist groups are so international in nature, an international approach to hinder funding, training, communicating, and operation of terrorists was necessary. The government's strategic goal therefore was to, "Reduce the risk from international terrorism so that our people can go about their business freely and with confidence" (p. 6). In other words, strike a balance between security and freedoms.

Paper Three: International Terrorism: Reconciling Liberty and Security—The Government's Strategy to Reduce the Threat

The purpose of this paper was to demonstrate how the prevention of terrorism and the pursuit of those who commit terrorist acts could handle the terrorism threats.

The emphasis was to be on prevention of extremist attitudes and of terrorist group recruitment while making the effort to understand the motivations behind joining and how organizations utilized propaganda, poor social and economic conditions, and a fostered hatred of the West to increase their ranks. Government programs were to be improved to reduce alienation and help to resolve conflict by supporting the Middle East peace processes. Also needed was improved hate crime legislation to better protect British Muslims and to further recognize the contributions of Islamic people in British society. In other words, the British government's goal was to attempt to counteract the propaganda of the recruiting terrorists and to build mutual understanding between the Islamic world and those in the United Kingdom.

The government's strategy in terms of pursuit was to "pursue, disrupt and, wherever possible, prosecute and convict networks of terrorists intent on mounting attacks against the UK" (Home Office, 2005c, p. 10). It was also stressed that this must be done without compromising civil liberties through improved gathering of intelligence, increased funding of counter-terrorism efforts, and when necessary, ". . . to take powers which enable (it) [sic. the government] to abridge in strictly limited circumstances the freedom of the individual in the interest of wider security" (p. 10). The government did stress in the paper that the latter will only be done when strictly required by the circumstances of the threat.

Paper Four: International Terrorism: Protect and Prepare

The government noted in this paper that it was making every effort to make it harder for terrorists to operate in the United Kingdom by tightening borders, introducing ID cards to make it more difficult for terrorists to forge identity, gaining international agreements to freeze terrorist assets, and reducing the ability for terrorists to attack specific targets and the United Kingdom's interests overseas.

Germane to our discussion, and which the government cited as preparation against terrorism, was updating the modernization of the legislative network for civil protection that would enable emergency powers to come into effect more quickly and efficiently should an emergency arise. Also included was a discussion concerning the preparedness of the government agencies and their roles and operations should a terrorist attack occur as well as the counter-terrorism exercises designed to manage a crisis should one arise. Thus, although not entirely convincing, the case was made that the threats were real, very serious, and on a national scale.

Security versus Civil Liberties

The bill for the act was introduced on February 22, 2005, and the House of Commons passed it with considerable opposition from members of the Labour Party. It was then sent to the House of Lords, who made several amendments including a sunset clause, meaning the act would expire in March 2006 unless renewed. Other amendments included a requirement whereby the Director of Public Prosecutions was to make a statement that a prosecution would be impossible before each control order on an individual could be issued, a requirement that a judge had to authorize each control order, a requirement for the review of the legislation by Privy Councilors, and a restoration of the burden of proof to such cases to prove beyond a reasonable doubt as opposed to the former and weaker "balance of the probabilities" (Prevention of Terrorism Act 2005: 1).

The House of Commons rejected most of the amendments on March 10, 2005, leading to an exchange of the bill several more times between houses; after the longest session of Parliament ever (over 30 hours), a compromise was reached whereby those who opposed the amendments would concede if the legislation could be reviewed later in the year. The bill was given Royal Assent on March 11, 2005. There were many critics of this bill in the media, in Parliament, and in the public.

The main objection was the perception that the new act would violate civil liberties on a grand scale. Moreover, although most would agree that anti-terrorism legislation was needed, rushing the process due to the imminent expiration of the previous legislation and implementing control orders was not necessarily the best route to gain security in the United Kingdom. Time was needed to properly pursue the legislation.

In exploring the Prevention of Terrorism Act of 2005, several aspects are commonly criticized. First, the definition of terrorism was essentially the same as in TACT 2000, but modified in the following ways:

> (8) *For the purposes of this Act involvement in terrorism-related activity is any one or more of the following—*
>
> *(a) the commission, preparation or instigation of acts of terrorism;*
>
> *(b) conduct which facilitates the commission, preparation or instigation of such acts, or which is intended to do so;*
>
> *(c) conduct which gives encouragement to the commission, preparation or instigation of such acts, or which is intended to do so;*
>
> *(d) conduct which gives support or assistance to individuals who are known or believed to be involved in terrorism-related activity;*

and for the purposes of this subsection it is immaterial whether the acts of terrorism in question are specific acts of terrorism or acts of terrorism generally." (Prevention of Terrorism Bill, 2005, p. 3)

Clearly, if original definitions from TACT 2000 and modifications from 2005 are compared, there was a profound shift to the prevention of terrorist acts. Many ordinary, daily activities would now fall under the category of the preparation to commit terrorist acts, generally or specifically. For example, the purchase of a cellular phone, which may be utilized to detonate a bomb, would have fallen under these new guidelines, as would the purchase of a book bag for the intention of carrying a bomb onto public transportation.

Also controversial were the control orders that may be issued by the Home Secretary or a court. They were designed to replace the controversial Article 4 of the Anti-Terrorism, Crime and Security Act of 2001, which was found illegal by the House of Lords, and to prevent individuals from participating in terrorism-related activities. Control orders, as found in the Prevention of Terrorism Act of 2005, were defined as and delineated as follows:

> (3) *The obligations that may be imposed by a control order made against an individual are any obligations that the Secretary of State or (as the case may be) the court considers necessary for purposes connected with preventing or restricting involvement by that individual in terrorism-related activity.*
>
> (4) *Those obligations may include, in particular-*
>
> *(a) a prohibition or restriction on his possession or use of specified articles or substances;*
>
> *(b) a prohibition or restriction on his use of specified services or specified facilities, or on his carrying on specified activities;*
>
> *(c) a restriction in respect of his work or other occupation, or in respect of his business;*
>
> *(d) a restriction on his association or communications with specified persons or with other persons generally;*
>
> *(e) a restriction in respect of his place of residence or on the persons to whom he gives access to his place of residence;*
>
> *(f) a prohibition on his being at specified places or within a specified area at specified times or on specified days;*
>
> *(g) a prohibition or restriction on his movements to, from or within the United Kingdom, a specified part of the United Kingdom or a specified place or area within the United Kingdom;*
>
> *(h) a requirement on him to comply with such other prohibitions or restrictions on his movements as may be imposed, for a period not exceeding 24 hours, by directions given to him in the specified manner, by a specified person and for the purpose of securing compliance with other obligations imposed by or under the order;*

(i) a requirement on him to surrender his passport, or anything in his possession to which a prohibition or restriction imposed by the order relates, to a specified person for a period not exceeding the period for which the order remains in force;

(j) a requirement on him to give access to specified persons to his place of residence or to other premises to which he has power to grant access;

(k) a requirement on him to allow specified persons to search that place or any such premises for the purpose of ascertaining whether obligations imposed by or under the order have been, are being or are about to be contravened;

(l) a requirement on him to allow specified persons, either for that purpose or for the purpose of securing that the order is complied with, to remove anything found in that place or on any such premises and to subject it to tests or to retain it for a period not exceeding the period for which the order remains in force;

(m) a requirement on him to allow himself to be photographed;

(n) a requirement on him to co-operate with specified arrangements for enabling his movements, communications or other activities to be monitored by electronic or other means;

(o) a requirement on him to comply with a demand made in the specified manner to provide information to a specified person in accordance with the demand;

(p) a requirement on him to report to a specified person at specified times and places.

(5) Power by or under a control order to prohibit or restrict the controlled person's movements includes, in particular, power to impose a requirement on him to remain at or within a particular place or area (whether for a particular period or at particular times or generally). (Prevention of Terrorism Act, 2005, Ch. 2, Section 1)

After much debate and opposition to the control orders, and to address the civil liberties concerns, the Home Office put forth its own information on control orders, which included the following:

1. *Control orders enable the authorities to impose conditions upon individuals ranging from prohibitions on access to specific items or services (such as the Internet), and restrictions on association with named individuals, to the imposition of restrictions on movement or curfews. A control order does not mean "house arrest."*

2. *Specific conditions imposed under a control order are tailored to each case to ensure effective disruption and prevention of terrorist activity.*

3. *The Home Secretary must normally apply to the courts to impose a control order based on an assessment of the intelligence information. If the court allows the order to be made, the case*

will be automatically referred to the court for a judicial review of the decision.

4. *In emergency cases the Home Secretary may impose a provisional order which must then be reviewed by the court within 7 days.*

5. *A court may consider the case in open or closed session—depending on the nature and sensitivity of the information under consideration. Special Advocates will be used to represent the interests of the controlled individuals in closed sessions.*

6. *Control orders will be time limited and may be imposed for a period of up to 12 months at a time. A fresh application for renewal has to be made thereafter.*

7. *A control order and its conditions can be challenged.*

8. *Breach of any of the obligations of the control order without reasonable excuse is a criminal offense punishable with a prison sentence of up to five years and/or an unlimited fine.*

9. *Individuals who are subject to control order provisions have the option of applying for an anonymity order.*

10. *To date the Government has not sought to make a control order requiring derogation from Article 5 of the European Convention on Human Rights. (Home Office, 2005c, p. 1)*

Critics of this act were numerous and quite vocal. Many noted that in this act, a person was not necessarily punished for committing certain acts but was punished for assisting persons who were suspected of committing terrorist acts. This was vague for several reasons. First was the subjectivity of the determination. What type of activities may be linked to terrorism? How was this verified and determined, and by whom? Further, the Secretary of State wielded a great deal of power in that he or she may impose "any obligation" on an individual he or she deemed necessary, "to prevent or restrict further involvement by that individual in terrorism activity" (Clause 1[2]). There were also criticisms that control orders were so general that a person would have no way of knowing the extent of the interference by the state, the extent of search and seizure during the 12 months of the control order, or what was required for "cooperation." Nor would any provision be made concerning the impact on the families or other pertinent people in the life of the person under the order.

Concerning freedom of movement and liberty issues, it is highly unlikely these restrictions will pass the same legal challenges put forth by Article 5(1) of the European Convention on Human Rights that declared Article 4 of the Anti-terrorism, Crime and Security Act of 2001 illegal. The problem again is likely to be that these restrictions placed on individual liberty will be demonstrably necessary, immediate, and threatening the life of the nation enough to warrant their use.

Terrorism Act of 2006

In the wake of the July bombings in London, new legislation was proposed on October 12, 2005, and passed into law by an Act of Parliament on March 30, 2006.

Known as the Terrorism Act of 2006, it consisted of 39 clauses and 3 schedules. The act proposed to outlaw encouragement or glorification of terrorism, create a new offense to tackle extremist bookshops that disseminate radical material, make it illegal to give or receive terrorist training or attend a "terrorist training camp," create a new offense to catch those planning or preparing to commit terrorist acts, extend the maximum limit of precharge detention in terrorist cases to 3 months, and widen the grounds for proscription to include groups that glorify terrorism (Home Office, 2005a, p. 1).

"Glorifying" terrorism would carry a maximum 5-year term; publishing, disseminating, or selling literature encouraging terrorism would bring a maximum 7-year jail term, and the law was retroactive for 20 years. The Home Secretary also proposed further measures in the Immigration and Nationality Bill to "strengthen the UK's ability to deny asylum to terrorists; make it easier to remove British citizenship from those whose presence in the UK is not conducive to the public good; and speed up the removal of those being deported on national security grounds by removing an in-country right of appeal except on human rights grounds" (Home Office, 2005a, p. 1).

Also added were the criminalization of acts that may be deemed as preparatory or harboring the intention of committing terrorist acts (maximum term: life in prison); indirect incitement to terrorism, which meant inciting others to commit terrorist acts (maximum term: 7 years); and training others to use hazardous substances to commit terrorist acts (maximum term: 10 years) (Peck, 2005, p. 24).

In addition, Prime Minister Tony Blair proposed the following on August 5, 2005:

- Creating an offense of glorifying terrorism, in the United Kingdom or abroad
- Examining calls for police to be able to hold terror suspects for longer prior to charging
- Proscribing the group Hizb ut Tahrir and the successor organization of Al-Muhajiroun, and looking at whether the grounds for proscription need to be widened
- Consulting on creating new powers to close places of worship used to foment extremism (Peck, 2005, pp. 13–14)

Since the bill's inception, and particularly after the passage of the law, there was much debate in Parliament and in the media concerning the violation of civil liberties in reference to the glorification of terrorism and detainment of suspects without charge. Critics cited the former as too vague and broad, for example, allowing the Home Secretary to determine which historical figures were freedom fighters and which were terrorists. The latter was controversial in that terrorist suspects could be held for extended periods without charging while a case was being built against them. Further, an amendment proposed in November 2005 would have extended the time a suspect could be held from 14 days, as was the standard in the Terrorist Act of 2000, to 90 days. The House of Commons settled on a 28-day period of detention on July 26, 2006, although advocates, particularly the police, are still lobbying for the 90-day period to enable a better and more thorough investigation of the suspects.

This and future legislation is very likely to continue the debate between civil libertarians and crime control advocates against the very serious backdrop of future potential terrorist acts. Time will tell how the balance between these two camps will be struck.

■ Conclusion

Earlier in the chapter, we discussed how the ideals of legislation were to protect the public from criminal acts, not to encroach upon the liberties of the individual; to be fair to the accused; and to have a deterrent effect on future criminal acts. It is clear the latter aspect, deterrence, is not a going concern because there is very little that any legislation will do to deter today's terrorists. Even death sentences for proscribed terrorist acts would not be effective because death, whether it occurs during the terrorist act or as a punishment, is viewed by these terrorists as a martyring act and thus desirable when fighting one's enemies. And given that technology has enabled terrorists to do a great deal of damage and to take many lives with what may be held in a backpack, the necessary shift for law enforcement and legislation has been to the prevention of these acts. This is not without its costs for everyone, because such an approach leads inevitably to an encroachment on civil liberties. Therefore, the first step in avoiding this problem is to clearly define our legislative terms.

From "terrorism" to "glorification," there has been little clarity in defining terrorist terminology. Couple this with a need for general definitions that will encompass most terrorist acts, and it is not surprising that there has been confusion in law enforcement, in the courts, and in the public. The retooling of the laws has also spurred debates concerning the unintended consequences of the legislation in matters of privacy and potential civil liberties violations. Both the British and American governments have been criticized for overemphasizing the terrorist threat to justify legislation that allows domestic spying on citizens and for holding foreign nationals without a trial with little prima facie evidence. Given the fact that terrorist acts evolve more quickly in deed, imagination, and scale and that legislation is not likely to keep pace, it is unlikely clear definitions involving terrorism will ever evolve or encompass what they are designed for and intended to cover. Place this in a world environment where there is constant potential for devastating terrorism and where there are politicians who might use this threat to foster fear in the public to advance invasive legislation, and it is unlikely we will ever see an increase in our civil liberties. Thus, there is a hotly debated conundrum in that the legislation designed to protect ordinary people from terrorists and that has been aggressive towards terrorists may encroach where it was never intended, thus giving one the impression that power is shifting away from the people and to the police and government. The result is that civil liberties are being taken and/or bartered away to gain security in this new age.

CHAPTER SPOTLIGHT

- The United Kingdom is no stranger to terrorism, particularly since 1969 when the Northern Ireland "troubles" began. However, the attacks of September 11, 2001, in the United States and the subway and bus bombings in London on July 7, 2005, placed both countries in unfamiliar territory, dealing with suicide bombings and large scale, mass casualty attacks.

- This led to changes in the way both countries addressed terrorism, new legislation, and a reorientation of the balance between individual rights and security.

- Historically, pre-9/11 the United States relied on ordinary criminal law to deal with terrorists. The United Kingdom, with its experience of sustained, organized terror campaigns, developed legislation to deal specifically with terrorism.

- Post-2001, both the United States and United Kingdom have shifted the emphasis of their legislation to the prevention of future terrorist attacks rather than investigating and prosecuting those involved in attacks that have happened.

- This impacts on basic civil liberties because it necessarily undermines rights to freedom of association, expression, and belief. The authorities may seek to listen to telephone calls, view emails, and restrict communications.

- Legislating against terrorism is inherently difficult because it is difficult to define what a terrorist or terrorist act is. Terrorist can be synonymous with freedom fighter or patriot, and some people may be viewed as patriots at one time but terrorists later, while all the time committing the same acts.

- Anti-terrorism legislation is necessarily broad to cover more eventualities, but this is also likely to lead to greater intrusion into individual liberties.

- Britain fought an intermittent war of independence or insurgency war in Ireland and with the Irish Republican Army since its occupation of Ireland in the seventeenth century. In the twentieth century, modern terrorism developed, first in the insurgency that followed the Easter Uprising of 1916, then after the subsequent partition of Ireland and the establishment of Northern Ireland and the rise of the Provisional IRA in 1969.

- Between 1969 and 1996 the Provisional IRA, along with smaller groups such as the INLA, perpetrated hundreds of terrorist attacks in Northern Ireland, in the Irish Republic, on the British mainland, and against British targets elsewhere, particularly British military personnel in mainland Europe. Targets included royalty, politicians, economic targets, and the general public.

- The IRA, along with some other Republican groups and the Protestant so-called "Loyalists" terror groups, have been in a cease-fire, and the levels of violence and terrorist activities associated with Northern Ireland have declined dramatically.

- Since 1996, public support and tolerance of terrorism in Northern Ireland has declined. In particular the bombing of the town of Omagh in 1998 and the deaths of 29 civilians in the process were seen as a turning point. The worldwide condemnation of those involved and even the condemnation by some Republicans was significant. Shortly afterwards, the IRA announced it would lay down its arms and commit to the peace process.

- This decline has been reflected in English crime statistics. In 1998, for the first time, detentions of international terrorists outnumbered those of "domestic" terrorists.
- Al-Qaeda is a worldwide fundamentalist Islamic organization that has been widely held to be responsible for the 9/11 attacks and the attacks on the East African U.S. embassies, the U.S. military in Saudi Arabia, and the *U.S.S. Cole*. However, its roots are in the U.S.-sponsored insurgency against the Soviet invasion and occupation of Afghanistan in 1979.
- The United Kingdom has a long history of anti-terrorism legislation that has been altered, updated, or refined to meet what is perceived as the changing threat. Most of the legislation was entitled the "Prevention of Terrorism Act" but all were repealed by the Terrorism Act of 2000. This act was updated following the 9/11 attacks by the Anti-Terrorism, Crime and Security Act of 2001 and again by the Prevention of Terrorism Act of 2005.
- The first Prevention of Terrorism Act was enacted on November 29, 1974, as an emergency response to the bombings of pubs in the city of Birmingham earlier that month. Among other things, it created the power to "prohibit" organizations and to exclude from the United Kingdom people suspected of involvement in terrorism.
- The Prevention of Terrorism (Supplemental Temporary Provisions) Act replaced the 1974 act in March 1976. This made it an offense to fundraise for terrorist groups and to withhold information relating to acts of terrorism. It prohibited additional organizations but provided a mechanism for those subject to exclusion orders to seek review of them.
- The 1976 act was replaced by the Prevention of Terrorism (Temporary Provisions) Act of 1984 and the Prevention of Terrorism (Supplemental Temporary Provisions) Order of 1984. These came into force on March 22 and 27, 1984, respectively. They empowered the Home Secretary to authorize detention without trial for up to 5 days and limited exclusion orders to 3 years, after which they would expire, although they were eligible to be renewed.
- The Prevention of Terrorism Act of 1989 added further organizations to the "proscribed" list and made it an arrestable offense to be a member of, show support for, solicit funds for, or associate with such organizations. Also, persons could be arrested without warrant if the police had "reasonable suspicion" that they were guilty of offenses or concerned with the commission of acts of terrorism.
- All of this previous legislation was repealed by the Terrorism Act of 2000 (TACT 2000), which became effective on February 19, 2001. It followed the development of the Northern Ireland "Peace Process" and was in response to the perceived threat of international terrorism. This was the most comprehensive of the terrorism acts, and was designed to broaden, streamline, and reform the previous legislation.

- TACT 2000 specifically outlawed 25 international terrorist groups that had previously operated lawfully in the United Kingdom. Fourteen Northern Irish groups, Republican and Loyalist, were also identified. The act also made it a duty for persons to notify authorities when, in the normal course of business or employment, they discover information about a person who is involved in terrorism.

- TACT 2000 gave the police increased powers to stop, search, seize, and arrest persons they reasonably believed to be involved in terrorism. The act also made offenses of inciting terrorist acts, seeking or providing terrorist training for purposes in the United Kingdom or abroad, and providing instruction to terrorists in activities such as use of firearms, bomb-making, or the use of nuclear or biological weapons.

- Neither internment without trial nor exclusion orders were in TACT 2000, thus repealing those provisions.

- Despite the intent to be comprehensive, TACT 2000 was insufficient, and the Anti-Terrorism, Crime and Security Act of 2001 was passed swiftly in response to the attacks of September 11, 2001. It provided for financial institutions to disclose information pertaining to criminal investigations in Britain and overseas and provided procedures for forfeiture of assets.

- The most controversial element of the 2001 act dealt with immigration and asylum issues. It provided that resident foreign nationals could be detained, without trial, for indefinite periods on the grounds of national security. This part of the act was challenged in the English courts under the European Convention on Human Rights (ECHR), and was found to be "inconsistent" with (i.e., in violation of) the convention and the Human Rights Act (HRA) of 1998.

- However, the government was caught between two elements of the ECHR. It could not deport foreign nationals to places where they would likely be tortured and killed, but they could not allow them to be at liberty in Britain if they posed a threat, and they could not jail them without adequate evidence to obtain a conviction.

- A separate, but equally controversial section made it an offense to "incite religious hatred."

- Exclusion orders were reinstated.

- The Prevention of Terrorism Act of 2005 amended some of the provisions of the 2001 act to make them compliant with the HRA of 1998 and the ECHR. This included the introduction of "control orders." Control orders were made by the Home Secretary and designed to restrict a specific individual's accommodation, associations, possession of items, or daily routine.

- The 2005 act expanded on the 2000 and 2001 acts and included many ordinary daily activities within the definition of "preparation to commit terrorist acts."

- Following the July 7, 2005, bombings in London, a further terrorism bill was introduced in October 2005 and became law in March 2006. It included further restrictions on attending terrorist training camps, a 28-day period for detention without charge pending, and investigation and other issues.

KEY TERMS

Al-Qaeda: An armed, Islamic, decentralized international organization with loosely affiliated, independent cells that carry out attacks and bombings in order to disrupt the economies and influence of Western nations, to rid Muslim countries of Western influence, and to advance Islamic fundamentalism.

Easter Uprising: Resulted from the instability created by the Government of Ireland Act of 1914, which stated that Ireland would gain regional self-government after World War I. The act was never implemented for fear that civil war might be sparked if the act came to fruition.

Irish Republican Army (IRA): A paramilitary organization of Irish nationalists who utilized terrorist tactics and guerrilla warfare to force English armed forces out of Northern Ireland, to achieve a united and an independent Ireland, and to be constitutionally free from English influence. The group is still in existence today.

Irish Republican Brotherhood (IRB): A group that gradually formed, organized, and fought for Irish independence from Britain. After home rule was again defeated in the British Parliament, the conflict came to a head in the 1916 Easter Uprising. The organization disbanded in 1924 after it perceived its goals had been achieved.

Officials: The larger of the two IRA groups that advocated a socialist Ireland that would be accomplished without the use of violence

Plantation of Ulster: The best agrarian tracts of land that Elizabeth I seized and gave to her subjects to colonize Ireland. This displaced many Irish and caused much friction between the Irish and the English.

Prevention of Terrorism Acts: A series of progressive acts started in November 29, 1974, that bestowed emergency powers upon the government and the police in times when terrorism was suspected and deemed a threat. The successive acts built upon each other, with legislators retaining the facets of previous acts that had been effective and rejecting those that had not. The acts were initially created in response to IRA bombing campaigns in the early 1970s after increases in IRA activities. The powers bestowed by the acts were broadened, debated, and reworked to balance civil rights and national security after the 9/11 and 7/11 terrorist attacks.

Provisionals: The smaller of the two IRA groups that advocated terrorism as a necessary evil that would bring about Irish reunification.

Siege of Drogheda: In 1649, Oliver Cromwell clearly demonstrated his hatred of the Irish by not only killing all soldiers after their surrender, but also by killing the civilians.

PUTTING IT ALL TOGETHER

1. Why has the focus of terrorist legislation shifted towards a preventive paradigm?

2. What were the basic issues behind the problems in Ireland?

3. How was religion an issue in the Irish troubles?

4. Why has the IRA "withdrawn" from the world of terrorism?

5. What were the issues that turned Osama bin Laden to attack the United States, the United Kingdom, and the Western world?

6. Why are definitions of terrorism not very specific?

7. Why was the Anti-terrorism, Crime and Security Act of 2001 so controversial?

8. What was the justification of the United Kingdom in holding foreign nationals suspected of terrorism indefinitely?

9. Why were control orders so controversial in the 2005 and 2006 legislation?

ENDNOTES

1. At the time of this writing, there have been no prosecutions under this section (Privy Counsellor Review Committee, 2003, p. 38).

2. This definition was adopted for the Anti-terrorism, Crime and Security Act of 2001, which updated the Terrorism Act of 2000 after the 9/11 attacks. It was also adopted and modified for the Prevention of Terrorism Act of 2005.

REFERENCES

Anti-terrorism, Crime and Security Act 2001. (2001). Retrieved June 4, 2006, from http://www.opsi.gov.uk/ACTS/acts2001/20010024.htm.

Chahal v. The United Kingdom. (November 15, 1996). 22414/93 [1996] ECHR 54.

Council of Europe. (1966). European convention on human rights and its five protocols. Retrieved June 23, 2006, from http://www.hri.org/docs/ECHR50.html.

Dingley, J. (2001). The bombing of Omagh, 15 August 1998: The bombers, their tactics, strategy and purpose behind the incident. Studies in Conflict and Terrorism, 24, 451–465.

Home Office. (2001). Statistics on the operation of prevention of terrorism legislation, Great Britain, 2000. Retrieved June 12, 2007, from http://www.homeoffice.gov.uk/rds/pdfs/hosb1601.pdf

Home Office. (2004). COUNTER-TERRORISM POWERS: Reconciling Security and Liberty in an Open Society: A Discussion Paper Retrieved June 12, 2007, from http://www.archive2.official-documents.co.uk/document/cm61/6147/6147.pdf.

Home Office. (2005a). New laws to combat terrorism. Retrieved July 3, 2006, from http://www.homeoffice.gov.uk/about-us/news/extending-detention-period.

Home Office. (2005b). Prevention of Terrorism Bill 2005, background briefing papers. Retrieved July 3, 2006, from http://cryptome.org/ptb2005-briefs.htm.

Home Office. (2005c). The Prevention of Terrorism Act 2005. Retrieved July 3, 2006, from http://www.homeoffice.gov.uk/security/terrorism-and-the-law/prevention-of-terrorism/?version=1

Katzman, K. (2005). *Al Qaeda: Profile and threat assessment.* CRS Report for Congress. Retrieved July 3, 2006, from http://www.mipt.org/pdf/CRS_RS22049 .pdf.

"7 July 2005 London bombings." Wikipedia. Retrieved June 11, 2007, from http://en.wikipedia.org/wiki/7_July_2005_London_bombings.

"Oliver Cromwell." Wikipedia. Retrieved July 3, 2006, from http://en.wikipedia.org/ wiki/Oliver_Cromwell.

Patterns of Global Terrorism 2001 (U.S. State Department) 2001. Retrieved on June 28, 2007, from http://fl1.findlaw.com/news.findlaw.com/hdocs/docs/dos/ trrpt2001/dostrrpt2001p10.pdf

Peck, M. (2005). The Terrorism Bill 2005–06, Bill 55 of 2005–06. Retrieved from http://www.parliament.uk/commons/lib/research/rp2005/rp05-066.pdf.

"Prevention of Terrorism Act." Wikipedia. Retrieved June 12, 2007, from http:// en.wikipedia.org/wiki/Prevention_of_Terrorism_Act_2005.

Prevention of Terrorism Bill: 2005. Retrieved June 12, 2006, from http://www. publications.parliament.uk/pa/cm200405/cmbills/061/2005061.pdf.

Prevention of Terrorism (Temporary Provisions) Act 1989. Retrieved June 12, 2006, from http://www.opsi.gov.uk/acts/acts1989/Ukpga_19890004_en_2.htm.

Privy Counsellor Review Committee. (2003). *Anti-terrorism, Crime and Security Act 2001 review: Report.* Retrieved on June 12, 2007, from http://www.statewatch .org/news/2003/dec/atcsReport.pdf

Simonsen, C., & Spindlove, J. (2000). *Terrorism today: The past, the players and the future.* NJ: Prentice Hall.

Terrorism Act 2000. Retrieved June 13, 2006, from http://www.opsi.gov.uk/ACTS/ acts2000/20000011.htm.

Terrorism Act 2006. Retrieved June 13, 2006, from http://www.opsi.gov.uk/acts/ acts2006/20060011.htm.

USA Patriot Act. (2001). Retrieved June 14, 2006, from http://www.epic.org/ privacy/terrorism/hr3162.html.

Walker, C. (2002a). *Blackstone's guide to the anti-terrorism legislation.* Oxford, England: Oxford University Press.

Walker, C. (2002b). 1.4.1 Terrorism Act 2000. Retrieved June 13, 2006 from http://www.oup.co.uk/pdf/1-84-174183-3.pdf.

White, J. R. (2002). *Terrorism: An introduction.* (3rd ed.): Wadsworth Publications. Belmont, CA.

Juvenile Justice in England

Chapter Objectives

After completing this chapter, you will be able to:

- Discuss the history and evolution of criminological thought concerning juveniles and juvenile justice, and pertinent legislation in England.
- Explain how attitudes towards age and criminal intent have affected the prosecution of juveniles and legislation for the past 200 years.
- Discuss how the English juvenile justice system has maintained its rehabilitative ideals in the face of outcries for tough punishment for juveniles.
- Discuss how parents have become more responsible for the acts of their children, particularly in recent years.
- Trace the evolution of legislation as its focus switches from placing responsibility for anti-social behavior on the individual, to society, to a mix of both, and then ultimately back to the individual.
- Discuss how juveniles' rights have increased in the past 200 years.
- Explain how the idea of structure and positive activities in a juvenile's life are important to the prevention of crime in England.
- Discuss how education and religion were instrumental in reforming the juvenile justice system in England.

■ Introduction

The debate over the origins of juvenile delinquency has waged for centuries, as have arguments about the most effective manner in which to deal with juvenile delinquents. Whether the approach is treatment or punishment or somewhere in between, the only consistency throughout the centuries concerning effective solutions is that no one has ever agreed what mix of these "solutions" is correct. Conclusions drawn about juvenile culpability, society's role in shaping the juvenile, parental involvement or a lack thereof, and ultimately where responsibility for the unacceptable behaviors lies have always been open to question and interpretation. Applied philosophies intended to understand juveniles and their wrongdoings and to guide legislation have evolved slowly but dramatically in the last 200 years and still reflect this tension between punishment and rehabilitation. Further, every society has opined that juvenile delinquency and crimes committed by juveniles have increased in number and seriousness, and therefore all have felt the urgency for action. In this chapter, we explore the history of the juvenile justice system in England with emphasis on the evolution of the juvenile justice system, the philosophies that shaped it, and the legislation that reflected both.

■ History of Juvenile Justice

In the 1800s England was experiencing a burgeoning urban population, particularly in London. Accompanying this were scarce employment opportunities, low wages, and poor living conditions resulting in a large number of unsupervised children, many of whom turned to begging, thievery, and other crimes to survive. In fact, children were often employed by adults to beg because they were better able to attract the compassion of strangers. During this time, crime was flourishing and juveniles constituted a large portion of the criminal population in England (Duckworth, 2002). Juveniles who violated the law received punishment that differed little from their adult counterparts (Wakefield & Hirschel, 1996, p. 91). Separate legal codes for children and adults did not exist in the 1800s, and beginning at the age of 7, the age of criminal responsibility in common law, children were tried for the same crimes in the same courts as adult criminals (Duckworth, p. 5), often with severe consequences. Juvenile law violators were also subject to the same punishments as adult criminals: death, imprisonment, and transportation (Duckworth, p. 40; Wakefield & Hirschel, p. 91). At this time, there were between 160 and 200 offenses for which juvenile offenders could be sentenced to death (Wakefield & Hirschel, p. 93). In fact, between 1801 and 1836, 103 children were sentenced to death at the Old Bailey, and although nearly all of these sentences were eventually commuted to imprisonment or transportation, there are records of juvenile offenders being executed as late as 1833 when a 9-year-old was hanged for taking merchandise from a shop window (Duckworth, p. 40). However, although the treatment of juveniles was still quite harsh and incidents such as these did occur, in practice, magistrates were more often lenient on younger offenders, and the execution of children was actually a rare event.

In the 1800s, the primary goal in using these extreme sanctions, regardless of age, was to prevent and deter future crimes. When sentenced to imprisonment, juveniles served their sentences in the same institutions as adult criminals with the hope that deterrence would be effective; obviously, however, this was not conducive to the reformation or rehabilitation of juvenile law violators. In reality, incarceration in the same institutions as adults seemed to increase the likelihood that children would continue to violate the law once released (Terrill, 2003, p. 99). Prisons of the day were schools for criminality and provided ample recruiting grounds for juveniles with criminal potential outside of prison.

Even in this time of harsh treatment of juveniles the law recognized that not all juveniles were as equally culpable as adults for their acts. Under common law there were two categories of juveniles, with differing presumptions with regard to their ability to possess the criminal intent required for prosecution and conviction as an adult. These presumptions applied to juveniles under the age of 14. At 14, a juvenile was considered an adult. Children up to the age of 7 were considered incapable of criminal intent (**_doli incapax_**), and therefore could not be held criminally responsible for their acts. Consequently, any criminal case against a child under 7 had to be dropped. The second category of children began at the age of 7 and extended up to the age of 14. The presumption was that children in this age group did not yet fully understand the notions of good and evil. The prosecution was, therefore, required to prove that a child between the ages of 7 and 14 knew the difference between good and evil (**_doli capax_**). In practice, however, the fact that the offense had been committed suggested to the court that the child knew the difference between right and wrong and had criminal intent in committing the act. Punishment, therefore, was usually justified for this age group, with rape as the notable exception. Boys under the age of 14 were presumed by law to be impotent, incapable of committing rape, and therefore presumed innocent (Duckworth, 2002, p. 41). Thus, even with the age categories and the criteria therein, the criminal intent of the child was established more or less on a case-by-case basis. This struggle with understanding criminal intent in juveniles continues to this day.

■ Changes in Juvenile Justice

By the mid-1800s, movement towards a more rehabilitative ideal became evident. **Ragged schools**, aimed at educating students who would otherwise be excluded from any educational opportunities because of their social class position, were created. The schools were named for the ragged appearance of the pupils. These schools were founded by philanthropic movements whose goal was to provide impoverished children schooling in the "three Rs" (reading, writing, and arithmetic), bolstered by Bible readings. Thus, as was the case in the United States, the early reform movements for juvenile justice found their roots partially in religion.

Concern regarding the brutal treatment of juveniles began as reformers in the early nineteenth century suggested that juvenile criminality was the result of poor living conditions, societal influences, and parental neglect, rather than poor moral character (Duckworth, 2002). With this premise, the hopes for children were twofold: (1) that children were malleable enough to change their ways if reached early enough in their "criminal careers," and (2) that if this was the case, then future criminality could be averted (Terrill, 2003, p. 99). As we shall see, this was a common philosophy in English criminal justice that waxed and waned in popularity and that exists to this day.

As discussed, religious thought and education played an instrumental role in these early changing ideologies and reforms. In the early 1800s, John Pounds, a Unitarian,[1] recognized the importance of education and began teaching a small group of children who could not afford the attendance fees and were considered unacceptable to British schools. Promises of free food were enticement enough for the malnourished children to attend on at least a semi-regular basis. Following Pound's death, the Portsmouth Unitarian Congregation continued his work with the children. Later, in 1844, the existing ragged schools were united within the Ragged School Union under the direction of Lord Shaftsbury. Although the educational standards in these ragged schools were quite low, they filled a necessary role, aided in altering the thinking on the treatment of juveniles, and continued to operate until the end of the nineteenth century.

Another influential person in the early history of British juvenile justice was Mary Carpenter. Carpenter, above all, was passionate in her concern for the welfare of children. The daughter of a Unitarian minister, Carpenter was responsible for the development of a ragged school and several reformatories (Manton, 1976). Her publications discussed the causes of juvenile delinquency, criticized the current treatment of juvenile delinquents, and suggested avenues for reform (Carpenter, 1970). Carpenter advocated for the continuation of the ragged schools and the development of free day schools and industrial feeding schools to handle juveniles before they became entrenched in the criminal justice system. She also promoted the development of reformatory schools as a substitute for the incarceration of juveniles in adult prisons (Carpenter, 1969).

However, to truly be effective, these reforms would require legislative action by Parliament, and they did not gain Parliamentary support until 1847 with the passage of the Juvenile Offenders Act and many other subsequent acts. With supporters like John Pounds and Mary Carpenter advocating for the development of ragged schools and reformatories, English Parliament officially established these institutions through the Youthful Offenders Act of 1854 and the Industrial Schools Act of 1857. These changes and the subsequent legislative acts did have their influence on the courts and were steps in a positive direction.

The Juvenile Offenders Act of 1847 allowed for summary trials for children under the age of 14 arrested for theft. In addition, trials were deemed a private matter and were quicker due to two magistrates conducting them rather than one. In 1854, Parliament further acknowledged the need to separate juvenile and adult criminals with the passing of the Youthful Offenders Act of 1854. This act provided for the development of reformatory schools for juvenile offenders, and industrial schools were later developed with the Industrial Schools Act of 1857 (Arthur, 2005). Industrial schools were another means beyond ragged schools to instill a work ethic so that a child learned a trade and developed into a responsible citizen. Both reform and industrial schools were regulated by the Home Office; however, most were operated by volunteer organizations. Although the development of reformatories and industrial schools prevented juveniles from being sanctioned as and incarcerated with adult criminals, juveniles continued to be processed through the same court system as adults (Terrill, 2003, p. 99).

Thus, the juvenile justice system was, in essence, feeling its way along in philosophy and practice.

The Summary Jurisdiction Act of 1879 was a first serious step in moving toward a separate juvenile justice system. This act provided for a summary trial for all indictable offenses except for homicide for individuals between the ages of 7 and 12 if the court agreed, and if the parent(s) did not object. In the case of a child between 12 and 16, the same standards applied if the child was informed of his or her right to a trial. Thus, a substantive step in recognizing intent as it is formed by juveniles of differing ages was applied legally and procedurally with this act. The act also differentiated punishments for juveniles and adults. For example, the maximum number of strokes a male child could be whipped by a birch cane was 6 whereas for an adult male it was 12 ("Summary jurisdiction"). Although these steps may appear insignificant, in reality, these slow changes in philosophy were important in delineating differences between juveniles and adults and how the criminal justice system should deal with each. These steps also heralded the creation of completely separate systems.

Finally, in 1908, the Children's Act not only provided children legal protection from harmful parents or guardians, but also developed a juvenile court system. The act required that all offenses committed by individuals under the age of 16, excluding murder, be adjudicated by these magistrate courts. Furthermore, this court had jurisdiction over issues of childcare for juveniles as young as 7 and for children considered to be beyond the control of their parents. However, unlike the juvenile courts developed in other parts of the world, the juvenile courts developed in England and Wales were little more than "modified criminal courts" (Bottoms & Dignan, 2004, p. 23) that mirrored their adult counterparts. Eventually, juvenile courts and legislation reflected deliberate intentions to move the criminal processing of juveniles away from the traditional and formal criminal procedures that adults were made to endure (Bottoms & Dignan, 2004).

Although the impetus for change was there, the movements may be regarded as somewhat piecemeal, and there were obstacles to efficiency. For example, the Children's Act of 1908 mandated a special sitting in Magistrates' Court to handle all juvenile matters, but many of these magistrates were not trained for work with juveniles (Wakefield & Hirschel, 1996, p. 94). In many courts, the primary difference between adult and juvenile proceedings was that juvenile proceedings usually occurred in a different location or at a different time than adult proceedings. Public access to juvenile court was also more restrictive than access to adult proceedings (Bottoms & Dignan, 2004). There was also criticism of the incarceration of juveniles by developing new discharge and fining options for sentencing authorities that somewhat "widened the net," bringing more juveniles into the system than otherwise would be there because of the new schemes. Therefore, although there were small steps towards a separate juvenile court system per se, the juvenile court continued to be a part of the normal adult court hierarchy.

There were also gains for juveniles in other areas of the criminal justice system. In the same year, 1908, the Prevention of Crime Act established a specialized juvenile institution called a "borstal," which provided another alternative to juveniles being sent to prison with adult populations. As discussed in Chapter 10, borstals housed delinquent boys between the ages of 16 and 21 and provided them with treatment designed to reform, educate, teach a work ethic, and instill strict discipline that, if violated, could bring the transgressor corporal punishment. Young boys were sent to borstals for an indeterminate length of time ranging from 6 months to 2 years. However, treatment did not end when the juvenile left the borstal; aftercare continued following release. Borstals remained a part of the English juvenile justice system until the 1982 Criminal Justice Act; following that act, borstals were renamed **Youth Custody Centers**. Sentences for juveniles also changed from being indeterminate in nature to determinate and shorter in length.

Two legislative acts later in the twentieth century, the Children and Young Persons Act of 1933 and the Children and Young Persons Act of 1969, further changed juvenile justice in England, focusing on the welfare of the child. First, the Children and Young Persons Act of 1933 made a number of small but important changes to the juvenile justice system. The age of criminal responsibility was raised from 7 to 8 years of age; individuals through the age of 16 (rather than up to the age of 16) were placed under the jurisdiction of the juvenile courts; and terms such as "convicted" and "sentenced," which could negatively label and affect juvenile offenders, were no longer used. Although the effects of this law may seem insignificant, this act reaffirmed the government's commitment to the concept of juveniles as separate from adults.

Some of the most significant changes to England's juvenile justice system did not occur until the 1960s. With the Labour Party gaining power in 1964, many social changes in Britain were championed by this party through legislation. For example, homosexuality was made "partially" legal; race relations and racial discrimination were encouraged and made illegal, respectively; the death penalty in Britain was abolished; and the Children and Young Persons Act of 1969 was enacted. The primary goal of this act was to de-emphasize the punishment aspect of juvenile justice and to place more of an emphasis on social services for juvenile offenders (Terrill, 2003, p. 100). It was believed that the etiology behind juveniles who happened to get into legal trouble was very similar to juveniles who were in less serious trouble. Therefore, using the court system to handle these "criminal" juveniles should be avoided until all other possibilities had been tried (Wakefield & Hirschel, 1996, p. 95). The thinking was that juvenile offenders should be offered social services in their own communities in an effort to avoid incarceration and the effects this would have on the child, so consequently, social workers were given significantly more control over the care and supervision of juvenile offenders (Joyce, 2006, p. 420; Terrill, 2003, p. 100). In fact, the incarceration of juveniles in borstals or detention centers was seen as a last resort and was strongly discouraged (Wakefield & Hirschel, p. 95).

The left wing and liberal Labour Party were in power when the Children and Young Persons Act of 1969 was originally passed. However, because of a rise in juvenile crime; a shift in government policy towards a more right wing, "law and order" philosophy in the 1970s; and a national economic crisis that hindered funding of social programs, the full potential of this piece of legislation was not realized (Terrill, 2003, p. 101). The Labour Party lost power after only a few of the provisions that the act contained had been implemented. The remainder of the provisions were not acted upon by the incoming Conservative government, nor any other subsequent governments (Bottoms & Dignan, 2004, p. 24). Thus, the debate as to whether the juvenile justice system was to be rehabilitative, punitive, or somewhere in between was again at the forefront, but leaning towards being punitive. The public was frustrated with repeat offenders and the seemingly more violent juvenile crime, but there was also a sense that exposing children at an early age to the criminal justice system might increase their chances of progressing through it (Joyce, 2006, p. 422). The English responded with strategies that were designed not only to handle these persistent offenders, but also to divert first-time offenders away from the formal criminal justice system through, for example, the increased use of cautions and other measures that were preventive in nature.

The importance of a juvenile justice system was recognized again and reaffirmed by Parliament with the Children's Act of 1989 and the Criminal Justice Act of 1991. Both of these acts emphasized the importance of limiting the amount of contact that juvenile offenders had in the criminal court system, as well as limiting the custodial sentences given to juvenile offenders. There was also recognition that although the juvenile is responsible for his or her acts, other influences such as poor parenting also caused the child's delinquency. Thus, the Children's Act of 1989 was one of the first acts to establish parental responsibility in a legislative act. It also amended the "care order," which was originally authorized by the Children and Young Person's Act of 1969. Care orders required that children be placed under the care of both the parent and the local authority if deemed necessary, meaning that the child was beyond the control of the parent or if there was a concern that harm could befall the child. The Family Proceedings Court was given jurisdiction over these matters.

Further expanding the jurisdiction of youth courts was the Criminal Justice Act of 1991. Courts were now to include 17-year-olds, individualized sentencing based on the maturity level of the offender was introduced, and even more emphasis on responsibility was placed on parents, in that parents were required to attend court at all proceedings where their children were involved, to potentially pay fines for the juvenile, and to maintain recognizance of the child (Sections 56–58). Youths who could be sentenced to detention were expanded from 14-year-olds to 15-year-olds, and the minimum length of detention was set at 2 months. This act also changed the name of juvenile courts to youth courts, which essentially was a special sitting of the Magistrates' Court. It is this court that had received criticism and debate as to the role of the juvenile court: as a punitive body or as a protector of a child's welfare (Terrill, 2003, p. 108). The debate continued and subsequent legislation reflected both philosophies.

The Crime and Disorder Act of 1998 ushered several key changes into the juvenile justice system. The act—designed to be national in scale, but very localized in operation—placed more emphasis on the prevention of offending rather than only focusing on the social factors that may have influenced criminality and harsh penalties for those who broke the law. In fact, the act explicitly stated that the primary goal of the juvenile justice system was to prevent juvenile offending. To this end, early intervention and intensive community supervision were utilized; for example, the act allowed local governments to impose curfews on juveniles age 10 and younger. In addition, the courts were empowered to place corrective requirements such as drug rehabilitation or counseling on juveniles who were at risk for criminal behavior, and parenting orders were implemented with counseling to help parents better care and gain control of their children.

An impetus for the act was the connection found in the criminological literature between insufficient parenting and juvenile crimes (Bottoms & Dignan, 2004, p. 88). Jack Straw, the Home Secretary at the time, suggested that this redirection of focus was an intentional movement away from the previous philosophy of juvenile justice that focused too heavily on a juvenile's social circumstance as an excuse for criminal behavior (Bottoms & Dignan, 2004, p. 43). There was also public and political pressure from the Labour Party to be tougher on juvenile crime. Given that these new implementations were focused more directly on the juvenile, their parents, and the behaviors by both that fostered the delinquency, one must notice that the state was shifting the emphasis to prevention in the early stages of a child's delinquency and narrowing responsibility. Also noteworthy was that although there were calls for "law and order" where juveniles were concerned, the effort to focus services where needed while maintaining a strategy to deal with repeat offenders with a community focus was still the order of the day. Thus, both rehabilitative and punitive philosophies were utilized.

Also developed by the Crime and Disorder Act of 1998 was an entity to monitor the multi-agency approach to the juvenile justice system (Joyce, 2006, p. 429). The Youth Justice Board was tasked with overseeing the youth justice system in England and Wales (Bottoms & Dignan, 2004, p. 42). Specifically, the 10 to 12 members of the Youth Justice Board monitor and support the efficient operation of the youth justice system and advise the Home Secretary regarding the operation of the youth justice system (Youth Justice Board). The Youth Justice Board is also responsible for overseeing the Youth Offending Teams and for finding suitable secure facilities for those youths remanded and sentenced by courts. As discussed in Chapter 6, Youth Offending Teams (YOTs) were developed to provide local coordination of much-needed services, such as social services, housing, drug and alcohol treatment, and health services, for youth ages 10 to 17. The teams consist of representatives from the police, social, health, housing, and probation services and are instrumental in assessing juveniles' needs, identifying risk factors, determining appropriate programming, providing court-ordered supervision, and acting as counselors for those in secure settings.

Another key change resulting from the Crime and Disorder Act of 1998 was a redirected focus on reparation and a modification of old standards. As discussed in a previous chapter, the repeated cautioning of young offenders was replaced by reprimand and final warning schemes designed to get younger offenders, and if necessary their parents, to deal with the youth's delinquent behavior earlier in the process and to make reparations where possible. Final warnings can activate a referral to a Youth Offending Team. In this case, the YOT is charged with determining the direct causes of the youth's delinquency and preparing a rehabilitative change program, or as it is also known, an action plan. The targets of the reprimands are first-time offenders who have committed minor offenses. There are, in general, three goals: (1) for the offender to make amends to the victim and/or the community for the harm that has been caused, (2) for the offender to learn empathy for the victim(s), and (3) for the Youth Offending Team to help the juvenile understand the causes of the offense itself in order to recognize what aspects of the juvenile's life place him or her at risk for future offending.

The act also modified the law of criminal intent by abolishing the common law presumption that a child under 14 does not know the difference between right and wrong. As mentioned previously, the practice and assumption in actual court operation was that the juvenile knew the difference between right and wrong because he or she committed the offense and was thus treated as such in court proceedings. Therefore, the act simply embraced and reflected court practice.

Subsequent legislation continued to advance the goal of diverting first-time minor offenders away from the court system while dealing with repeat and serious offenders in court. The Youth Justice and Criminal Evidence Act of 1999 continued the reformation of the juvenile justice system started by the Crime and Disorder Act of 1998, and utilized the restorative justice principles of restoration or restitution to the victim, reintegration into the community, and offender/parental responsibility. The notion that there was a wider range of issues causing a juvenile's offending behavior remained the philosophy advanced by the juvenile justice system, and therefore, according to the act, a conviction for first-time offenders in a youth court might prompt much closer examination by a youth offender panel (Section 6).[2] The panels, which have control of the juvenile for 3 to 12 months, can require both the parents and the juvenile to take responsibility for their respective behaviors as they pertain to the child's delinquency. Additionally, a contract may be negotiated by the child and the youth offender panel with specific conditions aimed at changing the juvenile's behavior and discouraging future offenses (Section 8). Stipulations in the contract are specific to the juvenile and his or her situation and can be in the form of reparations to victims, drug or alcohol treatment, consistent school attendance, or the like. Failure to sign the contract or to live up to the terms of the contract can trigger a return to court for sentencing (Section 11).

Further legislation concerning juveniles and their relationship with adults, particularly parents, was the Licensing (Young Persons) Act of 2000. This updated Section 169 of the Licensing Act of 1964 delineated the purchase of alcohol in situations when juveniles might be involved. Specifically, it was illegal to sell alcohol to those under 18 (Sections 169-A1) and it was an offense to purchase alcohol from off-license or licensed premises for a child (Sections 169C-2). However, parents could purchase alcohol for themselves and then give it to their own children, because the parent was not the child's agent for the purchasing of the alcohol. This is in stark contrast to the American legal drinking age of 21, and in the case of the English, may go a long way to reduce tension between parental and juvenile responsibility, because alcohol is a frequent and substantial variable in criminal activity. The English have the legal opportunity to teach responsible drinking to their children that may reduce future criminal activity. Other provisions in the act were that a 16-year-old may consume beer, porter, or hard cider provided it was part of a meal and not consumed in a bar area (Section 169D). Finally, a defense for adults existed if they had no reason to believe that the child was not less than 18 years of age when they purchased the alcohol for said child (Sections 169A, pp. 2–3).

Just as the Licensing (Young Persons) Act of 2000 provided the opportunity for parents to monitor certain aspects of their children's lives, the Criminal Justice and Police Act of 2001 greatly broadened the range of juveniles over whom control could be exerted by the government. The act raised the maximum age at which child curfew orders could be imposed from under 10 years to under 16 years of age (Section 48). It also allowed for 15- and 16-year-olds to be remanded to secure accommodations only after an inquiry into other remand options had been explored and if it was deemed that the juvenile in question constituted a harm to the general public and/or if he or she would commit further crimes (Sections 130–136). In addition, a court could impose upon a child electronic monitoring for remand or bail: (1) if the child was 12 years of age or older, (2) if the child was charged with or had been convicted of a violent or sexual offense or an offense whereby the punishment was greater than one year, (3) if the Youth Offending Team deemed it appropriate, and (4) if electronic monitoring was available (Sections 131–132). Thus, if necessary, the government could step in to closely monitor from first-time to more serious offenders, and although more responsibility was being imposed on juveniles, the Anti-Social and Behaviour Act of 2003 placed it squarely on parents.

The Anti-Social Behaviour Act of 2003 amended the parenting orders found in the Crime and Disorder Act of 1998 by requiring a parent to cooperate with specified orders concerning the care given to their child (i.e., guidance or counseling) for a period not exceeding 12 months (Section 18), provided these orders were germane to the prevention of future delinquent acts by the juvenile. The act also widened the circumstances in which they may be utilized and provided that parental contracts could also be imposed upon parents when their children have been truant or excluded from school. In this case, the parent enters into a contract with the local governing body for the school whereby both parties agree to aid the other in helping the child. The parent must also agree with the terms (i.e., guidance, a residential course, and/or counseling), and the length of the contract. The act also specifies how parenting contracts and orders should be conducted pursuant to criminal conduct and anti-social behavior of children (Sections 18–29). In this case, the child is referred to a Youth Offending Team, and the purpose of the contract is for the parent and the Youth Offending Team to aid each other in preventing the youth from committing further illegal acts or anti-social behavior, just as in the case of truancy and exclusion. If the parent fails to fulfill the terms of the contract, a fine of up to £1,000 could be imposed and the parent may be taken to court for another noncustodial penalty (Joyce, 2006, p. 435; National Association for the Care and Resettlement of Offenders [NACRO], 2006, p. 67).

The Children Act of 2004 received Royal assent on November 15, 2004, and represents yet another legislative work designed to improve the coordination and distribution of services to children and families. The overall aim of the act was to promote the health and safety of children and to enable them to make positive contributions to society. This act specifically moved the focus of attention from the national to the local level. The act created the position of a Children's Commissioner whose role is to learn and promote children's issues and concerns and to hold inquiries on policy, integrated planning, commissioning, the delivery of services, and accountability for agencies under his or her charge (Children's Act 2004: Part 1, Schedule 1). The act also required that those who look after children in need take responsibility for their educational needs (Children's Act 2004: Part 2, Section 10, Subsection 2). However, although the intent of this act was to improve the delivery of services to children and families, there was a lack of specificity as to exactly how this should be done.

The last work discussed here that attempted to answer what the needs of youth were was a green paper entitled, "Youth Matters: Next Steps" published in March 2006. This paper reported the results of a survey of 19,000 respondents and 1,000 organizations, professionals, and parents as to what their needs were, with the goal that the results would ultimately guide future legislation and proposals by the agencies that work with youth. Generally, respondents identified two common problems for British youth. First, there was a lack of positive activities in which youth may engage, and second, there were not enough safe places in the community for them to congregate. In reference to the former, a proposal by the organization Youth Matters was that a national standard be created that would define these "positive activities" in order to appropriately and effectively allocate resources, funding, and other financial opportunities for youth to engage in these defined activities. The survey disclosed that young people wanted empowerment and a say as to how local governments and services spent money on these yet to be defined activities, services, and projects (Department for Education and Skills, 2006, p. 6). Youth also wanted greater recognition for the positive activities they did. It was felt that more youth would be inclined to participate in these activities if there was some type of recognition, such as the provision of awards, certificates, or some kudos that might aid in gaining future employment (p. 11). In addition, adults supported high qualitative standards in reference to impartiality, challenging stereotypes, raising aspirations, and measuring the quality of the interventions for those government representatives who were responsible for information, advice, and guidance (IAG) for youth (p. 11). In terms of a timetable, the goal was set forth in the green paper that the proposals that stemmed from the findings of this study should be implemented by April 2008.

The "Youth Matters: Next Steps" report and the pending plans for the future are a logical culmination of what we have found in our exploration of the evolution of criminological thought in England concerning juveniles and the resulting key legislation over the last 200 years. Just as we have seen with the terrorism legislation designed to thwart those types of crimes (see Chapter 11), the scheme has been the same with juvenile delinquency: prevention of criminality at the individual level through a national effort. The report operates from the perspective we have seen many times: that there are multiple causes of juvenile crime, and whether these causes are found at the societal level is not so much the issue anymore, though the solution is. There is a national, governmental approach, localized in nature and operated through grassroots organizations in providing services, treatments, and the like, focused ultimately on the individual and his or her family. Thus, although the juvenile justice system has always been prepared to handle serious offenders through harsh sentences, a substantial investment is being made to prevent criminality before it begins or gets too serious.

■ Juvenile Courts

Following the implementation of the Crime and Disorder Act of 1998, the juvenile justice system in England and Wales received a significant overhaul in procedures and process, the first of which discussed here is defining the various categories of responsibility for children and young adults.

As was noted earlier in the chapter, what constituted a "juvenile" according to English law focused on the notion of what a child understood relevant to committing criminal acts and/or forming criminal intent. This has been clarified dramatically. What started as children being classified in roughly the same class as adults is now fairly categorized based on age, although juveniles are still very much processed on a case-by-case basis, as we shall see. Currently, anyone 14 to 17 years old is considered a **young person**, whereas those 13 and under are considered **children**. Within this latter category, no one under the age of 10 may be prosecuted for any offense; those between the ages of 10 and 13 may be, but the prosecution must prove that the child knew the difference between right and wrong. Thus, the *doli capax* aspect of criminal intent has not totally been eradicated. In reference to crimes such as homicide or manslaughter, a child between 10 and 13 may be tried on indictment if it can be proven he or she knew the difference between right and wrong. Young persons may also be tried on indictment; however, young persons may also be tried as adults for crimes that, if committed by an adult, would warrant 14 years or more in prison. Lastly, the youth courts have jurisdiction over those under the age of 17, with the exception of murder, unless *doli capax* is proved by the prosecutor. Ultimately, the goal in utilizing these age categories is for courts to deal with the age and development of youth on a case-by-case basis (Home Office, 2005b, p. 154).

Procedurally, a case involving an alleged juvenile offender begins with the police. Once a crime involving a juvenile offender has been reported, procedures are guided by the Police and Criminal Evidence Act of 1984. The police have several diversionary options that may be utilized prior to officially involving the youth in the criminal justice system. Provided the youth admits responsibility for the offense, police officers may informally warn the youth, reprimand the youth, or give the youth a final warning.[3] As noted previously (see Chapter 6), the Crime and Disorder Act of 1998 prohibited the use of cautions for juvenile offenders because it was found that some young people received multiple cautions for a series of offenses, were often reconvicted within 2 years, and prosecution of youth was more effective than cautioning (Audit Commission, 2004, p. 3). In addition, the act proscribed that reprimands and final warnings be used as a replacement. Reprimands are for minor offenses committed by first-time juvenile offenders under the age of 18. A second offense results in a final warning, which triggers a referral of the juvenile to a Youth Offending Team. Subsequent offenses typically result in charges being filed (Bottoms & Dignan, 2004, p. 84). This fairly quick and direct path to prosecution is the reverse of what occurred prior to the 1998 Crime and Disorder Act, and consequently it may be argued the system has been streamlined for persistent offenders. These changes also have vocal critics among those who subscribe to the "minimum intervention" philosophy (see Bottoms

& Dignan, 2005). In the end, the police have extraordinary discretion in deciding whether to send a case for prosecution or to divert the juvenile offender.

Once the police decide that prosecution is necessary, the juvenile case must be sent within 72 hours to the **Crown Prosecution Service (CPS)** magistrate for the final determination as to whether to continue with prosecution or to discontinue the case. If the juvenile case is discontinued, the police may be instructed by the Crown Prosecution Service to reprimand the juvenile. If the case is to be continued, the juvenile may be remanded to appear at either the Crown Court for treatment as an adult or the youth court. Most juvenile offenders are sent to the youth court, which is a specialized Magistrates' Court where the case may be heard by a magistrate or a District Judge. A typical circumstance where a juvenile would go to Crown Court occurs when she or he has been charged with an indictable crime with an adult. In this case, the crime must be a serious or a "grave crime," such as indecent assault or dangerous driving, with a high probability of conviction, and the sentence required must be such that only the Crown Court may impose it. Generally, the magistrates who sit in youth court are experienced in legal matters and are initially appointed by the Lord Chancellor or the Chancellor of the Duchy of Lancaster as local magistrates for adults (NACRO, 2006, p. 36). They are then selected by their peers to sit on the youth court. In court, magistrates sit as a panel of three judges with one acting as the chair. District judges are paid for their participation in the youth court, are members of the legal profession, and hear the more difficult and lengthier cases (Youth Justice Board); magistrates in youth court are not paid, but receive payment for expenses. Youth court differs from adult court, primarily in that its proceedings are closed to the public and are less formal than adult courts. They also interact with the juvenile and their family more. Only members of the court, witnesses, the parties involved in the case and their legal representatives, and those specifically authorized by the court (i.e., the media) are permitted to view the proceedings.

Typically, in the case where the juvenile offender pleads guilty and it is a first offense, four outcomes are possible: custody, an absolute discharge, a hospital order, or a referral order. Most cases result in a referral order, which can last anywhere from 3 to 12 months and is intended to reflect the seriousness of the offense. A referral order requires the juvenile to attend the youth offender panel, which is part of the Youth Offending Team. The victim may also attend the proceedings. As previously discussed, the goal of the youth offender panel is to develop a contract outlining a course of action to change the behavior of the juvenile to prevent future offending. If the juvenile refuses to agree to the referral order or later violates the terms of the referral order contract, the juvenile is returned to the youth court for sentencing. If the terms of the contract are fulfilled, the conviction is "spent" (Audit Commission, 2004, p. 23).

If a juvenile pleads not guilty and the case moves towards trial, a pretrial review convenes whereby the Crown Prosecution and the defense's solicitor meet to discuss the case, agree on a witness list, and set a trial date. In addition, a child has the same right as an adult to adequate representation and to have counsel present, and as it is in the United States, if the juvenile's funds are insufficient to pay for legal counsel, legal aid will provide this service. A parent's role in these procedures is to be present at trial, which is a requirement, and the parent may aid in the child's defense if legal aid is not utilized. It should be noted that the procedures are substantively the same as an adult court whether the case is summary or indictment in type.

During the trial, the prosecution begins the case against the juvenile by describing the offense, bringing forth any witnesses and other evidence; the defense has the opportunity to cross-examine. After the prosecution has completed their case, the magistrates may ask questions. From there, the defense brings forth their case, including witnesses and any exonerating evidence. The prosecution may ask questions of the defense's witnesses. When both sides have completed their cases, the magistrates retire to chambers to weigh the evidence to determine guilt. If found guilty, there are a wide array of sentencing options for the magistrates.

In 2004, for cases where juveniles were sentenced for indictable offenses, parents were required to pay fines for their children ages 10 to 17 11% of the time. For those 14 and under, parents paid the fine 44% of the time, which is an increasing trend since 1994 (Home Office, 2005b, p. 66). If sentencing for the youth becomes the outcome, a number of other options exist: discharges, fines, referral orders (ages 10–17), reparation orders, community orders (ages 16–17), community rehabilitation orders (ages 16–17), curfew orders, action plan orders, supervision orders (which can last up to 3 years), parenting orders, intensive supervision and surveillance programs, detention and training orders (ages 12–17 and lasting for 4 months to 2 years), and long-term custody, many of which are available pretrial or as a form of diversion (**Figure 12-1**).

In the case of murder, the juvenile may also be sentenced to what may be a life sentence. This is known as **<u>detainment during Her Majesty's Pleasure</u>**, which is a sentence with a minimum of 12 years with the parole board having the discretion to release the person when they determine he or she is no longer a threat to society. Upon release, the person will remain subject to license for their remaining life (NACRO, 2006, p. 53). In all cases, the seriousness of the crime and a prior record are most often the deciding factors in length of sentence.

Although cases will end up as either custodial or noncustodial in nature, the vast majority are noncustodial. Seventy-one percent of those cases that end up in court are fined as their sentence, followed by 13% receiving a community sentence, 8% being discharged, and 7% ending up in prison (Home Office, 2005a, p. 5). Thus, at every step of the process, particularly before trial, there are many opportunities for a juvenile to avoid going to court for a trial, and there are many sentencing options for those who are adjudicated as guilty.

If a juvenile wishes to appeal the decision of a youth court, this may be accomplished in Crown Court. This court will determine if the law was applied correctly and if facts in the case were rightly found. In the appeal, the judge sits with two magistrate judges and retries the case with the same evidence and witnesses, with the option to hear more evidence or call more witnesses. No jury is used. The Crown Court may overturn the youth court's sentence, hand down another sentence, or sentence the juvenile more harshly. In the event satisfaction is not received in this court, the juvenile may appeal to the Queen's Bench Division of the High Court for procedural issues, if the lower court acted irrationally, or if the court allowed or disallowed evidence that should have been

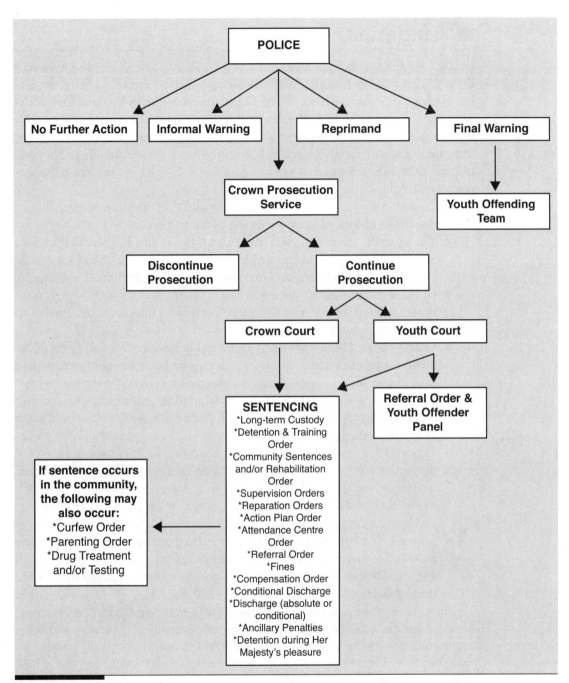

Figure 12-1 Juvenile Case Processing.

excluded or included (NACRO, 2006, p. 39). From here, the appeal may be heard by the Court of Appeal, the House of Lords, and ultimately, the European Court of Human Rights if the case may be in violation of human rights under the Human Rights Act of 1998.

■ Conclusion

In 2005, the Home Office published the report, "Young People, Crime and Antisocial Behaviour: Findings from the 2003 Crime and Justice Survey." The report was based on a 2003 survey that queried 12,000 respondents in England and Wales between the ages of 10 and 25 as to their experiences with anti-social behavior and the social dynamics behind those behaviors, because the literature has strongly indicated that anti-social behaviors lead to future criminal offending. The total number of respondents in that age category was 4,574. Key points from the survey were as follows:

- Twenty-nine percent of young people said they had committed at least one act of antisocial behavior in the previous year.
- The most common antisocial behavior was causing a public disturbance (15%), followed by causing "neighbor complaints" (13%). More serious incidents such as joyriding and carrying a weapon were much less common.
- Of those responsible for antisocial behavior, the majority (68%) only committed one type of behavior. Only 9% (2.4% of the sample) committed three or more different types.
- Males reported higher levels of antisocial behavior across all types of behavior. A third of males admitted at least one behavior, compared with a fifth of females.
- 14- to 16-year-olds were more likely to commit antisocial behavior than other age groups. Two fifths of them reported at least one act of antisocial behavior in the last 12 months.
- The following factors were strongly associated with antisocial behavior: disruptive school environment; delinquent peers; drug use; risky alcohol use; negative relationship with parents; "delinquent's personality traits"; living in a household in financial difficulties; living in a high disorder area; and being a victim of crime.
- About 17% of young people had committed antisocial behavior but no more serious offense. Twelve percent of young people had committed both antisocial behavior and offenses. Nine percent had committed offenses, but no antisocial behavior. (Home Office, 2005b, p. 1)

In reviewing the information above, and as noted in Chapter 4, we must bear in mind the pitfalls and problems inherent in self-report surveys and other crime statistics. However, as is the case with other crime statistics, trends or a "general idea" concerning the results can be reliable and conclusions on which to base policy may be gleaned.

Overall, it appears that crimes or anti-social behaviors by young people in England are fairly innocuous (i.e., public disturbance, disrupting the neighbors, etc.) and occur fairly infrequently. In addition, as is the case around the world, older youth (14 to 17) and males are typically the perpetrators of such behaviors, with this behavior most recently peaking at ages 17 for boys and 15 for girls (NACRO, 2006, p. 4). Social factors such as negative experiences with schools, peers, and parents combined with other factors such as drugs and financial problems in the home seem to be the impetus for these behaviors.

Thus, the conclusion that has been repeatedly reached in our exploration of the English juvenile justice system is that the English solution for anti-social behavior is social as well. Programs designed to educate and alter young people's behaviors have been in existence for hundreds of years and, in spite of politicians' and the public's desire to crack down on juveniles over the centuries, there have always been preventive, reparative, and rehabilitative ideals in operation. This is again evidenced in the Home Office report in 2005. Therefore, as previously mentioned, this scheme for juveniles is the same as for dealing with terrorism: prevention.

However, this is not to state that the system has become "soft" on juveniles, because the English criminal justice system has always had mechanisms to handle the toughest young criminals. And although it appears there is compassion for juveniles and their circumstances, the "no excuses" approach touted by Home Secretary Jack Straw is evident and is apparent in the statistics. The number of juveniles sent to court for formal prosecution has increased from 27% in 1992 to 42% in 2004, and the number remanded to custody has increased almost 60% in the same period (NACRO, 2006, p. 4). This is of great concern for those who embrace the idea that courts should be used as a last resort and that the courts will have a negative impact on juveniles.

Responsibility for anti-social behaviors and crimes has been placed squarely on the shoulders of the juveniles themselves and their parents, an issue that has been controversial in the United States. Parents are required to be in court during all stages of the proceedings, to participate in contracts designed to improve their children's behaviors, to pay fines for their children's acts, and to attend counseling, and the list increases in reference to what their child has done. Juveniles must also actively participate in their own "rehabilitation," and there are consequences for not succeeding. If they fail to fulfill terms of their contract they will end up back in court. If they have received their final warning and are still offending, they will be returned to court for sentencing.

In this rehabilitative vein, other aspects are clearly found in today's juvenile justice system. For example, youths are handled more on a case-by-case basis, and this is done for several reasons. First, there is an effort by juvenile justice agencies to determine what the causes of the anti-social behavior are so a corrective regimen may be implemented. As we have witnessed, the juvenile justice system has struggled with this issue for hundreds of years. Finding the right balance between punitive sanctions and rehabilitative services is key to a juvenile changing for the better and requires a great deal of effort in information gathering concerning each juvenile.

Second, a more victim-oriented approach to this rehabilitation of juveniles is apparent in reparations, victim-offender confrontations, contracts, community service, and the like, all with the goal of preventing future anti-social acts and making amends with the community, which is also instrumental in rehabilitation. Third, the age and maturity of the juvenile has always been an issue in how a given case will be processed. As we have seen, there has been a struggle by policymakers in determining what juveniles know about right and wrong at a given age, which varies from child to child. Therefore, cases must be assessed individually to identify levels of maturity, positive and negative influences, opportunities for channeling energy in positive directions, and so on, with the goal of bringing the juvenile and his or her family to the right services and the right level of contact with the system. This, in turn, will educate them about the causes of the delinquent behavior so negative behaviors do not repeat and thrust the child further into the criminal justice system.

Lastly, in exploring this evolution, what is apparent is an echo in history from John Pounds and Mary Carpenter, who advocated education and structure as means of reducing delinquency. What is being implemented today is their basic strategy on a grander scale for reducing juvenile delinquency. It is hoped that given our advanced knowledge, our technology, and our ability to learn lessons from the past we will have better luck in reducing juvenile delinquency and crime than did those in the past. Whether that will be the case we do not know, but we are certain that the pendulum that moves between punishment and rehabilitative ideologies will continue to swing.

CHAPTER SPOTLIGHT

- There has been a centuries-old debate as to whether juvenile delinquents should be dealt with by way of treatment, rehabilitation, or punishment, and if a mix of the three is used, where the balance should be among them. Policies have developed slowly but dramatically over the last 200 years.

- In the 1800s the population of urban England was growing rapidly with scarce employment opportunities, poor wages, and poor living conditions. This led to large numbers of unsupervised children living in cities like London, many of whom relied on crimes such as theft and begging to survive. As a result, juveniles constituted a large part of the criminal population of England at that time.

- In the 1800s, there were no separate criminal codes for children, and the age of responsibility was 7. Juveniles were subjected to the same punishments as adults, including imprisonment, death, and transportation. In 1833 a 9-year-old was hanged for theft. Juveniles also served sentences in the same prisons as adults.

- Under the common law, there were two categories of juveniles with different presumptions as to their ability to possess criminal intent. Children up to 7 years old were *doli incapax* or incapable of criminal intent. Children aged 7 to 14 were *doli capax*; they were capable of criminal intent but it was for the prosecution to show that the individual knew the difference between good and evil. An exception was that a boy under age 14 was regarded as impotent and therefore incapable of rape. Children over age 14 were treated as adults.

- In the mid-1800s there was a movement towards a more rehabilitative system. "Ragged schools" developed to educate some of those children excluded from education by social and economic circumstances. Philanthropic movements to provide education in reading, writing, arithmetic, and the Bible founded these schools.

- In the early nineteenth century, some social reformers suggested that juvenile criminality was the product of societal problems and poor parenting rather than poor moral character. They further suggested that through early intervention, future criminality could be averted. Some recognized the importance of education and began teaching children for free and with the promise of free food.

- Supporters of these schools included John Pounds and Mary Carpenter. They were also instrumental in the introduction of the Juvenile Offenders Act of 1847, the Youthful Offenders Act of 1854, and the Industrial Schools Act of 1857.

- These legislative changes changed the manner of trials for children under 14, acknowledged the need to separate juvenile offenders from adults, and began the development of reformatory schools and industrial schools to "instill a work ethic" in children and train them in a trade. These acts did not create a separate juvenile justice system.

- The Summary Jurisdiction Act of 1879 created the first separate juvenile justice system with summary trials on indictable offenses for all offenders ages 7 to 12 and similarly for children ages 12 to 16. Punishments for minors were also differentiated from those applied to adults.

- The Children's Act of 1908 provided children with legal protections from harmful parents or guardians, and required that all cases involving defendants under 16, excluding murder, be adjudicated by a magistrate. Cases were heard at special sittings of the Magistrates' Court, separate from adult sittings. However, the magistrates did not receive special training.

- The 1908 Prevention of Crime Act created "borstals," secure institutions for boys ages 16–21 where they could learn a trade. They were an alternative to jail and imposed strict discipline, enforced by corporal punishment.

- The Children and Young Persons Acts of 1933 and 1969 further changed juvenile justice to focus more on the welfare of the child. The 1933 act raised the age of criminal responsibility to 8 from 7, and children through the age of 16 were placed under the jurisdiction of the juvenile courts. The 1969 act further shifted the emphasis from punishment to the provision of services to juvenile offenders. Incarceration and borstals were seen as the last resort. However, the change of government in 1970 meant that much of the 1969 act was not brought into force.

- In 1970, the change of government led to further debate on the balance between rehabilitative and punitive approaches, with policy leaning toward the punitive. In particular, there was public frustration with persistent, repeat offenders. However, the English also devised strategies for diverting first-time offenders away from the formal criminal justice system.

- The next legislative steps were the Children Act of 1989 and the Criminal Justice Act of 1991. Both emphasized the importance of limiting contact between juvenile offenders and the criminal court system and limiting custodial sentences.

- The new acts recognized that juveniles are responsible for their actions but also that other influences, including poor parenting, played a role. Therefore the Children Act of 1989 established parental responsibility and amended "care orders" from the 1969 act to provide that the child could be placed under the care of both the parents and the local authority if deemed necessary. The Family Proceedings Court was given jurisdiction over these matters.

- The Criminal Justice Act of 1991 expanded the jurisdiction of the youth courts and provided for individualized sentencing of juvenile offenders based on the maturity of the offender. Even more emphasis was placed on the parents, who were required to attend all court proceedings, potentially pay fines for the child, and maintain recognizance of the child. Youths who were 14 or 15 years old could be sentenced to periods of detention, with 2 months set as the minimum detention period. Juvenile courts were renamed youth courts and were a special sitting of the Magistrates' Court.

- The Crime and Disorder Act of 1998 brought in several changes and was national in scale but intended to be localized in operation. It specifically stated that the role of the juvenile justice system was to *prevent* juvenile offending. It utilized early intervention and intensive community supervision and empowered local authorities to impose curfews on children under 10.

- The 1998 act also empowered courts to order "corrective" requirements such as drug rehabilitation, counseling for those at risk of committing offenses, and parenting orders and support for parents. The system deliberately moved away from the notion of a juvenile's social circumstances being responsible for their offending.

- The 1998 act also created the Youth Justice Board to oversee the youth justice system in England and Wales. The board is a multi-agency entity of 10 to 12 members who monitor the youth justice system and advise the Home Secretary. The board also oversees the work of Youth Offending Teams, which in turn coordinate local services for youths ages 10 to 17.

- The 1998 act replaced cautions with reprimands and final warnings for first-time minor- offenses, as well as schemes to include parents earlier in the process and to make reparations where possible. Final warnings trigger a referral to the Youth Offending Team. Subsequent offenses result in charges being filed and thus provide a swift path to prosecution. The act also abolished the common law notion that a child under 14 does not know the difference between right and wrong.

- This process continued with the Youth Justice and Criminal Evidence Act of 1999, which was based on notions of restorative justice, restitution to the victim, reintegration of the offender into the community, and offender and parental responsibility.

- Youth offender panels were also created, which can take control of the offender for 3 to 12 months and require both the juvenile and their parents to take responsibility for their respective behaviors as they pertain to the offense.

- Youth offender panels can enter into contracts with the juvenile with specific conditions related to their circumstances. These can include reparations, drug and alcohol counseling, or consistent school attendance.

- Licensing laws in 2000 also amended the law regarding juveniles and alcohol. A 16-year-old may consume beer, porter, or cider as part of a meal so long as it is not in the bar area.

- The Criminal Justice and Police Act of 2001 broadened the range of juveniles over whom the state could exercise control. Child curfews were extended from those ages 10 and under to those ages 16 and under. It also provided that 15- and 16-year-olds could be remanded to secure facilities if they were deemed a threat to the general public and likely to commit further crimes or, if age 12 or older, children on bail could be required to wear electronic tags.

- The Anti-Social Behavior Act of 2003 placed responsibility for the juvenile's delinquency squarely on the parents, requiring them to cooperate with specified orders given to the child. The concept of "parenting contracts" was also introduced, for example, forcing parents to tackle the juvenile's being truant. Failure to comply with the contract could result in a fine of up to £1,000.

- The Children Act of 2004 was another attempt to coordinate the distribution of services to children and families. It intended to promote the health and safety of children and created the position of Children's Commissioner.
- Further legislation is anticipated and a green paper, "Youth Matters: Next Steps" was published in March 2006. This reported the result of a survey of individuals and groups. Two major themes were identified—the lack of positive activities for young people to engage in and the lack of safe places for them to congregate. The goal is to implement the findings of this survey by April 2008.
- Since 1998 the law has been that anyone ages 14 to 17 is considered a "young person" and those 13 and under are "children." Children under 10 may not be prosecuted; those ages 10 to 13 can only be prosecuted if the state can show they knew the difference between right and wrong.
- Young people may be tried as adults for offenses that, if committed by an adult, would carry a sentence of 14 years or more in prison. Otherwise, youth courts have jurisdiction over those under age 17.
- When the police decide a prosecution is necessary, the file must be sent to the Crown Prosecution Service within 72 hours for determination as to whether a prosecution is necessary.
- Youth court judges are magistrates, selected by their peers to sit on the youth court bench. They typically sit in threes and are laypersons. District Judges are paid judges and members of the legal profession. They sit alone and deal with lengthier and more difficult cases. Only members of the court, witnesses, and the parties involved and their lawyers are permitted to view the proceedings. If the juvenile lacks sufficient funds to pay for a lawyer, legal aid will be granted.
- In the case of murder, a child may be sentenced to life in prison, known as being detained at Her Majesty's Pleasure. In reality, this is detention for at least 12 years with the parole board determining the release date based on the individual's potential threat to society.
- Appeals from the youth court are to the Crown Court where the judge will sit with two magistrates and without a jury. They may quash a conviction, confirm it, reduce a sentence, or impose a harsher sentence. The next level of appeal is to the Queen's Bench Division of the High Court, then to the House of Lords and, potentially, to the European Court of Human Rights.

KEY TERMS

Children: Legally, those who are age 13 or under in England and Wales.

Crown Prosecution Service (CPS): Responsible for the prosecution of crimes that were investigated by police in England and Wales. The CPS makes the final determination as to whether to continue with prosecution or to discontinue a criminal case. If proceedings are warranted, the CPS will prepare and carry out the court proceedings and the prosecution of the case to its conclusion.

Detainment during Her Majesty's Pleasure: A sentence with a minimum of 12 years, after which the parole board has the discretion to release the person when they determine the person in question is no longer a threat to society.

Doli capax: Latin meaning "capable of criminal intent."

Doli incapax: Latin meaning "incapable of criminal intent."

Ragged schools: A nineteenth-century charity school for destitute children, where they were taught, clothed, and sometimes fed and lodged. There was little to no government support for the schools. The name originated from the common and poor appearance of the pupils.

Young person: A person who is 14 to 17 years old in England and Wales.

Youth Custody Centers: Formerly known as borstals, they were renamed by the Criminal Justice Act of 1982.

PUTTING IT ALL TOGETHER

1. Is it fair that parents have such a large responsibility for the acts of their children? Should parents be jailed for what their children do? If so, under what circumstances should that happen? Where is the line whereby a parent is no longer responsible for his/her child's acts?

2. In spite of a plethora of examples of juveniles committing violent crimes, why do you believe the English have maintained such rehabilitative ideals in dealing with juveniles for the past 200 years?

3. In reference to the treatment of juveniles in the justice system, which country, the United States or England, treats their juveniles in a more civilized manner? Why?

4. Why was the Crime and Disorder Act of 1998 so instrumental in shaping the juvenile justice system?

5. At what age do you believe children should be held criminally liable for their acts? Does it depend on the act? Can set standards, as found in the English juvenile justice system, concerning age and responsibility truly be effective or should cases always be handled on a case-by-case basis? What circumstances guide your logic?

6. Has the juvenile justice system in England become too bureaucratic? Can such a system really prevent criminality as it is trying to do?

7. How did religion and education reform the juvenile justice system in England? England does not have a separation of church and state as is found in the United States. Why do you think the religious side of this reformation is not apparent today in legislation and rehabilitative ideals?

8. Given what you have read about the English strategy to deal with juvenile crime, do you think it will be effective in the end? Why or why not? Can England maintain the expense of such programs or is its strategy just another in a long line that will be replaced over time?

ENDNOTES

1. Unitarianism is a religion that rejects the traditional Christian doctrine of the Trinity and believes in the moral authority of Jesus; however, it rejects the traditional Christian belief of Jesus Christ as a deity.

2. The court may also discharge or remand to custody the youth in question.

3. No formal record is kept of informal warnings and they cannot be mentioned in any subsequent court proceedings. The Home Office currently discourages the use of informal warnings.

REFERENCES

The Anti-Social Behaviour Act of 2003. Retrieved January 2, 2006, from http://www.opsi.gov.uk/acts/acts2003/20030038.htm.

Arthur, R. (2005). Punishing parents for the crimes of their children. *Howard Journal of Criminal Justice, 44*(3), 233–253.

Audit Commission. (2004). *Youth justice 2004: A review of the reformed youth justice system.* Retrieved January 23, 2006, from http://www.audit-commission.gov.uk/Products/NATIONAL-REPORT/7C75C6C3-DFAE-472d-A820-262DD49580BF/Youth%20Justice_report_web.pdf.

Bottoms, A., & Dignan, J. (2004). Youth justice in Great Britain. In M. Tonry & A. N. Doob (Eds.). *Youth crime and youth justice: Comparative and cross-national perspectives* (Crime and justice: A review of research). Chicago: University of Chicago Press.

Carpenter, M. (1969). *Reformatory schools, for the children of the perishing and dangerous classes, and for juvenile offenders.* London: Woburn Books Limited.

Carpenter, M. (1970). *Juvenile delinquents: Their conditions and treatment.* Montclair, NJ: Patterson Smith.

The Children's Act of 2004. Retrieved January 6, 2006, from http://www.opsi.gov.uk/acts/acts2004/20040031.htm.

The Children's Act of 1989. Retrieved January 18, 2006, from http://www.opsi.gov.uk/acts/acts1989/Ukpga_19890041_en_1.htm.

The Crime and Disorder Act of 1998. Retrieved January 29, 2006, from http://www.opsi.gov.uk/acts/acts1998/19980037.htm.

The Criminal Justice Act of 2003. Retrieved February 2, 2006, from http://www.opsi.gov.uk/acts/acts2003/30044-by.htm#sch30

Criminal Justice and Police Act of 2001. Retrieved February 2, 2006, from http://www.opsi.gov.uk/acts/acts2001/20010016.htm

Department for Education and Skills. (2006). Youth matters: Next steps. Something to do, somewhere to go, someone to talk to. Retrieved August, 24, 2006 from http://www.mayc.info/temp/NextspStepsspMarchsp2006.pdf

Duckworth, J. (2002). *Fagin's children: Criminal children in Victorian England.* London: Hambledon and London.

Home Office. (2005a). Crime, sentencing, and your community: Sentencing explained. London: Home Office Communication Directorate. Retrieved February 3, 2006, from http://www.homeoffice.gov.uk/documents/catching-up-with-crime-and-sente?view=Binary.

Home Office. (2005b). Sentencing statistics, 2004 England and Wales. Retrieved February 3, 2006, from http://www.homeoffice.gov.uk/rds/pdfs05/hosb1505.pdf.

Home Office. (2005c). Young people, crime and antisocial behaviour: Findings from the 2003 Crime and Justice Survey. Retrieved February 3, 2006, from http://www.homeoffice.gov.uk/rds/pdfs05/r245.pdf.

Joyce, P. (2006). *Criminal justice: An introduction to crime and the criminal justice system*. Devon, England: Willan.

Licensing (Young Persons) Act of 2000. Retrieved February 3, 2006, from http://www.opsi.gov.uk/ACTS/acts2000/20000030.htm.

Manton, J. (1976). *Mary Carpenter and the children of the streets*. London: Heinemann Educational.

National Association for the Care and Resettlement of Offenders. (2006). *Nacro guide to the youth justice system in England and Wales*. London: Author.

The Protection of Children Act of 1999. Retrieved February 2, 2006, from http://www.opsi.gov.uk/acts/acts1999/19990014.htm.

Sexual Offences (Amendment) Act of 2000. Retrieved on January 6, 2006, from http://www.opsi.gov.uk/ACTS/acts2000/20000044.htm.

"Summary jurisdiction." Wikipedia. Retrieved June 11, 2007, from http://en.wikipedia.org/wiki/Summary_jurisdiction.

Terrill, R. J. (2003). *World criminal justice systems: A survey*. (5th ed.). Cincinnati, OH: Anderson.

Wakefield, W., & Hirschel, J. D. (1996). England. In D. J. Shoemaker (Ed.). *International Handbook on Juvenile Justice* (pp. 90–109). Westport, CT: Greenwood Press.

The Youth Justice and Criminal Evidence Act of 1999. Retrieved January 5, 2006, from http://www.opsi.gov.uk/acts/acts1999/19990023.htm.

Youth Justice Board. Youth court: Courts. Retrieved January 2, 2006, from http://www.youth-justice-board.gov.uk/YouthJusticeBoard/Courts/YouthCourt.htm.

Youth Justice Board. Youth justice board for England and Wales. Retrieved January 5, 2006, from http://www.youth-justice-board.gov.uk/YouthJusticeBoard/.

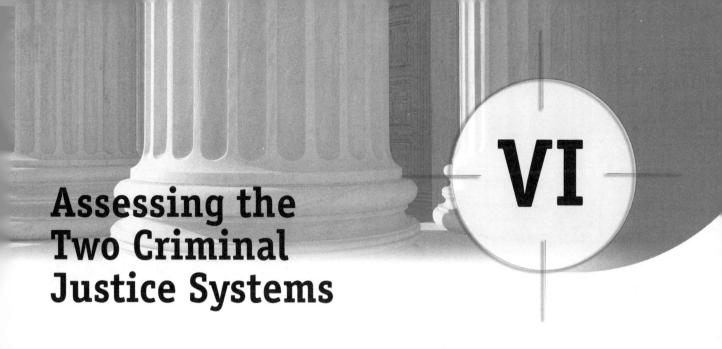

Assessing the Two Criminal Justice Systems

VI

Evaluating the English Criminal Justice System: Lessons to Be Learned

13

Chapter Objectives

After completing this chapter, you will be able to:

- Compare and contrast Packer's crime control and due process models of the criminal justice process.
- Compare and contrast how the criminal justice systems of England and the United States reflect Packer's crime control and due process models of the criminal justice process.
- Outline overall objectives for the criminal justice system.
- List changes derived from an examination of the English criminal justice system that should improve the operation of the U.S. criminal justice systems.
- Describe how the English have gone about assessing both the effectiveness and efficiency of the components of the criminal justice system: the police, courts, and corrections.

Introduction

We began our analysis of the English system of justice with an examination of England's history and culture, her socioeconomic conditions, and her system of government. As discussed in Chapter 2, this examination was undertaken with a specific purpose in mind. In describing any system of justice other than one's own, we considered that it was necessary at the outset to place the justice system in the cultural and political context in which it developed. This helps us to comprehend more completely why the various components of the criminal justice system of England developed the way they did, and why they operate as they currently do. In addition, a thorough discussion of the history and development of the laws, police, courts, and corrections is highly desirable if one is to assess properly their current operation in comparison to other systems of criminal justice—for our purposes, the criminal justice systems of the United States. The English system set the foundation for the development of the U.S. systems of criminal justice, and with a discussion of its operation, there are "lessons to be learned" for both the United States and England.

Despite her small geographical area, England has contributed greatly to the history of the world. England covers roughly 58,000 square miles, and her 52 million inhabitants have made significant contributions to many areas such as the sciences, the arts, philosophy, and literature. For our purposes, the development and legacy of the English system of laws and criminal justice have been the main focus of this book; however, without a rudimentary understanding of England's colorful history and culture, the description of the English criminal justice system would hold little meaning.

During the development of this book, the authors frequently heard comments questioning the need for such a work. What could possibly be gained from studying another system when we have enough problems with our own? Specifically, why examine a system that does not have the violent crime the United States has, or its large and multicultural population and great geographical diversity? The difference in government structure, the dual federal-state systems in the United States, and the absence of a formal written constitution in England were seen as barriers to comparing the two systems. Further, the fact that the English police "don't even carry guns" was sometimes offered as reason enough not to pursue such a comparison. Finally, a society that had abolished the death penalty, it was argued, would certainly have little in common with the United States and its federal and state systems of capital punishment laws, jurisdictions, and political diversities. However, it was the authors' intention to produce a work that maintained the tenets of "comparative criminal justice" research,[1] while simultaneously providing valuable insight into the rich culture of another important country. And more fundamentally, because the United States' systems of criminal justice find their origins mostly in the English system, a comparison of the two nations' histories, structures, and philosophies can yield significant information about each system's positive and negative aspects as we endeavor on both sides of the Atlantic to determine "what works best" in dealing with crime.

Since the publication of the first edition of this book a decade ago, the world has shrunk in many ways. The Internet and immediate media coverage of faraway events have combined to make incidents occurring in one location of potential concern to all. International crime and, in particular, international terrorism, have fostered the need for greater cooperation among nations, particularly those like England and the United States that possess a similar heritage and world view.

Against this backdrop, it is perhaps not surprising that some of the differences between the two nations' systems of justice have blurred. The differences in crime rates reported in England and the United States have narrowed considerably, and starting in the 1990s, crime trends have been to a large extent strikingly similar. Terrorism, and the consequent concern with improving homeland security, have highlighted the need for U.S. agencies at all government levels to work far more closely with each other and have spawned new national coordinating bodies. With regard to the processing of criminal cases, the systems have shown signs of convergence with England, for example, somewhat softening her approach to illegally obtained physical evidence and moving a little closer to formally recognizing plea-bargaining. In addition, there are even proposals to dispense with the judges' and barristers' robes and wigs. Despite all of this, differences between the countries do still exist that merit examination for the comparative analysis undertaken in this book.

We suggest that the primary goal of the criminal justice system should be to deter crime and to deliver justice in an efficient manner. We seek to discourage potential wrongdoers from committing offenses, and if that attempt fails and an offense is committed, we try to apprehend, charge, and convict the wrongdoer as efficiently as possible. Once convicted, our objective is to sentence the offender in a manner that metes out punishment commensurate with the offense, rehabilitates and possibly incapacitates the offender, and deters the offender and others from committing future offenses. In seeking to achieve this, we are aware that there is a danger that the wrong person may have been apprehended. There is also a concern that in seeking to deter and detect crime the state may unduly discourage law-abiding citizens from engaging in activities in which they desire to engage.

This tension between effective crime fighting and the protection of the rights of citizens is well captured by Packer's (1968) crime control and due process models of the criminal justice process. The **crime control model** is "based on the proposition that the repression of criminal conduct is by far the most important function to be performed by the criminal process" (p. 158). It "requires that primary attention be paid to the efficiency with which the criminal process operates to screen suspects, determine guilt, and secure appropriate dispositions of persons convicted of crime" (p. 158). It is a model of a system that evidences great faith in system officials and gives them great latitude in carrying out their duties. It stresses informal processes over formal processes and extra-judicial processes over judicial processes. It seeks the avoidance of "ceremonious rituals that do not advance the progress of the case" (p. 159), and advocates the uniform handling of cases. After disposing of cases in which it appears unlikely that the person apprehended is the offender, it "secures, as expeditiously as possible, the conviction of the rest, with the minimum of occasions for challenge, let alone post-audit" (p. 160).

The **due process model**, on the other hand, focuses on the rights of the accused and guards against errors in the fact-finding process. The accused is allowed to challenge not only the reliability of evidence, but also the manner in which it was obtained. As Packer (1968) notes: "If the Crime Control model resembles an assembly line, the Due Process model looks very much like an obstacle course" (p. 163).

In this chapter, we will once again consider differences noted in the previous chapters in the criminal justice systems of England and the United States. As we do so, we will examine the extent to which the respective systems reflect Packer's crime control and due process models. We will then consider what constitutes appropriate goals for the criminal justice system, and how the English have gone about evaluating efficiency and effectiveness in achieving system goals.

■ External Factors Affecting the Operation of the Criminal Justice System

As we have discussed in Chapters 2 through 4, there is a series of factors external to the criminal justice system that affect its operation. These factors include the geography and socioeconomic conditions in the country under study, the government framework in which that country's criminal justice system operates, the criminal laws, and the nature and volume of crime.

England's smaller geographical area and the fact that she is separated from mainland Europe by the sea make crime control measures easier to implement in England than in the United States. It is also far easier to respond to critical incidents at any geographic location in England than in the United States because geographic proximity is clearly a factor that helps promote the speed and efficiency of response. The opening of the channel tunnel and the relaxation of borders among countries in the European Union have, however, facilitated movement into and out of the country. This, in conjunction with the collapse of communism and the dissolution of the Iron Curtain, has had consequences for such activities as illegal immigration, drug trafficking, international prostitution, and terrorism.

Socioeconomic and cultural differences have been noted between the populations of England and the United States. These differences are likely to impact both the commission of crime and the response to crime. Thus, differences in cultural attitudes towards gun ownership are reflected in the strict gun control laws that exist in England, the far lower gun ownership rates, and the far lower rates of crimes of violence committed with a firearm. Conversely, we may surmise that the comparatively higher cost and lower rates of automobile ownership make automobiles attractive targets and play a part in the higher rates of automobile theft observed in England.

The association between cultural attitudes and the designation of acts as criminal offenses is manifested in a variety of ways. Traditionally, the English have possessed a more tolerant attitude toward diversity. This is reflected, for example, by the fact that they have not looked to the criminal law to control behaviors such as consensual sexual activity and gambling to the same extent as we have in the United States. Thus, culture influences what acts are designated as criminal offenses. Culture also impacts the response to acts that have been designated as criminal offenses. Whether a particular system of justice is imbued with more of a crime control or due process model of criminal justice is also influenced by the culture that is prevalent in that nation.

The framework of the government system clearly impacts the response to criminal activity. Their unitary system of government clearly provides the English with a major advantage in responding both efficiently and effectively to crime. Unlike the United States, which has local, state, and federal agencies operating under a wide range of sometimes conflicting laws and regulations, the English have a unified system with a clear organizational structure. Thus, they are able to set out national plans and objectives and to indicate how local initiatives are to fit into those national plans and objectives. And, with the concept of ministerial responsibility, they have a single person, in this case generally the Home

Secretary, answerable to Parliament, and through Parliament to the electorate, for the actions of those for whom he or she is responsible.

■ Differences in the Two Justice Systems

In this section we review differences noted in previous chapters between the various components of the criminal justice systems of England and the United States. We also reexamine what we have observed about the respective juvenile justice systems and the two nations' responses to terrorism.

The Police

As with all components of the English criminal justice system, the police have been the subject of legislation, procedural reform, and research throughout their history and development. As we have seen, Parliament has passed a large amount of legislation that directly or indirectly has affected the performance of their duties. Statutes, administrative directives, and case law have impacted the organization of the police and police powers to arrest, search, and seize evidence.

Generally, the English appear to have placed more trust in the police and given them more power than their counterparts in the United States. Thus, for example, the amount of evidence required for a police officer to arrest or search a suspect in England, "reasonable cause to believe," is less than the "probable cause" required to take such actions in the United States. The police also possess the legal authority to keep those suspected of committing serious criminal offenses from consulting with an attorney longer than those suspected of committing more minor offenses, an approach that runs counter to the law in the United States. The English orientation toward illegally obtained physical evidence has been to allow the inclusion of such evidence in court proceedings provided it is "relevant" and does not "unduly prejudice" the defendant, and not to let the criminal go free because the constable has blundered. Indeed, it might well be argued that suspects in the United States are afforded far more rights than those in England. Thus, the English system of policing appears more inclined toward Packer's (1968) crime control model with its inherent faith in the ability of officials to make the right decisions, while the systems of the United States appear to have adopted more of the features of the due process model with its focus on the possibility of error and the establishment of formal adversarial fact-finding processes, as evidenced, for example, by holding pretrial hearings to determine whether evidence has been legally or illegally obtained. This finding does not, however, completely mesh with the underlying philosophies that are often alleged to drive the operation of the respective justice systems. It may be argued that the United States has more often than not touted crime control as the primary focus. This has been most clearly evidenced by the utilization of the death penalty in the United States. The English, on the other hand, have traditionally placed more emphasis on prevention and rehabilitation.

The police in England have undoubtedly been aided in their crime control endeavors by the fact that they have a unitary system. The 43 forces enforce the same set of criminal laws, have the same hiring guidelines and training programs, and operate under the general guidance of the central government. The supervisory control exercised by the national government was strengthened in the 1990s and the early twenty-first century by the development of a national policing plan and national objectives, with local plans and objectives tied into the national plan and objectives. The English thus have a national police agenda that must be followed by local police departments. And, as indicated in Chapter 6, the national government sets performance targets for achieving policing objectives.

Because of the organizational structure, police officers in England do not encounter the same legal and jurisdictional problems faced by their counterparts in the United States. Given the far smaller number of police forces in England, they do not have the same degree of problems posed by the lack of shared information and cooperation prevalent in the United States, where as many as 20 or 30 law enforcement agencies may operate within a single county. The focus on local control of law enforcement in the United States has probably increased the influence of politics in the U.S. systems, though this may also be attributed to the official role of the police officer. In England, the police constable has traditionally been viewed as simply a "citizen in uniform" who is answerable to the Crown, as opposed to his or her counterpart in the United States who must answer to the city, county, state, or national government that hired the officer. While recognizing that a certain degree of responsiveness to local concerns might be lost through the amalgamation of police forces in the United States, it might be suggested that this could be more than offset by a potential increase in both the efficiency and effectiveness with which the larger police force might combat crime. Moreover, the current focus in the United States on community policing should allow a larger police force to be responsive to local community problems and needs.

The English police have also been required to take on a quasi-correctional role by administering cautions—formal stern warnings—to offenders. Despite concerns over the potential for net widening, the caution appears to be an effective mechanism for helping reduce the clogging of the English courts. However, the United States has not seemed too interested in adopting this formal practice for minor criminal offenses, though "warning" tickets have traditionally been issued for traffic offenses and "unofficial" warnings are often given to citizens. Perhaps this is an option that the United States, faced with overcrowded dockets in most of its lower courts, might examine more closely. However, it is perhaps unlikely the U.S. citizenry would accept a uniformed officer giving stern warnings in any setting, particularly in a home setting, no matter how effective the practice may be in reducing crime. This is most likely the result of the police being perceived as "crime fighters" in the United States as opposed to "public servants" in England.

Though the police in England still do not routinely carry firearms, and weapons' training is not a component of basic training, terrorism and the rising tide of violent crime have resulted in armed police becoming a more visible feature of policing in England. This development, coupled with public concern raised by the police role in various well-publicized miscarriages of justice cases and tension between the police and minority groups (both black and Muslim), threatens to distance the police from the public they serve. Attempts to make the police more accountable for their actions, as evidenced, for example, by the detailed documentation requirements instituted by the Police and Criminal Evidence Act of 1984, arguably moved English policing a little more toward the due process model. In addition, the Police Reform Act of 2002 created the **Independent Police Complaints Commission** to handle complaints against the police. This entirely independent nondepartmental public body, composed of 17 commissioners who can never have served as police officers, is intended to play a major role in raising the level of public confidence in the police, an issue that is monitored in the annual English crime surveys. However, despite these developments, the English policing system would still appear to be operating under a framework that incorporates more of the crime control model than is the case in the United States.

The Legal System

As we have seen, the English legal system provided the foundation for the systems of criminal justice in the United States. Much of what is in place in the legal systems in the United States today is a direct result of the English common law tradition; the laws, systems of criminal procedures, and specific functions of the courts are direct descendants of the English legal system.

During the course of the twentieth century the English did not look to the criminal law to promote moral values as much as was the case in the United States. Thus, for example, gambling and consensual sexual activity between adults were not regulated by the criminal law to the extent that they were in the United States. The U.S. orientation toward some types of consensual adult sexual activity has softened: consider, for example, the U.S. Supreme Court decision in *Lawrence v. Texas* (2003), which struck down as unconstitutional a Texas statute that criminalized sexual activity between same-sex partners. However, the English approach to other types of consensual sexual activity, in particular prostitution, and toward gambling, is still very different. The fact that prostitution is legal and regulated gambling is widely available has implications for both the police and the criminal courts, who are not asked to cover such a wide range of human activity, and consequently face smaller caseloads. Past experience indicates that the criminal law is not a very effective mechanism for changing human behavior, and it might be suggested that legislatures in the United States should not be so inclined to turn to the criminal law to promulgate particular moral positions.

The English court system provides, perhaps, the most visible differences between the United States and England. From the bifurcated legal profession, consisting of barristers and solicitors, to the stylistic formal attire of robes and wigs in the Crown Courts, the observer immediately notices the differences. The ritualized movements of the barristers as they plead their cases and the very formal responses of the judges reflect an atmosphere of conservative and restrictive parameters in the courtroom. Contrast this with the very informal and "showy" procedures in U.S. courtrooms, starkly evident in the televising of the court procedures surrounding the O. J. Simpson case in 1994, and one can easily reach the conclusion that the differences between courtroom demeanor in England and the United States are indeed vast. However, as we have indicated, these are differences more of style than of substance. Although the English have to deal with the issues of the antiquated and rigid nature of their courtroom procedures, and the possible fusion of the two branches of the legal profession, their far tighter control over media coverage of criminal cases that are sub judice results in the almost total elimination of problems of prejudicial pretrial publicity. Although recognizing that the First Amendment of the U.S. Constitution guarantees freedom of the press, it may be argued that some restrictions might still be imposed on the nature of the coverage the press gives to cases that are still being processed through the criminal justice system.

The English require that their barristers and solicitors serve an apprenticeship before full admission to the profession. Under the supervision of established lawyers, they learn the practice of law. Perhaps the legal profession in the United States might consider adopting a form of this practice for the legal training currently required of prospective attorneys. This might ultimately improve both the quality and flow of justice in the United States. At a minimum, it would assure newly trained law school graduates a greater familiarity with legal practice in the United States and might serve as a screening process for the less competent individuals seeking careers in law.

The English rely on the process of appointment for the selection of their prosecution and judicial officials. Concerns have been raised about the prevalence of older, white, conservative-oriented males of the upper socioeconomic classes among those appointed lay magistrates, and the highly select group from whom the higher court judges are selected. In many ways an appointment process can more readily meet these concerns than the election process so common in the United States. In addition, it would appear preferable to improve the current appointment procedures than move to selection by election with all its attendant problems, such as selection by an uninformed electorate, low voter turnout, and qualified candidates unwilling to run for office.

The systems in both countries operate under a presumption of pretrial release, and do in fact release the vast majority of defendants awaiting trial. In England, both the police and the courts are actively involved in the pretrial release process. The police may release the accused on police bail, and since 1994 have been allowed to place conditions on release. In England, unlike in the United States, defendants are released by both the police and the courts without requiring them to post money. As a consequence, there are no professional bail bondsmen working in the system. Concerns have long been voiced about the involvement of bail bondsmen in pretrial release in the United States, and the movement for unsecured release should take encouragement from the fact that the failure to appear rates in England compare favorably to those in the United States.

As in the United States, the bulk of English criminal proceedings take place in the lower courts. In England, these courts are the Magistrates' Courts, where one important difference is in the practice of using lay magistrates to hear cases. As noted, depending upon the geographical area, the cases are heard by District Judges, salaried legally qualified judges, or by a court having a minimum of two, but no more than seven, lay magistrates. The powers of both are equal; however, because the lay magistrates are not trained lawyers, the presence of a legally trained chief clerk is required during proceedings to advise on points of law. Low comparative costs and flexibility are two major advantages of the system. The United States does not use this system, but it appears to work effectively for England and her more than 25,000 lay magistrates.

The category of cases "triable-either-way" allows for certain cases to be officially resolved in either the lower or higher criminal courts. This adds a certain officially sanctioned flexibility to the court process in England. In providing incentives for defendants to opt for trial in Magistrates' Court, and more directly by reclassifying certain triable-either-way offenses as summary offenses, the English can be observed to be paying heed to the factors of cost and time. These are certainly factors that are of concern to court jurisdictions in the United States.

In England, the process of plea-bargaining is still not fully acknowledged. The fact that the prosecution is not involved in the sentencing process means that there is no direct sentence bargaining between prosecution and defense, though since 1994 the law has provided that, in passing sentence on a defendant who has pled guilty, the judge should take into account both the stage at which the defendant indicated the intention to plead guilty and the circumstances in which this indication was given. The United States has been utilizing plea-bargaining, some might say effectively and expeditiously, for some time. The English may want to experiment more openly with plea-bargaining as an option for decreasing crowded courtrooms.

The English criminal court system has functioned effectively without the grand jury since 1933. More recently the remaining filter on the strength of the prosecution's case, committal proceedings (comparable in nature to the United States' felony/probable cause hearings), came under question and in the late 1990s they were abolished for indictable offenses and reduced to paper proceedings for triable-either-way offenses. With no in-court challenging by opposing counsel of the strength of the prosecution's case, as can take place in a U.S. felony/probable cause hearing, the English have effectively removed committal proceedings as a filter of the strength of the prosecution's case. This would appear to constitute an example of an English focus on efficiency and expediency, and a lesser concern with defendant rights. Although we do not suggest that any of the states modify the open court format of the felony/probable cause hearing, the English experience would appear to provide strong support for those who argue that grand jury hearings should not be part of the routine processing of felony cases in the United States. However, it is still possible that the grand jury should be retained for its investigatory function.

The English practice of trial by jury and selection of jury members is of major importance. Although today more than 90% of all trials in England are summary trials without a jury, the jury process remains, as in the United States, a cornerstone of the criminal justice system. The English place great faith in the ability of random selection of jurors to achieve impartial juries. To ensure maximum representation, classes of individuals who had been previously automatically ineligible or excusable from jury duty (such as doctors, members of Parliament, and people involved in the administration of justice) now have to provide reasons not to serve.

The actual rules governing jury trials in England, however, seem to lean more in favor of the prosecution than they do in the United States. There is no detailed questioning (voir dire) of prospective jurors. Although there are more limited challenges for cause, and the defense no longer has the right to a single peremptory challenge, the prosecution may ask a prospective juror to "stand by for the Crown," and may in certain cases engage in jury vetting. Finally, all jury trials are by majority verdict. Many of these differences expedite the court process and save the state both time and money. However, it must be questioned whether this is not accomplished at the expense of defendant rights and the overall quality of justice.

The English sentencing schema traditionally has left a great deal of discretion to the judge or magistrate. However, there has been considerable opportunity for appellate review of sentencing. The Court of Appeals provides a broader overview of sentences than occurs in the United States, where sentences can generally only be challenged on the grounds that they infringe the Eighth Amendment's prohibition against the imposition of "cruel and unusual" punishment. Moreover, the court's jurisdiction to hear cases in which "unduly lenient" sentences have been imposed may well be envied by prosecutors in the United States.

Concerns about sentencing disparity in England led to a renewed legislative focus in the early 1990s on proportionality between the offense and the sentence and to the establishment in 2003 of a Sentencing Guidelines Council, which issues guidelines that must be consulted by all courts when sentencing defendants. Here again one can observe an advantage that accrues from the existence of a unitary system of justice.

No system of justice, finally, is immune from miscarriages of justice. The English criminal justice system has seen its share, and as a result established in 1995 a Criminal Cases Review Commission. This independent commission reviews suspected miscarriages of justice and decides whether they should be referred to an appellate court for review. The existence of an independent review authority can help alleviate the public unrest that often accompanies suspected injustice, and the establishment of such an authority should perhaps be considered in the United States.

Corrections

As demonstrated in the chapter concerning the history of corrections in England (Chapter 9), corrections has undergone an interesting and "colorful" development. Here again the English have provided the foundation for correctional systems in the United States. Yet, if there is any area in which the English might take a cue from the United States, it is perhaps in this component of the criminal justice system. Given the mushrooming of the prison population in England, the system appears to be in the position in which corrections in the United States has found itself for quite some time: overcrowded, understaffed, and underfunded. The United States has built new, modern prisons to keep up with the burgeoning population in the prisons; England, however, has opted to keep many of the old, outdated prisons with occasional remodeling and build a few new institutions. Both countries are currently wrestling with this problem of overcrowding, and how they attempt to solve it will be of critical importance to the future of both criminal justice systems.

The United States is beginning once again to recognize, at least theoretically, the importance of community alternatives to incarceration, and has begun to implement significant changes in the administration of probation and parole. The innovations include, inter alia, electronic monitoring and the utilization of private noninstitutional alternatives. England is also looking at community alternatives and experimenting with privatization.

The English are far more likely than their counterparts in the United States to fine a convicted offender, and it may be prudent for officials in the United

States to consider using this potential income-generating sentencing option more frequently than is currently the case. Conversely, the English are less likely to use probation, giving the offender a conditional discharge instead. Greater use of these sentencing options should also be contemplated in the United States where viable options need to be applied in the place of the much overused probation order.

Although the English have a reputation among Europeans for being imprisonment oriented, convicted offenders appear to be less likely to be imprisoned in England than in the United States. And if they are given prison sentences these are likely to be shorter. Concerns about burgeoning prison populations may prompt more of the states in the United States to emulate the current English orientation toward sending fewer nonviolent offenders to prison.

In contrast to the general situation in the United States, jails in England are part of a unified prison service and are not administered by locally elected officials, who may have no correctional background. Although we would not go as far as to suggest that the office of sheriff be abolished, more states may want to consider having jails be part of a centralized prison service. This could facilitate the development of a comprehensive state plan for the institutionalization of both pretrial and sentenced prisoners, decrease the wide discrepancies that currently exist in the conditions and operation of local jails, and lift the financial burden of administering these institutions from the shoulders of local government. The establishment of goals and key performance indicators for the prison service in England may be regarded as a welcome development, and one that should be emulated more widely in the United States. Although there may be disagreement about both the goals and the manner in which progress toward meeting them is measured, there is merit in clearly articulating public policy and measuring the degree to which it is being achieved.

Terrorism

Terrorism has become a very serious global concern since September 11, 2001, and July 7, 2005, with England and the United States thrust into the forefront both as targets of, and active combatants against, terrorist groups. In an effort to deal with the new and terrifying threats, and before the dust had settled after the New York and Washington attacks, both nations rushed out legislation designed to give government, law enforcement, and the courts power to deal with those who would perpetrate terrorist acts. Although such legislation was greatly needed on both sides of the Atlantic, the various anti-terrorist statutes in England and the Patriot Act in the United States have been greatly criticized for giving the respective governments and the police too much power to investigate the citizenry. At the time of this writing, legislators in both countries are struggling with the creation of legislation that not only is effective in dealing with a real and highly lethal threat, but also does not encroach too far on civil liberties; a tall order, indeed. The real change for both English and American citizens is that for the prevention of terrorism to be effective some invasion of civil liberties is necessary. Surveys in both countries have found that people are willing to give up some of their rights to be more secure, and trust their government to find the right balance.

A comparative analysis of the two nations' strategies for fighting terrorism is revealing. The English are making a concerted effort at preventing future acts by thwarting recruiting, striving to include in English society those minorities who may be viable candidates for terrorist groups, and strictly targeting and surveilling those individuals who may be terrorists. This scheme is more social than investigative, although the movement is towards the latter orientation with individual control orders being debated in the latest legislation. In the United States, the reverse appears to be the case. The current political climate is largely anti-immigrant, and there is little attempt to understand the cultures and philosophies that foster terrorism, activities that are crucial if one is pursuing preventive goals. It remains to be seen which approach is more effective. However, both countries appear to be moving toward more invasive investigatory methods to prevent future terrorist acts. Both the Patriot Act in the United States and the 2006 English legislation have been criticized for being indefinite and vague regarding civil liberties and personal privacy. In England there is also the concern that too much power will be given to the Home Secretary because it is the Home Secretary who will ultimately decide who is to be subject to a control order, the conditions under which a contol order is to be imposed, and the extent to which such an order may affect families and the basic rights to liberty. And, given that these orders may be imposed without any imminent threat, and that criminal punishment may occur for the violation of any order, the possibility of abuse of such a system is not out of the question. The widespread employment by the English of CCTV in public places and its utility in helping reveal the perpetrators of acts of terrorism suggests that greater use of this type of technology should be considered in the United States. This would, however, constitute an encroachment on civil liberties.

Clearly, large-scale terrorism is an unpleasant reality confronting both nations and the world at large. Although England has had more experience with terrorism because of groups such as the IRA, the devastating nature of today's terrorist acts is such that the United States and England are now on the same plane with regard to such threats. It is hoped that both nations can cooperate and learn from each other ways to effectively deal with this serious problem for their mutual benefit, and that clarity and narrowness of definition and purpose will prevail in future legislation.

Juvenile Justice

The state of juvenile justice in England and the United States is of primary concern to both countries. The United States' experience of increasing juvenile crime throughout the past three decades has led to an increased emphasis on prevention, control, and treatment of juveniles as they come into contact with the justice system. Boot camps and other rehabilitative measures have experienced moderate success, but the United States continues to struggle with the problems of drugs, gangs, and violence, and the country appears to be destined to continue fighting these problems well into the twenty-first century.

England is facing similar problems with her youth, although not on the scale of the United States, and is grappling with the dynamics of juvenile behavior and the proper and effective treatment for juvenile offenders. Prevention is the key in England and, as we have seen, the English have made great use of diversion for her youth. For a long time, societal attitudes have favored this approach. However, there have also been renewed calls for harsher penalties for juveniles who persist in criminal activities. Today, there are many opportunities for a juvenile who is in trouble to avoid being processed through the juvenile justice system. This is particularly true for a first-time offender. However, the opportunities are not innumerable and there is now a greater willingness to process a juvenile through the system if it is apparent that all other diversion programs and tactics have failed. And, as juvenile crime increases in both frequency and severity, and the youth committing crime become younger and more violent, the attitude of the public is shifting to a crime-control model. England must address this change, especially given her rehabilitative paradigm for dealing with juveniles.

■ Efficiency and Effectiveness in Achieving System Goals

Earlier in this chapter, we suggested that the primary goal of the criminal justice system should be to deter crime and to deliver justice in an efficient manner. Two major objectives are contained in this goal: (1) the deterrence of crime, and (2) the administration of justice. These two objectives are to be achieved in an "efficient manner," that is by using the minimum amount of energy or resources.

Deterrence is traditionally divided into two components: (1) specific deterrence, and (2) general deterrence. **Specific deterrence** focuses on whether a particular offender has been deterred from offending again. **General deterrence** focuses on whether people in general have been discouraged from engaging in criminal activity. Although assessing specific deterrence is relatively straightforward, measuring general deterrence is difficult because we cannot directly determine who has been deterred, and consequently we have to focus on those we know have not been deterred (i.e., those who have committed crimes). Thus, whether we are seeking to determine the effect of mandatory arrest laws on the commission of domestic violence offenses, or of saturation police patrols on the commission of crime in a neighborhood, we have to pay attention to the known failures, those who have committed domestic violence offenses and those who have committed crimes in the neighborhood. Consequently, an examination of the general deterrent effect of any law, policy, or practice is highly complex and the results of any such examination are often subject to intense debate. Fortunately for us, the focus of this book is not so much on deterring criminal activity as on the criminal justice system's response to crime and the manner in which the police, courts, and correctional system deal with detecting, apprehending, trying, convicting, sentencing, and both punishing and rehabilitating offenders. Of course, rehabilitation ties in with specific deterrence. If an offender has been successfully rehabilitated, then he or she will not offend again.

The objective of "administering justice" contains two aspects: (1) letting go those who have been wrongly suspected of committing crime and those who have been wrongly processed through the different stages of the criminal justice process (arrest, trial, conviction, and sentence); and (2) detecting, apprehending, convicting, and appropriately punishing the guilty. A system that concentrated solely on punishing the guilty without any concern about wrongfully punishing the innocent would approximate an absolute version of Packer's crime control model. Both the English and U.S. systems of justice, however, manifest grave concern about the danger of convicting an innocent person. It is to minimize this error of convicting an innocent person that the two systems require proof beyond a reasonable doubt to convict an accused of a criminal offense, a standard of proof that is far higher than the standard of preponderance of the evidence or balance of probabilities required to win a civil case. This orientation finds expression in the saying that, "It is better to let 10 guilty people go free than to convict an innocent person."

It is suggested that an ideal criminal justice system is one that sets out clear objectives and is both efficient and effective in meeting those objectives. The concept of effectiveness focuses on the extent to which a system is successful in attaining desired outcomes. The concept of efficiency focuses on the amount of energy expended on achieving those outcomes.

Efficiency

Efficiency is of concern both at the system and at the individual case level. At the system level, the English have clearly articulated a concern with efficiency. Performance objectives for police departments contain measures of efficiency and, it may be recalled, the Home Secretary can require a police authority to call upon a chief constable or an assistant chief constable to retire in the interests of efficiency. Moreover, in policing, effective communication is a great contributor to efficiency. Because the United States has over 36,000 law enforcement agencies, crimes committed by, say, a serial murderer or a terrorist who moves from state to state are more difficult to solve because of jurisdictional and "turf" issues and, quite simply, communication problems between agencies. In England and Europe, because the system(s) are nationalized and cooperate through INTERPOL, there is greater efficiency in police investigations of everything from murder to terrorist crimes and drug trafficking.

The court system likewise reflects a concern with efficiency. The category of cases "triable-either-way" and the effective removal of committal proceedings as a filter of the strength of the prosecution's case are designed to save the courts both time and money. National performance standards clearly show a concern for efficiency in operations. For example, the Crown Court has eight performance criteria to measure performance: timeliness or the number of trials in a given time, average waiting times for trials or hearings, percentage of committals outstanding after 16 weeks, the rate at which cases are completed, courtroom usage, "**cracked trial**" rate (resolving a case on the day a trial was to take place without actually having to hold a trial), ineffective trials (trials that were not able to start on their scheduled day), and days jurors sit as a percentage of attendance and non-attendance (Her Majesty's Courts Service [HMCS], 2001, p. 1). Results are as varied as the individual courts and their jurisdictions because of the types and levels of crimes found in their jurisdictions (e.g., urban versus rural), but courts at all levels in England report on these criteria.

In corrections, the English similarly show a concern for efficiency. In their assessment of probation agencies, for example, they include such performance measures as timeliness of court reports, timeliness of probation revocation action, and employee working days lost to sickness (National Probation Service [NPS], 2006).

Both the legal parameters within which the systems operate and the operating procedures themselves suggest that the English criminal justice system is more oriented than the U.S. systems to promote efficiency in processing individual cases. The lower standard of proof required to arrest a suspected criminal in England and the greater latitude afforded police when dealing with suspects, such as by denying an arrestee access to a lawyer for 36 hours in the case of serious arrestable offenses, seem designed to enable officers to get to the facts more expeditiously than in the United States. A suspect's knowledge that a court is entitled to draw inferences from his or her failure to answer police questions may likewise prompt a suspect to answer questions, thus enabling the police to resolve the case in a speedier fashion. In court the lack of detailed questioning (voir dire) of prospective jurors and the provision for majority verdicts help expedite the processing of court cases.

Effectiveness

In the United States, we have tended to examine the performance of the various components of the criminal justice system, asking, for example, how successful the police have been at solving reported crime or corrections in rehabilitating offenders. Consequently, measures of effectiveness have focused on such items as police clearance rates and recidivism rates. Although excellent research has been conducted on specific programs and initiatives, there has been no consistent systematic evaluation at either the state or federal level of the performance of the police, courts, and corrections in achieving either their own stated objectives or general system objectives.

Police

In England, however, as we have seen in earlier chapters in this book, there has been a concerted move to develop performance measures for the various components of the criminal justice system. In the police area, for example, the English require the Home Secretary to present to Parliament each year a 3-year national plan with strategic policing priorities clearly set out. Local police departments must develop local plans and objectives tied into the national plan and objectives, and the performance of these departments is assessed on a range of dimensions including citizen focus, reducing crime, investigating crime, promoting public safety, and providing assistance (Home Office, 2003, Annex C). Police agencies are evaluated on 7 key broad performance areas with 32 performance indicators assessing 26 narrower key areas of policing. The broad areas are reducing crime, investigating crime, promoting safety, providing assistance, citizen focus, resource use, and local policing. Two assessments are conducted. The first focuses on the force's performance in delivering services compared to peer agencies. The second assessment focuses on improvement, or a lack thereof, in reference to the previous year's performance. Both assessments are based on the seven criteria listed above, and each of the 43 police departments in England and Wales receives a "grade" (Home Office, 2006a, p. 1).

The 2004/2005 report was the first to use performance data and baseline grades together to assess police performance. Ultimately, the findings of the reports for the individual police agencies and at the national level will help identify those areas in policing that are working and those that are not. Specific areas of focus are:

- reductions in crime and disorder;
- improvements in crime investigation;
- bringing more offenders to justice;
- promoting community confidence and engagement;
- increasing satisfaction;
- reducing fear of crime; and
- improving the value-for-money (Home Office, 2006b, p. 1).

With performance indicators established to monitor performance and targets set for performance improvement, the English have put in place a system to assess on an ongoing basis how each police department in the country is performing. At present, for example, the Metropolitan Police use four broad categories to measure performance: public satisfaction, crime rates, detection rates, and violence against the person (Metropolitan Police, 2006, p. 1). In May 2006, the Metropolitan Police Service reported generally very positive findings. Overall public satisfaction was up 10% from the previous year, crime was down 3%, violence against the person was down 2.3%, and the detection rate was 18% (Metropolitan Police, 2006, p. 1). The report also noted specific trends in public satisfaction. Public approval for actions taken by police was at 78% (up 14%), with being kept informed about police progress and overall police satisfaction both up 11% at levels of 59% and 79%, respectively (Metropolitan Police, p. 4). Generally, public perception of the Metropolitan Police Service is very positive and is improving on all measures.

As we explored in Chapter 4, such measures are not infallible. There are measurement problems with each, and each has limited applicability. However, effectiveness is being measured and triangulated from different perspectives. As a result, what is gleaned is reasonably valid and useful for improving police services in England. The Home Office is also making an effort to compare English police agencies in major cities with comparable agencies in North America, Europe, Australasia, and Southeast Asia (Home Office, 2006a, p. 1). Data exchanges have already occurred on police performance and will continue in the future.

Courts

The English are also striving to assess the effectiveness of the courts in a number of different ways. One major area of concern focuses on citizen access to, and confidence in, the court system. As was stated in a 2006 report issued by Her Majesty's Courts Service (HMCS), the goal of HMCS is that:

> *All citizens according to their differing needs are entitled to access to justice, whether as victims of crime, defendants accused of crimes, consumers in debt, children in need of care, or business people in commercial disputes. Our aim is to ensure that access is provided as quickly as possible and at the lowest cost consistent with open justice and that citizens have greater confidence in, and respect for, the system of justice. (HMCS, 2006, p. 2)*

The strategies to achieve HMCS's goal are based on the following six principles: (1) putting the needs of citizens first; (2) increasing access to justice; (3) ensuring respect for, and confidence in, the courts; (4) transforming service delivery; (5) supporting the independence of the judiciary; and (6) being an employer of choice for those wishing to deliver first class public service (pp. 4–6). Strategies involve assessing citizens' needs with priority being given in criminal cases to the needs of victims, witnesses, and jurors; simplifying court procedures; and distributing quality information about the court processes as well as how to resolve disputes outside the courtroom. Access to justice is to be increased by allowing court users greater choice, more convenience, and easier access to information by using the Internet and other electronic means. In addition, efforts are to be made to ensure that the courts are reflective of the needs and priorities of the community in which they sit, and that customer standards are developed that build the trust of those who utilize English courts. In carrying out court operations, support for the judiciary is to be a priority. Well-trained staff will enable courts to have what is required in resources and operations to manage courts more effectively and efficiently (HMCS, 2006, pp. 4–16).

Another way in which the English are seeking to assess the effectiveness of the criminal court system, and indeed the effectiveness of the whole criminal justice system, is by focusing on an outcome measure that examines "the number of offenses brought to justice" (Home Office, 2005, Ch. 5). An offense is "considered to have been brought to justice when an offender has been cautioned, convicted or had the offense taken into consideration by the court" (Home Office, p. 5.2). The Home Office publication *Criminal Statistics* reports the number of offenses brought to justice each year (see, for example Home Office, 2005, Figure 5-1 for a listing for 12-month periods from March 1999 to March 2005). In addition, this annual report contains a flow chart that attempts to depict the flow of criminal offenses through the criminal justice system. After giving an estimate of the number of offenses committed based on the result of the British Crime Survey, the chart then gives criminal justice system case figures all the way from crimes recorded by the police through offenses resulting in custodial or community sentences (Home Office, Figure 1-1). Tracking the flow of cases through the criminal justice system represents a good way of assessing effectiveness. However, a number of problems exist with the way in which the English are currently showing the flow of cases. For example, the system figures they currently present do not include all offenses. Many summary offenses are omitted from all data counts. In addition, the figures for the different stages of the criminal justice process are not directly comparable. For example, although the figures on recorded offenses include some summary offenses, the figures on actions taken by the Crown Prosecution Service include only indictable offences. In addition, it should be noted that although arrest rates may constitute fairly reliable and valid indicators of police performance, conviction rates and "brought to justice" rates are very imprecise indicators of a court's performance. Courts are simply too removed from the ultimate measure of crime reduction to accurately determine their contribution.

Corrections

The underlying philosophy in both male and female correctional institutions is that the offender is a human being and that the punishment is the deprivation of liberty, not what corrections officers and officials can do to make the prison experience worse. People are sent to prison "as punishment," not "for punishment." Programs designed to improve the offender's lot in life such as reading proficiency classes, learning a trade, drug treatment, aiding in maintaining family ties, or simply learning the basics of life (such as how to apply for a job, how to write a resume, how to balance a checkbook, etc.) have been shown to be effective in reducing repeat offending.

The English system of corrections is focusing on performance indicators and measures that incorporate such programs, and this should lead them toward a more effective evaluation system. With regard to probation, they are using such outcome measures as "skills for life referrals," "employment gained," and "accredited programme completions," as well as a measure that taps into repeat offending: "proportion of cases that reach the six month stage without requiring breach action" (NPS, 2006, pp. 4, 8). In addition to such measures, correctional institutions are focusing on such dimensions as serious injuries inflicted within the prison and prison escapes (Her Majesty's Prison Service [HMPS], 2005). System data, findings from external inspections by HM Chief Inspector of Prison and independent Monitoring Boards, and surveys of Prison Service Area Managers and the Prison Service Management Board are being used to conduct these evaluations (HMPS, 2006).

The Criminal Justice System

So far we have, for the most part, talked about how the English are attempting to assess the effectiveness of the different components of the criminal justice system. How about the effectiveness of the system as a whole? Here again, the English are in the forefront with regard to setting overall goals for their criminal justice system and attempting to measure progress in terms of achieving those goals.

In July 2004, the **Office of Criminal Justice Reform (OCJR)** was established to coordinate the work of the different components of the criminal justice system in order to help realize overall system goals. As the OCJR states:

> The key goals for criminal justice are to help reduce crime by bringing more offenses to justice, and to raise public confidence that the system is fair and will deliver for the law-abiding citizen. . . . Together with other partners, the CJS works to prevent crime from happening in the first place, to meet the wider needs of victims, and to help turn offenders away from crime. (OCJR, 2006, p. 1)

The Office for Criminal Justice Reform has put forth a strategic plan for 2004 to 2008. In the preface to the plan, the three ministers responsible for criminal justice (the Home Secretary, the Secretary of State for Constitutional Affairs, and the Attorney General) stress that the "overriding principle" of the plan is "to deliver criminal justice that puts the victim of crime and the law abiding citizen first" (OCJR, 2004, p. 7). They also note that there are two "key themes" that underlie this principle. These are:

- "Unifying the system so that it is efficient, gets things right first time, and produces outcomes that are effective in protecting the innocent, deterring criminals and rehabilitating offenders"

- "Engaging with the community so that its concerns are reflected and people's confidence maintained" (OCJR, 2004, p. 7).

The plan stresses the two themes we have been discussing in this subsection: efficiency and effectiveness. The plan also presents objectives that we have already examined in this chapter, including reducing crime, increasing the police detection rate, bringing more offenders to justice, providing the public with greater knowledge and understanding of the criminal justice system, and treating victims with more respect and making the process they go through more understandable and user-friendly. Specific programs are set out, and some performance targets are set. Effectiveness is to be measured by examining crime and criminal justice system processing data and by conducting surveys. A vision is presented of what the criminal justice process should look like in 2008. It is that:

- "The public will have confidence that the Criminal Justice System is effective and it serves all communities fairly."
- "Victims and witnesses will receive a consistent high standard of service from all criminal justice agencies."
- ". . . More offenses (will be brought) to justice through a more modern and efficient justice process."
- "Rigorous enforcement will revolutionize compliance with sentences and orders of the court."
- "Criminal justice will be a joined up, modern and well run service, and an excellent place to work for people from all backgrounds" (OCJR, 2004, pp. 9–11).

In order to promote more effective coordination among the various components of the criminal justice system, and to facilitate "closer co-operation at the working level" (OCJR, 2004, p. 16), local criminal justice boards were created in 2003. A **National Criminal Justice Board** coordinates and supports their efforts. The National Criminal Justice Board does this by "removing barriers to joint working" and providing "strategic direction of resources to secure achievement of objectives" (National Criminal Justice Board, 2006).

From the above we may conclude that the English have established a varied set of measures to assess the efficiency and effectiveness of the criminal justice system. The measures they have selected are: (1) easily measured and understandable, and (2) quantitative in nature. The English are seeking to measure a broad range of issues concerning the operation of the criminal justice system. However, some aspects are far easier to measure than others. Assessing the level of public confidence in a particular component of the criminal justice system, or the system in general, is not a difficult task. However, assessing the contribution of a particular criminal justice initiative to an observed reduction in crime is a different matter. The leap to the conclusion that the criminal justice system, or a particular component of the system, reduced crime rates is difficult to make. The challenge in evaluating any individual program or an entire criminal justice system is to determine whether there is a causal linkage between the activities of the entity and the desired objectives. In this case, has the English government and the criminal justice system been responsible for the reduction of crime that we explored in Chapter 4, or is there some other phenomenon causing the reduction?

As we saw in Chapter 4, no matter what new and improved measure one chooses to use, crime has been on the decline in both England and the United

States. However, certain crimes are on the rise again. Can we honestly conclude that the respective systems' tactics, philosophies, and day-to-day operations are responsible for these reductions? There are many ways in which the criminal justice systems in the United States and England and the philosophies that underlie them differ. Yet both nations saw reductions in crime after the 1995 peak. Politicians would have us believe that their ideas and the implementation of those ideas were the real cause for this crime reduction. However, it is by trial and error on both sides of the Atlantic, and by examining through the academic study of programs and system changes the ensuing effects on crime, that we will be able to sort through the myriad of competing explanations for the occurrence of crime. Hopefully, this will ultimately help us understand the phenomenon we label "crime" and control, at least to some extent, its incidence.

■ "Quo Vadis" England? Where Do You Go from Here?

There are several challenges facing England as we begin to progress through the twenty-first century. Chief among these are questions concerning the need of the English government to balance the rights of suspects with the public's right to protection. This issue has been thrust to the forefront as a result of 9/11 and 7/7. The challenge is to provide the police with sufficient power to conduct their mission to safeguard the public without encroaching too much on the individual rights of citizens. This will entail more accountability on the part of police than in the past. Currently, both the form and extent of this accountability are being set out. The United States, historically more fastidious about protecting citizens against government intrusion, may provide some guidance for the English as they make accountability a top priority for the future.

A second challenge relates to the ability of the courts to operate in an effective and efficient manner while paying more attention to the rights of both suspects and innocent citizens, and overseeing the operation of the other components of the criminal justice system without unduly hampering their efficiency and effectiveness. In some ways, England may be disadvantaged by not having a written constitution, and this is an innovation, long advocated by some, that is being considered.

In corrections, there is a paramount need to use resources wisely. Just building more prisons is clearly not the answer to rising crime rates. More headway must be made in determining the appropriate candidates for incarceration, and in developing community alternatives for those for whom incarceration is deemed unwarranted. Rehabilitation programs geared toward specific offender needs must be continually developed and properly evaluated.

The English have set performance objectives for the criminal justice system and its components and have begun the arduous task of evaluating the extent to which these objectives are being achieved. Although there may be disagreement about both the objectives and the manner in which progress toward meeting them is measured, there is merit in clearly articulating public policy and measuring the degree to which it is being achieved.

Although the tone of this book has been primarily one of comparison and looking for ways in which the United States and England might share and exchange policies and practices, it appears to the authors that not all policies and practices are transferable. As stated at the outset, the cultural and political context in which a criminal justice system functions should remain uppermost in any comparative analysis. Thus, some practices simply do not lend themselves to exchange because they reflect the unique character of each country and desirable differences between the systems.

■ A Final Word

We would be remiss if we did not include as a part of this discussion a plea for a learned and academic approach to change through comparative analysis of the systems of England and the United States. Any type of policy change should come about as a result of properly designed research of the dynamics of any system. Political influence will always be a factor in any policy change; however, the rigorous pursuit of the tenets of sound research design can help to overcome the potential for bias in any change mechanism. It is our hope that the change agents in both systems of criminal justice will work together to bring about meaningful and effective results in their pursuit of criminal justice. Additionally, it is our desire through works such as this to promote the idea that comparative analysis is one avenue that can aid in this goal, whether the input comes from "inside" or "outside" the system.

CHAPTER SPOTLIGHT

- Packer (1968) illustrates the tension between effective crime fighting and protecting the individual rights of citizens with his comparison of the "crime control" and "due process" models of the criminal justice process. To some extent the English criminal justice system is an example of the former, and the U.S. systems are examples of the latter.

- The crime control model focuses on the repression of criminal conduct and the efficiency of the criminal justice process. The due process model focuses more on the rights of the accused and guards against errors in the fact-finding process.

- External factors can influence the criminal justice process. Socioeconomic and cultural differences exist between the United States and England. These influence both the commission of crime and the response to it. Examples include the lower rates of violent crime using firearms and higher rates of auto theft found in England.

- Cultural differences and England's more tolerant approach to diversity go some way to explaining why, unlike the United States, the English have not looked as much as inhabitants of the United States to the criminal law to promote moral values. In particular, the English do not prohibit gambling or most forms of consensual sexual activity. In the United States the opposite has been the case, although the Supreme Court decision in *Lawrence v. Texas* is indicative of a softening of attitudes toward consensual adult sexual relations.

- England and Wales have a unitary state criminal justice system with a clear organizational structure. The United States has a mixed federal, state, and local system of government with multiple agencies and laws that sometimes conflict. England's unitary system has made it easier to implement crime control strategies, and the English do not face the same jurisdictional problems as agencies in the United States.

- The English have placed more trust in the police and given them more powers than their American counterparts. The threshold for having grounds to arrest someone is lower in England (reasonable cause to believe) than in the United States (probable cause). English police also have broader powers to deny access to counsel and are less likely to be barred from relying on illegally obtained evidence.

- The English, with their highly centralized system, have been able to impose a "national police agenda" on police departments across the country. The government sets various performance targets and measures police departments and other criminal justice agencies (prisons, probation, and the courts) against these.

- The English police have taken on a quasi-correctional role, administering cautions and formal warnings to offenders as a method of saving court time and money. The United States has typically limited such activities to minor motoring offenses. Cultural differences may make the English system unacceptable in the United States.

- The English police do not routinely carry firearms or undergo routine weapons training. However, rising levels of violent and gun-related crimes, along with the threat of terrorism, has led to armed officers becoming more prevalent and more visible in England.

- The English system has moved more toward a due process model, with the introduction of the Police and Criminal Evidence Act in 1984 and the creation of the Independent Police Complaints Commission in 2002.

- The English court system is a very visible example of the differences between the two nations. England has two distinct and mutually exclusive legal professions—solicitors and barristers. Attorneys wear formal clothing, including gowns and, in the case of barristers and judges, wigs in court. The defendant does not sit next to counsel but apart from counsel in a separate part of the courtroom known as the "dock."

- English attorneys must serve a form of apprenticeship for at least one year prior to full admission to practice. They must work under the supervision of established lawyers and learn the practice of law. The United States might want to consider a similar system.

- The English have strict rules regarding the pretrial reporting of criminal cases by the media, and cameras or other recording devices are not permitted inside courthouses. Although this runs contrary to the First Amendment of the U.S. Constitution, such controls, it could be argued, enhance the fairness of trials by reducing the potential for prejudicing juries.

- English judiciary and prosecutors are appointed to their positions, ostensibly on merit, rather than being elected. This has provided the English a much more effective approach to tackling the issue of the judiciary being dominated by older white males.

- Both systems have a presumption of pretrial release, and release the vast majority of defendants before trial. However, in England, defendants are not required to post money for bail.

- In both systems, the bulk of cases are heard in the lower courts. England has a practice of using lay magistrates to hear cases, and this appears to work effectively.

- England does not have a formal system of plea-bargaining. However, judges may consider a guilty plea and when guilt was admitted when setting a sentence. The earlier a defendant pleads guilty, the more credit the defendant receives.

- England abolished grand juries in 1933 and has practically abolished committal proceedings (probable cause hearings in the United States). Again, this is an example of how the English have prioritized efficiency and expediency over concerns for defendants' rights.

- In England there is no voir dire of juries, and counsel have only minimal rights to question jurors at selection. The defense has no right to preemptory challenges and the state can, in certain circumstances, vet juries.

- The English have given judges more discretion in sentencing, and the appeals process provides a broader overview of sentences than is the case in the United States. However, the appeals court can impose harsher sentences when the court believes that the original sentence was "unduly lenient."

- The English also have an independent Criminal Cases Review Commission to review suspected miscarriages of justice and refer cases to the appellate court for further review. Such a body can enhance confidence in the criminal justice system and should be considered in the United States.

- Both England and the United States suffer from mushrooming prison populations, although England imprisons far fewer people. The United States has responded by building newer prisons; England has opted to keep many of its old and outdated institutions.

- English courts are far more likely to fine offenders than U.S. courts. The English are far less likely to place a convicted offender on probation or send him/her to prison.

- As with policing, prisons in England are part of a single, unified system, operated by appointed professionals. In the United States, elected officials at local, state, and federal levels operate prisons. The elected officials may have no background in corrections.

- Both the United States and England have been targets of global terrorism and have rushed to enact new legislation to combat this threat by giving the government, law enforcement, and the courts new tools, often at the expense of individual liberties. In both countries, surveys indicate that the public support giving up some of their rights in a fight against terrorism.

- Approaches do differ between the two countries. England has made a concerted effort to thwart recruitment of terrorists by striving to include minority groups and targeting individuals who may be terrorists. In the United States the reverse is the case, with anti-immigrant policies and little, if anything, in the way of attempts to understand diverse cultures.

- Both countries have moved and continue to move toward more invasive investigatory methods of preventing future terrorist attacks.

- Both countries have concerns about issues relating to juvenile crime. In the United States, increasing juvenile crime over the last three decades has led to greater emphasis on control, prevention, and treatment of juvenile defendants. In particular, the United States continues to deal with the problems of gang violence and drugs.

- England faces many of the same problems, but on a different scale. The English have made great use of diversion for youth. However, there have also been calls for harsher sentences for persistent juvenile offenders. There is now more willingness to process a juvenile through the criminal justice system if other programs have failed.

- We suggest that the two primary aims of a criminal justice system are to deter crime and deliver justice in an efficient manner. Deterrence falls into two categories: specific deterrence and general deterrence. The former relates to whether a specific offender has been discouraged from committing crime, and the latter to whether the general population has been discouraged.

- Both countries have grave concerns about convicting innocent people. As a consequence, both set the burden of proof for conviction in criminal cases as proof "beyond reasonable doubt."

- The English criminal justice system is more oriented than the U.S. ones toward promoting "efficiency" in processing individual cases. This is monitored and measured by the government with a concerted move toward "performance measures" and 3-year national plans. These apply across all areas of the system: police, corrections, and courts.
- This approach was enhanced by the 2004 creation of the Office of Criminal Justice Reform (OCJR), which coordinates the work of all the components of the criminal justice system.
- The OCJR oversees the National Criminal Justice Board and local criminal justice boards, which were established to promote closer cooperation at the working level within the criminal justice system.
- In the United States there is no systematic evaluation of efficiency at the state or federal level. Measures that are used are based on monitoring the various components of the system, rather than the whole.
- Whether the new English approach leads to a reduction in crime remains to be seen.

KEY TERMS

Cracked trial: A case that is resolved on the day a trial was to take place without the trial actually having to be held.

Crime control model: "Based on the proposition that the repression of criminal conduct is by far the most important function to be performed by the criminal process" (Packer, 1968, p. 158).

Due process model: Focuses on the rights of the accused and guards against errors in the fact-finding process.

General deterrence: Focuses on whether people in general have been discouraged from engaging in criminal activity by the penalties handed down to those who are caught committing that activity.

Independent Police Complaints Commission: An independent commission that deals with complaints against, and possible illegal action by, those serving with the police.

National Criminal Justice Board: A government organization responsible for promoting effective coordination among the various components of the criminal justice system by removing barriers to joint working and providing strategic direction of criminal justice resources to secure achievement of objectives. The national board is responsible for coordinating the work of local boards whose mission is to "facilitate closer co-operation at the working level" (OCJR, 2004, p. 16).

Office of Criminal Justice Reform (OCJR): An umbrella government office established under the direction of the three ministers responsible for criminal justice: the Home Secretary, the Secretary of State for Constitutional Affairs, and the Attorney General. It has overall responsibility for coordinating the work of the different components of the criminal justice system in order to help realize overall system goals and to improve service for the public. It sets forth multiple year strategic plans.

Specific deterrence: Focuses on whether a particular offender has been deterred from offending again by the imposition of a sanction.

PUTTING IT ALL TOGETHER

1. What is the ideal balance to be sought in a criminal justice system between Packer's crime control and due process models?
2. What, if any, restrictions should be placed on the freedom of the press to publish information about criminal cases that are being processed through the criminal justice system? What are your reasons for the position you are taking?
3. Is it better to let 10 guilty people go free than to convict one innocent person?
4. What three features of the English criminal justice system would you most like the U.S. systems of criminal justice to adopt?
5. What should be the overall objective of the criminal justice system?
6. What is the best way to measure the effectiveness of the criminal justice system?
7. What is the best way to measure the efficiency of the criminal justice system?
8. What is the most interesting fact you have learned about the English criminal justice system?

ENDNOTES

1. For a detailed discussion of the "science of comparative criminal justice research" versus "transnational" or "international criminal justice," see Chapter 19 of *Criminal Justice* by Adler, Mueller, and Laufer, 1994, pp. 527–548. In this work, the authors refer to comparative criminal justice research as incorporating the examination of two or more distinct systems of justice and evaluating the successes and failures in attempting to eradicate crime in their particular cultural context.

REFERENCES

Adler, F., Mueller, G., & Laufer, W. (1994). *Criminal justice*. New York: McGraw-Hill.

Her Majesty's Courts Services (HMCS). (2001). *Lord Chancellor launches drive to improve crown court performance*. Retrieved July 15, 2006, from http://www.hmcourts-service.gov.uk/news/press_notices/press_not_crwn_ct_performance.htm.

Her Majesty's Courts Services (HMCS). (2006). *HMCS business strategy*. Retrieved July 15, 2006, from http://www.hmcourts-service.gov.uk/docs/publications/hmcs_business_strategy.pdf.

Her Majesty's Prison Service (HMPS). (2005). *KPT performance targets and out-runs 2004–05.* Retrieved July 18, 2006, from http://www.hmprisonservice.gov.uk/assets/documents/10000E84EstablishmentOutturns2004-05.xls.

Her Majesty's Prison Service (HMPS). (2006). *Prison service performance rating system: Quarter 4—2005/06.* Retrieved July 18, 2006, from http://www.hmprisonservice.gov.uk/assets/documents/10001E06performance_ratings_qtr4_0506.pdf

Home Office. (2003). *The national policing plan 2004–2007.* London: H.M.S.O.

Home Office. (2005). *Criminal statistics 2004.* London: H.M.S.O.

Home Office. (2006a). *Performance and measurement, 2006.* Retrieved June 14, 2006, from http://police.homeoffice.gov.uk/performance-and-measurement/international-standards.html.

Home Office. (2006b). *Police performance assessments 2004/05—Frequently asked questions.* Retrieved June 14, 2006, from http://police.homeoffice.gov.uk/performance-and-measurement/performance-assessment/faqs/.

Lawrence v. Texas. (2003). 537 U.S. 1044.

Metropolitan Police. (2006). *London gets safer: Metropolitan Police Service performance 2005/2006.* Retrieved June 14, 2006, from http://www.met.police.uk/news/reports/performance0506.pdf.

National Criminal Justice Board. (2006). *National Criminal Justice Board.* Retrieved July 19, 2006, from http://lcjb.cjsonline.gov.uk/ncjb/1.html.

National Probation Service (NPS). (2006). *NPS performance targets and measures 2006–2007: Guidance.* Retrieved July 19, 2006, from http://www.probation.homeoffice.gov.uk/files/pdf/PC28%202006.pdf.

Office for Criminal Justice Reform. (2004). Cutting crime, delivering justice: A strategic plan for criminal justice 2004–2008. London: TSO.

Office for Criminal Justice Reform. (2006). *The CJS: Working together for justice.* Retrieved July 15, 2006, from http://www.cjsonline.gov.uk/the_cjs/aims_and_objectives/index.html.

Packer, H. L. (1968). *The limits of the criminal sanction.* Stanford, CA: Stanford University Press.

Index

Figures and tables are indicated by "f" and "t" following page numbers.

A

Abortion Act of 1967, 37
absolute discharge, 206, 272
Access to Justice Act of 1999, 168, 175
Act of Settlement of 1701, 21, 24, 163
Act of Settlement of Ireland of 1652, 295
Act of Union of 1536, 11
Act of Union of 1801, 295
adultery, 45
The Adult Offender (Home Office), 256
Afghanistan, 299
AFOs (Authorized Firearms Officers), 123
Albany Prison, 264
Al-Qaeda, 291–92, 299, 306–7
Anti-Social and Behaviour Act of 2003, 330
Anti-terrorism, Crime and Security Act of 2001 (ATCSA), 300, 304–6
Appellate Courts, 159–61, 160f, 163–65
apprenticeships for lawyers, 166, 176
arrestable offenses, 36
Arthur JS Hall and Co. v. Simons (2000), 167
Ashworth, A., 188
assaults, statistics on, 83–85t, 86, 88
association system of confinement, 236
ATCSA. *See* Anti-terrorism, Crime and Security Act of 2001
Attorney General, 170
attorneys. *See* lawyers
Australia, 229, 232
Authorized Firearms Officers (AFOs), 123

B

Bail Act of 1976, 188, 189
Baldwin, J., 199–200
Bar Council, 166
barristers, 165–67, 176
Beccaria, Cesare, 230
bellmen, 102
Bentham, Jeremy, 105, 230, 234
Betting, Gaming and Lotteries Act of 1963, 52, 53
Billam; R. v. (1986), 202
Bill of Rights (English), 21, 24
Bill of Rights (U.S.), 36
Bin-Laden, Osama, 299
Birmingham Six, 208
Blair, Tony, 312
Board of Customs and Excise, 53
Board of Prison Commissioners, 236
bobbies, 106
Borstal juvenile facilities, 248–49, 326
Bow Street Runners, 103–4

Bramshill Police Staff College. *See* Leadership Academy for Policing
Bridewell Palace, 228
bridewells, 228
Briggs, D., 200
British Crime Survey (BCS), 65, 71–73, 79–87
British Transport Police, 120
Brixton riots of 1981, 124, 139
buggery, 45, 82
Building Communities, Beating Crime: A Better Police Service for the 21st Century (Home Office), 140–41
burglaries, statistics on, 83–85t, 87
Burmah Oil Co. v. Lord Advocate (1965), 23

C

cabinet, 14
capital pledge, 101
capital punishment, 36–37, 237–38
Carpenter, Mary, 324
case law, 34
Catholics, 11, 12, 225, 266, 294, 296–97
cautions by police, 137
CCRC. *See* Criminal Cases Review Commission
Central Intelligence Agency (CIA, U.S.), 107, 299
Central Police Training and Development Authority (Centrex), 126
central prisons, 236
Chahal v. United Kingdom (1996), 304
Chamberlain, Neville, 15
Chancellor of the Exchequer, 14
Charles II, 102
Charlies, 102, 103
chief pledge, 101
children
    and prison maternity, 267
    in prisons, 233
    and prostitution, 47
    trafficking of, 16. *See also* Juvenile justice
Children Act of 1908, 237, 249, 325
Children Act of 1989, 47, 327
Children Act of 2004, 330
Children and Young Persons Act of 1933, 47, 326
Children and Young Persons Act of 1969, 137, 326
Church of England, 14, 266, 294
CIA (Central Intelligence Agency, U.S.), 107, 299
circuit judges, 163
City of London Police, 103
civil liberties and terrorism legislation, 308–11, 359–60
Civil Nuclear Constabulary, 120
class divisions as social issue, 17
Codes of Practice, 130–33
Collins, Michael, 296
Colquhoun, Patrick, 104, 105
committal proceedings, 192–93
common law, 34
community correctional alternatives, 206, 249–51, 269–70, 337
conditional cautions, 137

conditional discharge, 206, 272
Connelly, James, 296
consensual sexual activity. *See* sexual activity, regulation of
constables, 100, 101–2
Constitutional Reform Act of 2005, 24, 164
A Coordinated Prostitution Strategy (Home Office), 50
correctional system, 223–88
 alternatives to incarceration, 267–73
 and community alternatives, 249–51
 and death penalty, 237–38
 development of, 223–46
 future trends, 272–75
 historical background of, 224–28
 and hulks, 231–33
 and Independent Monitoring Boards, 255–56
 and institutional development, 249–51
 and medieval punishments, 229–30
 nonpunitive dispositions, 272
 organization and operations of, 247–88, 253*f*
 and Parole Board, 256–58
 and penal reformers, 230–31, 233, 235–37, 248–49
 philosophical foundations of, 251–52
 police role in, 136–38
 and Prison Service Agency, 253–54
 and public displays of punishment, 228
 and transportation as punishment, 231–33. *See also* prisons
Council of Legal Education, 165
County and Borough Police Act of 1856, 106
county courts, 34
County Police Act of 1839, 106
Court Act of 2003, 158
Court of Appeal, Criminal Division, 201–2
courts
 evaluation of, 364–65
 and juvenile justice, 332–35, 335*f*
 process in, 185–220. *See also* Judicial system; *specific court*
Courts and Legal Services Act of 1990, 163, 168
CPS. *See* Crown Prosecution Service
cracked trial rate, 362
Crawford, William, 234
crime
 control model, 351, 353, 362
 against morality, 40, 42*t*, 43
 against the person, 37–39, 38–39*t*
 against property, 39–40, 39*t*
 against the public order, 40, 41–42*t*
 against the State, 40, 41*t*. *See also* crime statistics; Property crimes; Violent crimes
Crime and Disorder Act of 1998, 37, 137, 193, 203, 327, 328, 329, 332
Crime in England and Wales Report, 71–73, 79–87
crime statistics, 63–96
 definitional issues in, 67–68
 in Great Britain, 71–73, 79–87
 methodological problems in, 65–66
 and police reports, 66–67, 89–90
 sources of, 64–65
 in U.S., 68–71, 69*t*, 74–79, 75–78*t*
 U.S. and U.K. comparison, 87–90, 90*t*

Criminal Appeal Act of 1995, 160, 209
Criminal Cases Review Commission (CCRC), 160–61, 209, 212, 358
Criminal Defence Service, 175–76
Criminal Justice Act of 1948, 250
Criminal Justice Act of 1967, 199, 271
Criminal Justice Act of 1972, 270
Criminal Justice Act of 1982, 207, 326
Criminal Justice Act of 1987, 193
Criminal Justice Act of 1988
 and Appellate Courts, 159
 and committal proceedings, 193
 and juries, 197, 198
 and pretrial release, 189
 and sentencing, 201
 summary offenses under, 191
Criminal Justice Act of 1991
 and financial penalties, 271
 and juvenile sentencing, 261, 327
 and parole, 257, 258
 and probation, 269
 and sentencing, 201, 202, 206, 207, 259, 273
 and suspended sentences, 271
Criminal Justice Act of 1993
 and financial penalties, 271
 and sentencing, 202, 206, 207
Criminal Justice Act of 2003
 and bail, 188
 and conditional cautions, 137
 and detention of suspects, 186
 and juries, 197, 199, 210–11
 and parole, 257
 and pretrial release, 189
 and sentencing, 159, 191, 196, 203
 and summary trials, 158
Criminal Justice and Police Act of 2001, 49, 126, 330
Criminal Justice and Public Order Act of 1994, 46, 132, 188, 189, 192, 196, 210
criminal law, 33–61
 categorization of offenses in, 36–44, 38–39*t*, 41–42*t*
 source and structure of, 34–36
 substantive vs. procedural, 33. *See also specific types of crime*
Criminal Law Act of 1967, 34–35, 36
Criminal Law Policy Unit, 35
Criminal Law Revision Committee, 35, 45, 47, 48, 49
Criminal Procedure and Investigations Act of 1996, 191, 193
Criminal Procedure (Insanity) Act of 1964, 43
Criminal Procedure (Insanity and Unfitness to Plead) Act of 1991, 43
Criminal Statistics: England and Wales report (CSEW), 65
Critchley, T. A., 101
Crofton, Walter, 235
Cromwell, Oliver, 11, 295
crown, 21–22
Crown Courts, 159, 160*f*, 163, 190–91
Crown Prosecution Service (CPS), 171, 172–73, 177, 195, 333
CSEW (Criminal Statistics: England and Wales), 65
Cultural and Communities Resource Unit, 127
cultural heritage of England, 10–12, 14–16

Current Sentencing Practice (Thomas), 202
Custody Plus program, 274

D

Davey Paxman & Co. v. Post Office (1954), 23
DCA (Department of Constitutional Affairs), 164
death penalty, 36–37, 237–38
debtors' prisons, 227–28
defense counsel, right to, 173–76
deferred sentences, 272
demographics, 10
Departmental Committee of Inquiry into Prisons, 237
Department of Constitutional Affairs (DCA), 164
Detainment during Her Majesty's Pleasure, 334
detention, PACE requirements for, 132
deterrence of crime, 361
deviant sexual activity, 45–47
Dingley, J., 298
Diploma in Probation Studies, 269
Director of Public Prosecutions (DPP), 133, 170
Directors of Convict Prisons, 236
discharges. *See* absolute discharge; conditional discharge
District Judges, 162
Dockyards Protection Act of 1772, 36–37
doli incapax, 323
DPP (Director of Public Prosecutions), 133, 170
Drogheda, siege of, 295
drug use and abuse, 40, 43
DuCane, Edmund, 236
due process model, 351

E

Easter Uprising, 296
economic issues, 16–21
Eden, William, 230–31
education, 18, 265
Edward VI, 228
Elizabeth I, 11, 295
England and Wales
  cultural heritage of, 10–12, 14–16
  demographics of, 10
entrapment, 43
Ethelbert I, 226
ethnic minorities. *See* minorities
European Commission of Human Rights, 24, 255
European Convention for the Protection of Human Rights and Fundamental Freedoms, 24, 25, 35, 188, 305
European Court of Human Rights, 24, 35, 247, 274, 304
European Union (EU), 16
exclusion of evidence, 193–95
exile as punishment method, 231–33

F

Federal Bureau of Investigation (FBI, U.S.), 65, 107
Fielding, Henry, 103–4, 105
Fielding, John, 105
financial penalties, 204–5*t,* 205, 225, 270–71

Financial Services Authority, 53
firearms
  police use of, 123
  and violent crimes, 15–16, 88
Foreign Secretary, 14
Forfeiture Act of 1870, 36
fornication, 45
Fox, Lionel, 250
frankpledge system, 100
Fry, Elizabeth, 233

G

gambling, 50–54
Gambling Act of 2005, 50, 51, 52, 53, 54
Gambling Commission, 53
Gambling Enforcement Act of 2006 (U.S.), 52–53
Gaming Act of 1968, 53, 54
Gaming Board, 53
gaols, 228
Gay Police Association, 128
Gladstone Committee, 237
governmental system, 9–31
  framework of, 21–25
  structure of, 12–14
Government of Ireland Act of 1914, 296
Greater London Authority Act of 1999, 118
guidance circulars, 119
Guildford Four, 208
guns. *See* firearms

H

Hancock, L., 200
hangings, 229
Hardwick, Nick, 136
health care, 17, 266
Health Care Service for Prisoners, 266
Hendon Police College, 107, 122
Henry II, 227
Henry VIII, 228, 294
Henry's Act of 1531, 228
Her Majesty's Chief Inspector of Constabulary Office, 118
Her Majesty's Government, 12
High Court judges, 163
Higher Rights of Audience (Law Society), 168
High Potential Development Scheme (HPD), 121–22
H.M. Prison Service, 252–54, 261
Holloway Prison, 267
Home Office Recorded Crime Statistics, 80
Home Office Review of Crime Statistics, 71
Home Secretary, 14, 118, 119
Homicide Act of 1957, 43
homicide, statistics on, 83–85*t,* 85–88
homosexuality, 45–47, 128
Horserace Betting Levy Board, 53
Horserace Totalisator Board, 53
House of Commons, 12–13, 21
House of Lords, 13, 21

House of Lords Act of 1999, 13
Howard Association, 251
Howard, John, 230, 231
Howard, Michael, 135
Howard League for Penal Reform, 237, 251
HPD (High Potential Development Scheme), 121–22
hue and cry system, 100
hulks, use of as punishment, 231–33
Human Rights Act of 1998, 25, 35
human trafficking, 16
hundred courts, 34
hundredmen, 101
hundreds, 100

I

Imbert, Peter, 139
immigration, 10, 16, 305
incarceration alternatives, 267–73
incompetence to stand trial, 43
Independent Monitoring Boards, 255–56
Independent Police Complaints Commission (IPCC), 136, 140, 355
An Independent Police Complaints Commission (Liberty), 135–36
indictable offenses, 37, 186–87
Industrial Schools Act of 1857, 324
infant mortality, 17
Inns of Court, 165
"An Inquiry into the Causes of the Late Increase of Robbers" (H. Fielding), 103
insanity as defense, 43
Intensive Development Scheme, 263
Internet gambling, 52
interrogation, PACE requirements for, 132
IPCC. *See* Independent Police Complaints Commission
Ireland, 11, 294–99
Irish marks system, 235
Irish National Liberation Army, 300
Irish Republican Army (IRA), 12, 292, 294, 295–97, 298
Irish Republican Brotherhood (IRB), 296

J

James I, 11
James II, 295
JCSB (Jury Central Summoning Bureau), 197–98
Jebb, Joshua, 234
Jeffrey v. Black (1977), 194
John Howard League. *See* Howard League for Penal Reform
judges
  in Appellate Courts, 163–65
  in Crown Courts, 163
Judicial Appointments Commission, 164
judicial system, 157–83
  and committal proceedings, 192–93
  and court process, 185–220
  and evidentiary rules, 193–95
  and indictable offenses, 186–87

judges in, 161–65
  and juries, 197–200
  and lawyers, 165–68
  and miscarriages of justice, 208–9
  and plea bargaining, 195–96
  and pretrial release, 188–90
  and sentencing, 200–208
  special features of, 187
  structure and jurisdiction of, 158–61
  and summary offenses, 186
  and triable-either-way cases, 190–92
junior counsels, 166
juries, 197–200, 357
jurisdiction, 158–61
Jury Act of 1974, 197, 198, 199
Jury Central Summoning Bureau (JCSB), 197–98
jury nobbling, 199
Justice Committee, 170
Justice of the Peace Act of 1979, 162
Justices of the Peace Act of 1361, 102
juvenile justice, 321–45
  changes in, 323–31
  and courts, 332–35, 335*f*
  history of, 322–23
  U.S. and U.K. comparison, 360–61
Juvenile Offenders Act of 1847, 324

K

Keir, D. L., 23
kerb crawling, 82
Khan; R. v. (1996), 194

L

Labour Party, 11
Ladies Newgate Committee, 233
law enforcement
  City of London Police, 103
  history of, 100–109
  Metropolitan Police Force, 103, 105–6
  recruitment and training, 120–28
  reformers of, 103–5
  structure and organization, 118–20, 119*f*
  twentieth-century changes in, 106–7. *See also* Police; *specific agency*
Law of the Ten Tables, 224–25
Law Society, 167–68
Lawson, F. H., 23
lawyers
  for defense, 173–76
  for prosecution, 169–73
lay magistrates, 161, 177, 356
Leadership Academy for Policing, 107, 125–26
Lee, Melville, 100
Legal Aid Act of 1982, 174, 175
Legal Aid Act of 1988, 174
Legal Aid and Advice Act of 1949, 174
Legal Aid Board, 174

Legal Services Commission, 175
Liberty (civil rights group), 135–36
Licensing (Young Person) Act of 2000, 329
local prisons, 236, 260
London
  barristers in, 16
  bombings in, 123, 141, 291–92
  crime in, 100–101
  hangings in, 229
  minorities in, 10
  as Olympics host city, 141
  police in, 102–5, 108, 118, 133
  prisons in, 228–32, 234, 266, 267
Loosely; R. v. (2001), 44
Lord Chancellor, 24, 158
Lord Justices of Appeal, 164
Lords of Appeal in Ordinary, 13, 164
Lotteries Act of 1976, 53

M

Maconochie, Alexander, 235
magistrates, 161–63
Magistrates' Association, 201, 207
Magistrates' Courts, 158–59, 160*f*, 190–91, 327, 333, 356
Magistrates' Courts Act of 1952, 159
Magistrates' Courts Act of 1980, 158, 190, 193
Magna Carta, 21, 24
Maguire Seven, 208
Maitland, F. W., 34
Making Punishments Work: Report of a Review of the Sentencing Framework for England and Wales (Home Office), 202
manor courts, 34
Marine Police, 104
Martinson, R., 251–52
maternity in prisons, 267
Matthews, R., 200
Mayne, Richard, 105
McConville, M., 199–200
McGuinness, Martin, 12
McPherson Report, 135
media coverage. *See* news media and courts
medical care. *See* health care
medieval punishments, 229–30
Members of Parliament (MPs), 12–13
mental illness of defendants, 43
Met Careers Team, 127
Metropolitan Police Commissioner; R. v. (1986), 169
Metropolitan Police Force, 103, 105–6, 118
Millbank Prison, 234
Ministry of Defense Police and Guarding Agency, 120
minorities
  in England, 10
  law enforcement recruitment of, 126–28
  in prisons, 260
  as prison staff, 262
miscarriages of justice, 208–9
misconduct investigations, police, 132–33
M'Naghten rules, 43

monarchy, 21–22
morality, legislation of, 355
MPs, 12–13
Municipal Corporations Act of 1835, 106
Murder (Abolition of the Death Penalty) Act of 1965, 36–37, 238

N

Narey, Martin, 254
Nathaniel; R. v. (1995), 194
National Association for the Care and Resettlement of Offenders (NACRO), 272–73
National Campaign for the Abolition of Capital Punishment, 238
National Crime Recording Standard (NCRS), 71, 82
National Crime Victimization Survey (NCVS, U.S.), 65, 70–71, 79, 80*t*, 81*t*
National Criminal Justice Board, 367
National Duty Solicitor Scheme, 175
National Health Service, 17
National Incident-Based Reporting System (NIBRS, U.S.), 65, 68–70, 69*t*
National Lottery Act of 1993, 51
National Lottery Commission, 53
National Offender Management System (NOMS), 35, 252–54, 267–68
National Police Library, 126
National Policing Improvement Agency (NPIA), 126
National Policing Plan, 120, 140
National Probation Service, 267–68
NCVS. *See* National Crime Victimization Survey
Newbold Revel, 263
Newgate Prison, 228, 233
New Scotland Yard, 105
news media and courts, 187, 355–56
New York City, 108, 239
NIBRS. *See* National Incident-Based Reporting System
NOMS. *See* National Offender Management System
nonarrestable offenses, 36
nonpunitive dispositions, 272
Northern Ireland, 11–12, 294–99
Northern Ireland Act of 1996, 301
Northern Ireland Act of 1998, 12
"Nothing Works" report (Martinson), 251–52

O

Obscene Publications Act of 1959, 49
Observations on the Visiting, Superintendence, and Government of Female Prisoners (Fry), 233
Offences against the Person Act of 1861, 34, 37, 158
Office of Criminal Justice Reform (OCJR), 366
Office of Fair Trading, 167
Official Secrets Act of 1911, 163, 255
Officials (IRA), 297
Old Bailey, 10, 236
Organization for Economic Co-operation and Development (OECD), 17–18
oubliettes, 227

P

Packer, H. L., 351, 353, 362
Paisley, Ian, 12
paper committals, 192
parents and juvenile justice system, 329
parish constables. See constables
Parliament, 12–13, 21, 35. See also specific members or
    legislation
Parliament Act of 1911, 13
Parole Board, 256–58
Patriot Act. See USA Patriot Act
Paying the Price: A Consultation Paper on Prostitution (Home
    Office), 49
PCA (Police Complaints Authority), 133
PCTs (Primary Care Trusts), 266
Pearse, Patrick, 296
Peel, Robert, 104, 105–6, 231
Peel Centre Police Training Facility, 120, 122–25
Peelers, 106
penal reform. See reform
Penal Reform League, 251
Penal Servitude Act of 1853, 234
Penal Servitude Act of 1857, 232, 234
Pentonville Prison, 234
philosophical foundations of correctional system, 251–52
Piracy Act of 1837, 36
Plantation of Ulster, 295
plea bargaining, 195–96, 357
plea before venue procedures, 191
police
    and constables, 101–2
    correctional role of, 136–38
    evaluation of, 363–64
    future trends for, 140–41
    misconduct investigations of, 132–33
    powers and procedures of, 129–36
    public image of, 138–39
    reports of offenses to, 66–67, 89–90, 90t
    U.S. and U.K. comparison of, 353–55
Police Act of 1857 (U.S.), 108
Police Act of 1964, 129–30
Police Act of 1976, 133
Police Act of 1996, 119
Police and Criminal Evidence Act of 1984 (PACE), 130–33, 139
    arrestable offense under, 36
    and detention of suspects, 186
    and evidenciary issues, 194
    and juvenile justice, 332
    and police accountability, 354–55
    and pretrial release, 188
    and right to counsel, 173–74, 175
Police and Magistrates' Courts Act of 1994, 119, 140, 162
Police Complaints Authority (PCA), 133
Police Complaints Board, 133
Police Federation, 135, 139
Police Reform: A Police Service for the Twenty-First Century
    (Home Office), 134–35
Police Reform Act of 2002, 120, 136, 140, 355

Poor Prisoners' Defence Act of 1903, 174
Poor Prisoners' Defence Act of 1930, 174
pornography, 43
Positive Action Team, 127
posse comitatus, 100
POST (Prison Officer Selection Test), 263–64
Pounds, John, 324
Powers of Criminal Courts Act of 1973, 271, 272
Powers of Criminal Courts (Sentencing) Act of 2000, 159, 190
pretrial release, 188–90, 356
Prevention of Crime Act of 1908, 250, 325
Prevention of Terrorism Act of 1974, 300
Prevention of Terrorism Act of 1976, 300
Prevention of Terrorism Act of 1984, 301
Prevention of Terrorism Act of 1989, 301
Prevention of Terrorism Act of 2005, 25, 300, 306–11
prima facie cases, 187
Primary Care Trusts (PCTs), 266
Prime Minister, 14
Prison Act of 1877, 236
Prison Act of 1898, 237, 248
Prison Officer Selection Test (POST), 263–64
prisons, 258–67
    alternatives to, 267–72
    design of, 234, 235
    development of, 234–35, 249–51
    evaluation of, 365–66
    and maternity, 267
    minorities in, 260
    programs in, 264–66
    staffing of, 262–64
    types of, 260–62
    women in, 266–67. See also correctional system
Prison Service Agency, 253–54
Prison Service Management Board, 254
Prison Staff College at Wakefield, 263, 264
probation, 204–5t, 206, 268–69, 366
Probation of Offenders Act of 1907, 250
Probation Service, 189, 250
Professional Conduct and Complaints Committee, 167
property crimes
    in Great Britain, 84t, 85t, 87
    statistics on, 88
    in U.S., 77–78t, 78–79
prosecution lawyers, 169–73
Prosecution of Offences Act of 1985, 133–34, 169, 171–72, 175
prostitution, 47–50
Protestants, 11, 12, 296, 297
Provisionals (IRA), 297
public displays of punishment, 228
public image of police, 138–39
Public Order Act of 1986, 40
punishment methods
    death penalty, 36–37, 237–38
    exile, 231–33
    history of, 224–28
    medieval examples of, 229–30
    public displays of, 228
    U.S. and U.K. comparison, 239

Q

Queen's Counsels (QCs), 166–67
questioning of suspects, PACE requirements for, 132

R

Race Relations Act of 2000, 127
ragged schools, 323
Ragged School Union, 324
rape, statistics on, 83–85t, 86
recorders, 163
recruitment of police officers, 120–28
Redmayne, M., 188
Reducing Crime—Changing Lives (Home Office), 252
Reduction for Guilty Plea (Sentencing Guidelines Council), 196
reform
    of law enforcement, 99, 103–5
    penal, 230–31, 233, 235–37, 248–49
rehabilitation programs, 206, 264–66, 337, 360
religion, 11, 14
    and juvenile justice, 324
    in prisons, 266. *See also* Catholics; Protestants
reparations, 337. *See also* financial penalties
reprimands, 137, 332
right to assistance of counsel, 173–76
roadchecks, PACE requirements for, 130–31
Road Traffic Act of 1972, 40
robberies, statistics on, 83–85t, 86, 88
Robin Redbreasts, 104–5
Romilly, Samuel, 230–31
Rowan, Charles, 105
Royal Commission on Betting, Lotteries, and Gaming, 51
Royal Commission on Capital Punishment, 237
Royal Commission on Criminal Procedure, 170
Royal Commission on Gambling, 51, 52, 53
Royal Commission on Legal Services, 168
Royal Irish Constabulary, 104
Ruggles-Briese, Evelyn, 248
Runciman Commission, 187, 192, 196, 197, 199, 209
Rushcliffe Committee, 174
R. v. See name of opposing party

S

Sang; R. v. (1979), 43
Saudi Arabia, 299
Scarman report, 139
Scotland, 11
Scotland Yard, 105
search warrants, PACE requirements for, 132
sector policing, 139
secure children's homes, 261
secure training centers (STCs), 261
Select Committee on Capital Punishment, 237
sentencing, 200–208
    patterns in, 203–8, 204–5t
    suspended sentences, 271–72
    U.S. and U.K. comparison, 358
Sentencing Advisory Panel, 203

Sentencing Guidelines Council, 196, 203, 211, 358
separation of powers, 22, 23–24
separation system of confinement, 236–37
Serious Organized Crime Agency (SOCA), 16, 120
sexual activity, regulation of, 39, 44–50
    deviant, 45–47
    prostitution, 47–50
Sexual Offences Act of 1956, 39, 46, 47–48
Sexual Offences Act of 1967, 45–46, 47, 48–49
Sexual Offences Act of 1985, 49
Sexual Offences Act of 2000, 46
Sexual Offences Act of 2003, 39, 45, 47, 49
Sheehy Report, 135
Shipman, Harold, 85–86
ships as prisons, 232–33
shire reeve, 101
shires, 100
Silverman, Sidney, 238
simple cautions by police, 137
Skogan, W., 139
SOCA. *See* Serious Organized Crime Agency
social issues, 16–21
    class divisions, 17
    educational system, 18
    underprivileged, aid for, 17–18
Society of Gentlemen Practisers in the Courts of Law and Equity. *See* Law Society
solicitors, 165, 167–68, 176
solitary confinement system, 236
Soviet occupation of Afghanistan, 299
staff of prisons, 262–64
standard of living, 20–21
star prisoners, 249
State of Prisons in England 1777 (Howard), 230
Statute of Acton Burnell of 1283, 227
Statute of Merchants of 1285, 227
Statute of Winchester of 1285, 101
statutes and criminal law, 34
STCs (Secure training centers), 261
Stephen Lawrence Inquiry, 127, 139
stop and search, PACE requirements for, 130–31
Strangeways Prison, 264
Straw, Jack, 328, 337
Street Offences Act of 1959, 48
strikes and lockouts, 19
Sudan, 299
Summary Jurisdiction Act of 1879, 325
summary offenses, 37, 186
Supremacy of Parliament doctrine, 21–23
Supreme Court (U.S.), 24
Supreme Court Act of 1981, 163
suspended sentences, 271–72

T

TACT. *See* Terrorist Act of 2000
Terrill, R., 256
terrorism, 291–319
    in England, 294–99

legislation to combat, 292–94, 299–313
security vs. civil liberties, 308–11, 359–60
U.S. and U.K. comparison, 359–60
Terrorism Act of 2006, 311–13
Terrorist Act of 2000 (TACT), 40, 298, 300, 301–4
Texas Rangers (U.S.), 108
Thames River Police, 104
Theft Act of 1968, 34, 39–40
thefts, statistics on, 83–85t, 87
Thomas, David, 202
Thorpe, K., 74
ticket of leave system, 232, 235
tithings and tithingmen, 100
trade unions, 19
trading justice practices, 102
trafficking in humans, 16
training of police officers, 120–26
training prisons, 260–61
Transport Act of 1981, 40
transportation of prisoners, 190–91, 231–33
Treason Act of 1351, 40
Treason Act of 1814, 36
Trenchard Police College, 122
triable-either-way cases, 186, 190–92
Turner; R. v. (1970), 195–96

U

underprivileged, aid for, 17–18
unemployment rate, 20
Uniform Crime Report (UCR, U.S.), 65, 68–70, 69t, 75–78t
unions, 19
unitary governmental system, 21
United Kingdom, 10–14. *See also* England and Wales;
    Northern Ireland
United Nations, 24
United States
  and Bill of Rights, 36, 211
  and crime statistics, 65, 68–70, 69t, 75–78t
  criminals exiled to, 229
  demographics of, 10
  and firearms, 15–16
  gambling legislation in, 52–53
  juvenile justice in, 360–61
  legal system and process in, 355–58, 364–65
  police in, 108, 353–55, 363–64
  prisons in, 359–60, 365–66
  and terrorism legislation, 293, 359–60
USA Patriot Act, 293, 303, 359, 360

V

Vennard, J., 200
vetting of juries, 198, 357
victim compensation, 270–71
victim-offender confrontations, 337
village constables. *See* constables
violent crimes
  and firearms, 15–16, 88
  in Great Britain, 82–87, 83t, 85t
  in U.S., 74–77, 75–76t
voir dire and jury selection, 198, 357

W

Wadham, John, 136
Wales. *See* England and Wales
Walker, C., 303
warnings by police, 137, 332
warrantless arrest, PACE requirements for, 130–31
watch system, 101
welfare system, 17–18
Widgery Committee, 174
William of Orange, 295
William the Conqueror, 227
Wireless Telegraph Act of 1954, 23
Wolfenden Committee, 44, 45, 47
women
  law enforcement recruitment of, 126–28
  in prisons, 233, 266–67
  as prison staff, 262
  as probation officers, 269
  and prostitution, 47–48
  trafficking of, 16
Woodcock, John, 139

Y

Young offender institutions (YOIs), 261
"Young People, Crime and Antisocial Behaviour"
    (Home Office), 336
Youth Custody Centers, 326
Youthful Offenders Act of 1854, 324
Youth Justice and Criminal Evidence Act of 1999, 329
Youth Justice Board, 328, 333
Youth Offending Team (YOT), 137, 328–29, 330, 332